Clean Energy for Low-Income Communities

Other related titles:

You may also like

- PBPO1590 | Probst | Transforming the Grid Towards Fully Renewable Energy | 2020
- PBPO160 | Chauhan | Microgrids for Rural Areas: Research and case studies | 2020
- PBPO193 | Chattopadhyay | Overhead Electric Power Lines: Theory and practice | 2021
- PBPO101 | Ting *et al.* | Methane and Hydrogen for Energy Storage (previous work by EiC of proposed book) | 2016
- PBPO130 | Ting *et al.* | Wind and Solar Based Energy Systems for Communities (previous work by EiC of proposed book) | 2018

We also publish a wide range of books on the following topics:
Computing and Networks
Control, Robotics and Sensors
Electrical Regulations
Electromagnetics and Radar
Energy Engineering
Healthcare Technologies
History and Management of Technology
IET Codes and Guidance
Materials, Circuits and Devices
Model Forms
Nanomaterials and Nanotechnologies
Optics, Photonics and Lasers
Production, Design and Manufacturing
Security
Telecommunications
Transportation

All books are available in print via https://shop.theiet.org or as eBooks via our Digital Library
https://digital-library.theiet.org.

IET ENERGY ENGINEERING SERIES 251

Clean Energy for Low-Income Communities

Technology, deployment and challenges

Edited by
David S-K. Ting and Jacqueline A. Stagner

The Institution of Engineering and Technology

About the IET

This book is published by the Institution of Engineering and Technology (The IET).

We inspire, inform and influence the global engineering community to engineer a better world. As a diverse home across engineering and technology, we share knowledge that helps make better sense of the world, to accelerate innovation and solve the global challenges that matter.

The IET is a not-for-profit organisation. The surplus we make from our books is used to support activities and products for the engineering community and promote the positive role of science, engineering and technology in the world. This includes education resources and outreach, scholarships and awards, events and courses, publications, professional development and mentoring, and advocacy to governments.

To discover more about the IET please visit https://www.theiet.org/

About IET books

The IET publishes books across many engineering and technology disciplines. Our authors and editors offer fresh perspectives from universities and industry. Within our subject areas, we have several book series steered by editorial boards made up of leading subject experts.

We peer review each book at the proposal stage to ensure the quality and relevance of our publications.

Get involved

If you are interested in becoming an author, editor, series advisor, or peer reviewer please visit https://www.theiet.org/publishing/publishing-with-iet-books/ or contact author_support@theiet.org.

Discovering our electronic content

All of our books are available online via the IET's Digital Library. Our Digital Library is the home of technical documents, eBooks, conference publications, real-life case studies and journal articles. To find out more, please visit https://digital-library.theiet.org.

In collaboration with the United Nations and the International Publishers Association, the IET is a Signatory member of the SDG Publishers Compact. The Compact aims to accelerate progress to achieve the Sustainable Development Goals (SDGs) by 2030. Signatories aspire to develop sustainable practices and act as champions of the SDGs during the Decade of Action (2020–30), publishing books and journals that will help inform, develop, and inspire action in that direction.

In line with our sustainable goals, our UK printing partner has FSC accreditation, which is reducing our environmental impact to the planet. We use a print-on-demand model to further reduce our carbon footprint.

Published by The Institution of Engineering and Technology, London, United Kingdom

The Institution of Engineering and Technology (the "**Publisher**") is registered as a Charity in England & Wales (no. 211014) and Scotland (no. SC038698).

Copyright © The Institution of Engineering and Technology and its licensors 2024

First published 2024

All intellectual property rights (including copyright) in and to this publication are owned by the Publisher and/or its licensors. All such rights are hereby reserved by their owners and are protected under the Copyright, Designs and Patents Act 1988 ("**CDPA**"), the Berne Convention and the Universal Copyright Convention.

With the exception of:

(i) any use of the publication solely to the extent as permitted under:

   a. the CDPA (including fair dealing for the purposes of research, private study, criticism or review); or
   b. the terms of a licence granted by the Copyright Licensing Agency ("**CLA**") (only applicable where the publication is represented by the CLA); and/or

(ii) any use of those parts of the publication which are identified within this publication as being reproduced by the Publisher under a Creative Commons licence, Open Government Licence or other open source licence (if any) in accordance with the terms of such licence,

no part of this publication, including any article, illustration, trade mark or other content whatsoever, may be used, reproduced, stored in a retrieval system, distributed or transmitted in any form or by any means (including electronically) without the prior permission in writing of the Publisher and/or its licensors (as applicable).

The commission of any unauthorised activity may give rise to civil or criminal liability.

Please visit https://digital-library.theiet.org/copyrights-and-permissions for information regarding seeking permission to reuse material from this and/or other publications published by the Publisher. Enquiries relating to the use, including any distribution, of this publication (or any part thereof) should be sent to the Publisher at the address below:

The Institution of Engineering and Technology
Futures Place
Kings Way, Stevenage
Herts, SG1 2UA, United Kingdom

www.theiet.org

While the Publisher and/or its licensors believe that the information and guidance given in this publication are correct, an individual must rely upon their own skill and judgement when performing any action or omitting to perform any action as a result of any statement, opinion or view expressed in the publication and neither the Publisher nor its licensors assume and hereby expressly disclaim any and all liability to anyone for any loss or damage caused by any action or omission of an action made in reliance on the publication and/or any error or omission in the publication, whether or not such an error or omission is the result of negligence or any other cause. Without limiting or otherwise affecting the generality of this statement and the disclaimer, whilst all URLs cited in the publication are correct at the time of press, the Publisher has no responsibility for the persistence or accuracy of URLs for external or third-party internet websites and does not guarantee that any content on such websites is, or will remain, accurate or appropriate.

Whilst every reasonable effort has been undertaken by the Publisher and its licensors to acknowledge copyright on material reproduced, if there has been an oversight, please contact the Publisher and we will endeavour to correct this upon a reprint.

Trade mark notice: Product or corporate names referred to within this publication may be trade marks or registered trade marks and are used only for identification and explanation without intent to infringe.

Where an author and/or contributor is identified in this publication by name, such author and/or contributor asserts their moral right under the CPDA to be identified as the author and/or contributor of this work.

British Library Cataloguing in Publication Data

A catalogue record for this product is available from the British Library

ISBN 978-1-83953-849-0 (hardback)
ISBN 978-1-83953-850-6 (PDF)

Typeset in India by MPS Limited

Cover image: Philippe Lissac – Godong/Stone via Getty Images

To those who practice cleaner living using clean energy and share this with their less-privileged neighbors.

Contents

Preface	**xv**
Acknowledgments	**xvii**
About the editors	**xix**

1 Sharing clean energy with the poor   **1**
David S-K. Ting and Jacqueline A. Stagner

1.1	Why share clean energy with the poor?	1
1.2	A complex challenge	2
1.3	A step in the right direction	2
References		4

2 Providing clean and affordable energy for all: possible, practical or propaganda?   **5**
Graham T. Reader

2.1	Introductory remarks		6
2.2	Affordable energy, poverty and fuel poverty		9
	2.2.1	Fuel poverty benchmarks	11
	2.2.2	Poverty measures and sustainable development	18
	2.2.3	Poverty and energy challenges – Least Developed Countries (LDC)	23
2.3	Clean energy		26
2.4	Access to electricity		29
	2.4.1	Why electricity?	29
	2.4.2	Energy measurement and data gathering	32
	2.4.3	Electricity versus energy	37
	2.4.4	Household electricity access benchmarks redefined	45
2.5	Clean cooking fuels		47
2.6	Space heating and warm households		50
2.7	Affordable energy for other end-use sectors		51
2.8	Discussion possible, practical, propaganda		53
2.9	Concluding remarks		58
References			60

x   *Clean energy for low-income communities*

3   Low-cost and energy-efficient housing design: a review on research trends   **79**
Abbas Shadmand and Semra Arslan Selçuk
3.1   Introduction   79
3.2   Method   82
3.3   Findings   84
   3.3.1   Publication and citation numbers by year   84
   3.3.2   Research areas   84
   3.3.3   Publication type   86
   3.3.4   Sources of publications   86
   3.3.5   Publication numbers by country   87
   3.3.6   Authorship and co-authorship analysis   87
   3.3.7   Publication and citation numbers by institutions   89
   3.3.8   Keyword analysis   90
   3.3.9   Most cited studies published between 2000 and 2023   93
3.4   Results and potential areas for future studies   97
References   98

4   Enabling solar energy production for low-income communities   **103**
Lutfu S. Sua and Figen Balo
4.1   Introduction   104
4.2   Solar power at low-income communities   107
   4.2.1   Solar energy implementation challenges in low-income communities 107
   4.2.2   Potential funding resources for cities   109
4.3   Methodology   109
4.4   Results and discussion   110
4.5   Conclusions   114
References   115

5   Building energy efficiency improvements and solar PV systems integration   **119**
Sogo Mayokun Abolarin, Manasseh Babale Shitta, Olanrewaju G. Oluwasanya, Charles Asirra Eguma, Azizat Olusola Gbadegesin, Emmanuel O. Ogedengbe and Louis Lagrange
5.1   Introduction   122
5.2   Data collection   125
5.3   Data reduction   128
   5.3.1   Energy consumption analysis   128
   5.3.2   Energy efficiency improvements   129
   5.3.3   Solar sizing analysis   130
   5.3.4   Solar PV sizing   130
   5.3.5   Energy yield of the solar PV array   131
   5.3.6   Battery sizing   132
   5.3.7   Inverter sizing   132

5.4	Model validation		132
5.5	Results and discussion		134
	5.5.1	Present case	134
	5.5.2	Alternative case—energy-efficient case	140
5.6	Conclusions		153
Acknowledgments			154
References			154

6 Sustainable energy solutions for rural electrification in a low-income community
163

Reza Babaei, David S-K. Ting and Rupp Carriveau

6.1	Introduction		164
6.2	Methods and materials		168
	6.2.1	Study area	168
6.3	System overview		169
	6.3.1	Renewable resources	169
	6.3.2	Desalinaion unit	169
	6.3.3	Load profile	170
	6.3.4	CHP plant	172
	6.3.5	PV module	172
	6.3.6	Wind turbine	173
	6.3.7	Battery storage	173
	6.3.8	Converter	174
6.4	Results and discussion		176
6.5	Conclusion		180
References			181

7 An introduction to the electrification in remote communities located in ecologically sensitive areas: from planning to implementation experience
185

Cresencio-Silvio Segura-Salas

7.1	Socio-environmental concerns in remote electrification programs		186
	7.1.1	Socio-environmental characterization	187
	7.1.2	Socio-environmental analysis	188
7.2	Multicriteria electrification approaches		189
	7.2.1	Input data—socio-environmental criteria characterization	192
	7.2.2	Heuristic approach for the expansion of the conventional distribution system in remote areas	193
	7.2.3	Microgrid formation methodology	196
	7.2.4	Stand-alone systems	197
	7.2.5	Application example in the case of the Pantanal, Mato Grosso do Sul, Brazil	204
7.3	Electrification model and experiences in prototype implementation		216
	7.3.1	Experiences in electrochemical battery application	217
	7.3.2	On the implementation stage	219

xii *Clean energy for low-income communities*

	7.3.3 Regarding the operations and maintenance plan	222
7.4	Summary	224
References		226

8 Enhancing solar insolation in agricultural greenhouses by adjusting its orientation and shape — **231**
Gurpreet Khanuja, Rajeev Ruparathna and David S-K. Ting

8.1	Introduction	232
8.2	Literature review	234
8.3	Methodology	236
	8.3.1 Model development	236
	8.3.2 The TRNSYS-18 model	238
8.4	Results	241
	8.4.1 Impact of greenhouse orientations on solar radiation availability	241
	8.4.2 Impact of roof inclinations on solar radiation availability	242
	8.4.3 Impact of greenhouse shapes on solar radiation availability	244
8.5	Conclusions	245
References		246

9 Recent advances in biofuels production: industrial applications — **249**
Kang Kang, Sophia Quan He and Yulin Hu

9.1	Introduction	249
9.2	Thermochemical conversion of biomass	251
	9.2.1 Combustion	251
	9.2.2 Torrefaction	253
	9.2.3 Pyrolysis	255
	9.2.4 Gasification	257
9.3	Anaerobic digestion of biomass	259
9.4	Industrial applications of biofuels	263
	9.4.1 Biomass-based energy for heating	264
	9.4.2 Use of bio-oil as fuel oil	266
	9.4.3 Use of biogas	269
9.5	Conclusions	270
Acknowledgments		270
References		270

10 Modelling and forecasting the energy mix scenarios for Türkiye via LEAP analysis — **281**
Fazıl Gökgöz and Fahrettin Filiz

10.1	Introduction	281
10.2	Türkiye's power system overview	283
10.3	Methodology	284
10.4	Scenario development	286
	10.4.1 Low-demand-BAU	288

Contents   xiii

10.4.2 Low-demand-renewable	289
10.4.3 Low-demand-renewable-nuclear	289
10.4.4 Base-demand-BAU	290
10.4.5 Base-demand-renewable	291
10.4.6 Base-demand-renewable-nuclear	292
10.4.7 High-demand-BAU	292
10.4.8 High-demand-renewable	292
10.4.9 High-demand-renewable-nuclear	293
10.5 Results	294
10.6 Conclusion	295
References	296

11 Realizing clean energy for every earthling   **299**
David S-K. Ting and Jacqueline A. Stagner

11.1 A tomorrow for every earthling	299
11.2 Cleaner cooking	299
11.3 More solar	300
11.4 Microgrid	300
11.5 Green and white hydrogen	300
11.6 Other considerations	301
11.7 Moving forward	301
References	303

Index   **307**

Preface

This volume serves as a catalyst to further cleaner energy, especially in less-privileged communities. The editors, **D.S-K. Ting** and **J.A. Stagner**, introduce the chapters making up the volume in Chapter 1, "Sharing Clean Energy with the Poor." There is no doubt that realizing the United Nations Sustainable Development Goal SDG #7, "Affordable and Clean Energy," is a complex challenge. **G.T. Reader** enlightens us on this challenge comprehensively in a somewhat philosophical tone in Chapter 2, "Providing Clean and Affordable Energy for All: Possible, Practical, or Propaganda?" In addition to pointing out that SDG#7 is a prerequisite for SDG #1, "No Poverty," he highlights the dilemma that for some of our less privileged fellow human beings, affordable energy may not be clean, and clean energy may not be affordable. As it happens, many of the green energy incentives end up benefiting the upper class at the cost of the lower class. **A. Shadmand** and **S. Arslan Selçuk** furnish a review on affordable, energy-efficient housing design in Chapter 3, "Low-Cost and Energy-Efficient Housing Design: A Review on Research Trends." Based on 195 articles, they conclude that it is feasible for low-income groups to have comfortable energy-efficient housing. Energy justice, in terms of mitigating inequalities, must be dealt with to enable clean energy for all. For this, energy-efficient housing can be an enabling tool. Tapping into freely available solar energy is essential, whether we are dealing with high-, medium-, or low-income communities. **L.S. Sua** and **F. Balo** expound on a quantitative technique for picking the most appropriate commercial photovoltaic system in Chapter 4, "Enabling Solar Energy Production for Low-Income Communities." Both subjective and quantitative factors, including those unique to low-income communities, factor into the decision-making process using the Analytic Hierarchy Process among Multi-Criteria Decision-Making techniques. A solar energy case for Nigeria is presented by **S.M. Abolarin, M.B. Shitta, O.G. Oluwasanya, C.A. Eguma, A.O. Gbadegesin, E.O. Ogedengbe**, and **L. Lagrange** in Chapter 5, "Building energy efficiency improvements and solar PV systems integration." A five-bedroom duplex residential building in Lagos State is being considered for an electronic energy audit. A combination of solar photovoltaic and battery storage systems, along with adopting more energy-efficient appliances, are recommended. Many places require electricity, heat, and water simultaneously and the addition of a combined heat and power unit may be appropriate. Such a case study is performed by **R. Babaei, D.S-K. Ting**, and **R. Carriveau** in Chapter 6, "Sustainable Energy Solutions for Rural Electrification in a Low-Income Community." A sensitivity analysis that considers solar, wind, grid breakeven distance, environmental

xvi *Clean energy for low-income communities*

impact, technical performance, etc. is carried out. For the studied case, increasing solar irradiation and wind speed can lower both the cost of electricity and net present cost. Additional challenges are expected when dealing with ecologically sensitive areas. **C-S. Segura-Salas** addresses these in Chapter 7, "An Introduction to the Electrification in Remote Communities located in Ecologically Sensitive Areas: From Planning to Implementation Experience." The feasibility of clean energy implementation varies around the world and ecological differences are a factor. Experience gathered from electrification studies in a vast, ecologically sensitive remote region of approximately 90,000 km^2 of Brazil is shared. Controlled-environment greenhouses are a means to improve food security, especially for remote communities located in unfavorable climate regions. Solar energy can empower this energy-intensive operation, mitigating the usage of dirty energy. Exploiting more solar insolation can be as easy as adjusting the greenhouse orientation and the shape of the roof, as disclosed by **G. Khanuja**, **R. Ruparathna**, and **D.S-K. Ting** in Chapter 8, "Enhancing Solar Insolation in Agricultural Greenhouses by Adjusting its Orientation and Shape." To realize clean energy for all, we must tap into biofuels. **K. Kang**, **S.Q. He**, and **Y. Hu** enlighten us on recent advances in biofuel production in Chapter 9, "Recent Advances in Biofuels Production: Industrial Applications." To further biomass as a renewable energy source, a better understanding of thermochemical and biological pathways from biomass to biofuels, along with practical challenges, is essential. Insights into the production and application of biofuels across different industries are provided. In Chapter 10, "Modeling and Forecasting the Energy Mix Scenarios for Türkiye via LEAP Analysis," **F. Gökgöz** and **F. Filiz** employ the Long Emissions Analysis Platform (LEAP) tool to quantify and forecast the composition of the electricity generation mix in Türkiye. Nine scenarios of varying future electricity demand are modeled and analyzed. These are: Low-Demand-Business as Usual (BAU), Low-Demand-Renewable, Low-Demand-Renewable-Nuclear, Base-Demand-BAU, Base-Demand-Renewable, Base-Demand-Renewable-Nuclear, High-Demand-BAU, High-Demand-Renewable, and High-Demand-Renewable-Nuclear scenarios. The volume wraps up with Chapter 11, "Realizing Clean Energy for Every Earthling," by **D.S-K. Ting** and **J.A. Stagner**. In addition to the many revelations disclosed in the preceding chapters, additional considerations, including hydrogen, possible impacts on other SDGs when cleaning energy, retaining some amount of fossil energy for the near term, and further community and cultural involvements are also highlighted.

Acknowledgments

This book could not have materialized without the unfailing striving of many punctilious individuals. The editors are forever indebted to the outstanding experts who fashioned the first-class chapters. A heartfelt "Thank You" goes to the reviewers who executed quality control. The editors truly enjoyed working with the fantastic IET publishing team. Above all, providence from above made this endeavor possible.

About the editors

David S-K. Ting is a professor in the Mechanical, Automotive and Materials Engineering Department at the University of Windsor, Canada. His research interests include flow turbulence, aerodynamics, energy conversion and management, and energy storage.

Jacqueline A. Stagner is an associate professor in the Mechanical, Automotive and Materials Engineering Department at the University of Windsor, Canada. Besides her research on renewable energy, energy storage, and climate change mitigation, she also coordinates the undergraduate programs.

Chapter 1

Sharing clean energy with the poor

David S-K. Ting[1] and Jacqueline A. Stagner[1]

The challenges confronting humanity must be overcome with solutions that work for the entire human race, or they will not be efficacious. As more advanced countries continue to gain ground in expanding renewable energy, efforts should also be invested in helping our underprivileged neighbors clean up the energy they use. Only with such care and collaboration can we clean up our habitation, Earth. This volume aims to make a step forward in furthering the furnishing of clean energy to poor, remote, and/or isolated communities. Solar energy, biofuels, wind, integrated energy systems with batteries, and energy-efficient housing are all enablers. These technologies on their own would not work unless the haves were willing to share their clean energy with the have-nots.

Keywords: Energy; Poor; Clean; Renewable; Remote; Isolated

1.1 Why share clean energy with the poor?

We are all in the same boat, and that boat is called Earth. To save the boat from sinking, we must continue to strive for cleaner energy for ourselves and, as importantly, for less privileged earthlings. It is time to put more effort into helping our poorer neighbors by providing them with life-supporting energy and replacing unclean energy with less polluting alternatives. Pollution knows no borders, and thus, there will not be a clean tomorrow as long as a significant portion of the population has no access to clean energy. Some of our neighbors may not be financially poor, but they are isolated from power grid networks and, thus, are clean-energy poor. For example, many small Mediterranean islands in well-developed Europe are neither connected to the central grid nor furnished with clean energy locally. That being the case, they hinder the European Union's clean energy transition [1]. Not helping these and other neighbors of ours will enable the damage brought about by unclean energy usage to linger, spread, or escalate, especially in developing countries where populations are growing rapidly.

---

[1]Turbulence & Energy Laboratory, University of Windsor, Canada

2 *Clean energy for low-income communities*

1.2 A complex challenge

Without a doubt, clean energy is a complex challenge because energy is not an independent entity. There are many factors associated with energy usage. Corsini *et al.* [2] present the situation in Nigeria, where important steps have been established by policymakers and measures executed to switch over to clean energy. The rapid growth in population and the consequential food insecurity, however, seriously hamper the realization of energy cleaning. It is suggested that environmentally friendly agricultural practices should go hand in hand with the transition from fossil fuels to renewable energy. This is but one example illustrating the complexity of the problem.

1.3 A step in the right direction

This volume contains, albeit non-exhaustive, the state of the art and many facets that we should consider and focus on in furnishing green energy for less privileged, remote, and/or isolated communities. It is comforting to see significant advancements in green and clean energy in various places. This progress in advanced nations ultimately translates into improved, higher-efficiency renewable energy technologies. Despite what preceded, the advancement cannot be duplicated universally. In other words, greening the energy systems of privileged populations is a big challenge. The challenge is much greater to do the same for the less privileged communities. To say the least, a strong local currency (economy) and political will are essential ingredients for delivering clean energy to a community.

G.T. Reader cuts to the chase and confronts us with the million-dollar question in Chapter 2, "Providing Clean and Affordable Energy for All: Possible, Practical or Propaganda?" Reader argues that the United Nations Sustainable Development Goal (SDG) to eventuate universal access to affordable and clean energy, SDG #7, is a prerequisite for SDG #1, alleviating poverty. He points out the dilemma that for some of our less privileged earthlings, affordable energy may not be clean, and clean energy may not be affordable. We are reminded of heavily subsidized solar panel installations, affordable only to the haves, leaving the have-nots in developed countries struggling with the heightened electricity cost, paying for the tariff handed over to the well-off. Arguably, this kind of political intervention has worsened energy poverty in advanced grid-connected communities, let alone remote, isolated, or poor communities. After more than 50 years of striving, there remain questions regarding the definition of "poverty," "clean," "affordable," and other terms. Reader ends the chapter by quoting Bob Iger, "What I have learned over time is that optimism is a very, very important part of leadership. However, you need a dose of realism."

Everyone needs shelter. If there is such a thing as human rights, an affordable house for every living soul must rank higher than clean energy. **A. Shadmand** and **S. Arslan Selçuk** present "Low-Cost and Energy-Efficient Housing Design: A Review on Research Trends" in Chapter 3. They highlight the sizable share of energy usage in the built environment. To that end, there is plenty of room for energy conservation via proper construction. Much can be done on the building aspect to provide thermally comfortable housing that requires minimal energy usage. They identify areas for further development by drawing from 195 articles.

While the amount varies, every corner of the Earth is bestowed with solar energy. Because of that, solar power is a priority for generating clean energy, and solar energy is particularly well suited for isolated communities where sunshine is abundant. Proper matching between solar resources, power demand, and photovoltaic systems is required to reduce waste of energy and capital. These are expounded in Chapter 4 by **L.S. Sua** and **F. Balo**, "Enabling Solar Energy Production for Low-income Communities." Characteristics of various commercial photovoltaic panels/ systems are evaluated using the Multi-Criteria Decision-Making technique to deduce the system with the maximum warranted energy output.

A case study invoking a five-bedroom duplex residential building in Nigeria is performed by **S.M. Abolarin**, **M.B. Shitta**, **O.G. Oluwasanya**, **C.A. Eguma**, **A.O. Gbadegesin**, **E.O.B. Ogedengbe**, and **L. Lagrange** in Chapter 5, "Building Energy Efficiency Improvements and Solar PV Systems Integration." A solar photovoltaic system plus storage batteries is employed to power lighting, air conditioning, and appliances. Based on an energy audit and data analysis, measures such as utilizing more energy-efficient appliances and switching them off when not in use are recommended. In addition to substantial energy savings, these measures also aid in the proper sizing of the required energy generation systems.

To optimize economic performance, more than solar photovoltaics should be considered in the energy mix. This is especially the case when the simultaneous provision of electricity, heat, and water is needed. **R. Babaei**, **D.S-K. Ting**, and **R. Carriveau** present such a case in Chapter 6, "Sustainable Energy Solutions for Rural Electrification in a Low-Income Community." Photovoltaic, wind turbine, combined heat and power units, battery, and brackish water reverse osmosis desalination technologies are integrated to meet the needs of Sar Goli village, Iran. The grid breakeven distance, environmental impact, and technical performance are concurrently considered. Ultimately, the cost of electricity and the net present cost are central in deciding the most viable combination. Increases in solar irradiance and wind speed can appreciably lower these costs.

The challenge of electrification becomes more severe in ecologically sensitive areas. This is the topic of Chapter 7, "An Introduction to the Electrification in Remote Communities Located in Ecologically Sensitive Areas: From Planning to Implementation Experience." In this chapter, **C-S. Segura-Salas** details the added challenges associated with providing electricity within continental zones, which are ecologically fragile. A conventional grid system is compared with microgrids and standalone systems via a multi-criteria heuristic method. Also disclosed are implementation, operation, and maintenance challenges, and these are discussed in terms of consumer satisfaction and financial stability. The evaluation makes use of field data and geospatial analysis of key decision variables.

There is no question that more affordable food can help alleviate a wide range of issues associated with low income. Controlled agriculture greenhouses are crucial in providing produce to communities located in cold climate regions.

Consequential transport-related emissions can be mitigated, in addition to promoting local employment, economics, and production. Maximizing the utilization of solar insolation is key to lowering greenhouse energy usage and simultaneously promoting photosynthesis. This leads to more delicious and affordable fresh local produce. **G. Khanuja**, **R. Ruparathna**, and **D.S-K. Ting** delineate "Enhancing Solar Insolation in Agricultural Greenhouses by Adjusting its Orientation and Shape" in Chapter 8. For southwestern Ontario, Canada, aligning the longer dimension of the greenhouse at 30° north of east is best for capturing the maximum solar radiation. A southerly roof inclined at 60° can harness the maximum solar radiation during the winter.

To maintain a good standard of living, the industrial sector must be sustainably fueled. In Chapter 9, "Recent Advances in Biofuels Production: Industrial Applications," **K. Kang, S.Q. He**, and **Y. Hu** enlighten us regarding the important role of biofuels in the global transition to clean energy. They reveal the advantages and challenges associated with combustion, torrefaction, pyrolysis, and gasification processes, especially in low-income communities. Anaerobic digestion, a viable means to take care of organic waste, including food, has been proven to be a viable way to produce biofuels. This chapter clearly demonstrates that biofuels can foster a sustainable and decarbonized economy.

The Long Emissions Analysis Platform (LEAP) tool is utilized by **F. Gökgöz** and **F. Filiz** to quantify and forecast the composition of the electricity generation mix. Nine diverse scenarios are modeled and analyzed in Chapter 10, "Modelling and Forecasting the Energy Mix Scenarios for Türkiye via LEAP Analysis." The versatility of LEAP is illustrated in the state of Türkiye. This study provides a roadmap for developing a low-carbon energy future. It highlights the importance of integrating quantitative methodologies into energy planning and policy formulation.

These chapters disclose the status of clean energy, with an emphasis on sharing it with underprivileged, low-income, and remote communities. While there remain outstanding challenges, significant progress has been realized in recent years. More importantly, lessons learned from both failures and successes enable us to review our position and tread forward more effectively. We are bestowed with abundant renewable energy. There is more than enough clean energy for every earthling to savor. It makes sense for the haves to share cleaner energy with the have-nots in low-income communities. Only when all neighborhoods are clean can there be a bright tomorrow.

References

[1] G. O. Atedhor, "Greenhouse gases emissions and their reduction strategies: Perspectives of Africa's largest economy," *Scientific African*, 20: e01705, 2023.

[2] A. Corsini, G. Delibra, I. Pizzuti, and E. Tajalli-Ardekani, "Challenges of renewable energy communities on small Mediterranean islands: A case study on Ponza island," *Renewable Energy*, 215: 118986, 2023.

Chapter 2

Providing clean and affordable energy for all: possible, practical or propaganda?

Graham T. Reader[1]

The overarching goal of the 2015 United Nations Sustainable Development Agenda is global poverty eradication. There is general international agreement that a necessary prerequisite for alleviating poverty is universal access to *clean and affordable energy*, which is also an Agenda goal. Affording clean energy requires consumers to have the financial means to purchase sufficient amounts for the intended purpose. In the case of the general population, this means heating or cooling their living space, providing some lighting after sunset, and cooking food. Therefore, energy affordability and poverty are closely associated, in a similar fashion to the relationships between general poverty and income, and between affordability and cost. But, for some people, affordable energy may not be clean, while clean energy may not be affordable. Presently, unclean energy sources supply at least three-quarters of global energy consumption. Is it economically and technically plausible that they can be wholly replaced by clean sources by 2030?

Moreover, even if the colossal amount of capital investment required for this energy source transition were forthcoming, could the necessary clean energy infrastructure be built in a few short years? Could such investments ensure that the somewhat hyped promise of cheap, clean energy is accessible for all future consumers sooner, rather than later? Is mitigating anthropogenic climate change compatible with sustainable development? If not, what is more important: the cleanliness of energy or its cost, or the temperature of a household or the surface temperature of the planet? Exactly what is meant by eliminating energy poverty, and providing access to clean and affordable energy, is discussed in this chapter, along with the plausibility of ensuring that both are achieved. The discussions are focused largely on household energy needs and the impacts of poverty, population growth, energy costs, and energy source transitions, with particular regard to access to electricity.

Keywords: Affordability; Clean energy; Electricity access; Poverty

---

[1]Department of Mechanical, Automotive and Materials Engineering, Faculty of Engineering, University of Windsor, Canada

6 *Clean energy for low-income communities*

2.1 Introductory remarks

The title of this chapter poses a question for which, arguably, precise definitions of 'clean' and particularly 'affordable' should prove prudent, along with established measures of their determination. The terms are prevalent in matters dealing with sustainable development, anthropogenic climate change[1], and many other societal issues impacting the financially disadvantaged [1–3]. Despite the common use of the terms, there is a frequent lack of rigour in their application, coupled with somewhat ambiguous explanations of their meaning, as discussed in Section 2.2 with respect to affordable energy and poverty, and in Section 2.3 with regard to clean energy. Without any doubt, for those individuals living in poverty, it is inevitable that not all the acknowledged essentials of life, i.e., sufficient food, clean water, and the right to adequate shelter, will be affordable to the extent necessary. Arguably, only the elimination of poverty can resolve these dilemmas. Many national governments have sought to tackle the poverty problem by defining and legislating a minimum wage, while others, especially the United States (US), have developed many welfare and anti-poverty programs [4–6]. Poverty is usually associated with individuals, families, and households rather than public and private institutions, industries, and commercial enterprises, although it could be argued that corporate bankruptcy is a kind of poverty and corporate subsidies are a form of government welfare. However, the focus of the discussions in this chapter concentrates on households.

All such poverty reduction endeavours have met with varying degrees of success and failure, as gauged, by the opinions of politicians, government analysts, and 'independent' experts. For example, as shown in Figure 2.1, by 2022, the US had over 130 social welfare anti-poverty programs, with annual Federal and State spending of approximately US\$1.8 trillion[2], yet the official poverty rate was about the same as it had been over 50 years earlier [5]. This considerable financial outlay, which questionably has had little real impact on a per capita basis, is equivalent to the total combined Gross Domestic Product of Indonesia and South Africa which, taken together, have a comparable population to the US [7]. Moreover, the amount the US spends on trying to reduce the share of its population living in poverty is greater than the national GDPs of over 160 global countries [7]. However, as global incomes vary substantially, so does the 'cost-of-living', thus, what may be considered poverty in one country will be different from another. The World Bank, for example, defines 'extreme poverty' as having an income of less than \$1.90 per day, based on 2011 prices, or \$2.15 per day using 2017 prices, but a country like Sweden sets its poverty line at \$30 per day, and many similarly rich countries, such as the US, have approximately the same defined level [8]. Yet, while the world has

---

[1]In the remainder of this chapter the term 'anthropogenic climate change' is regularly shortened to 'climate change' in harmony with the 1992 United Nations Framework Convention On Climate Change [UNFCCC] [1].

[2]When dollar figures are given in this chapter, they refer to what are called International United States dollars.

Providing clean and affordable energy for all 7

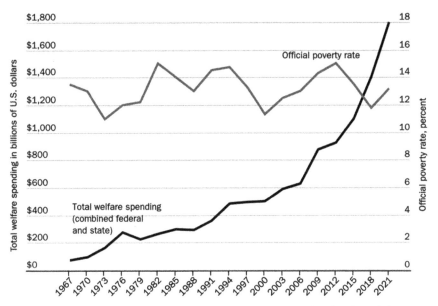

Sources: U.S. Census Bureau, Number in Poverty and Poverty Rate: 1959 to 2020 § (2021); Michael D. Tanner,

Figure 2.1   Welfare spending compared to poverty rate in the US 1967–2021 [5]

historically managed, almost since the 1st industrial revolution at least in percentage terms, to hugely reduce extreme poverty, the trends to achieve the Swedish-type levels have been far less impressive given that only 15% globally have per capita incomes of over $30 per day, i.e., $10,950 annually or just over £8,660[3], as illustrated on Figure 2.2 [8].

Does this situation suggest that it is easier to tackle poverty in poorer countries? Based on the $30 per day yardstick, there are few people in rich countries living in general poverty, so why are many people in these rich countries suffering from fuel poverty? Is the frequently mentioned adage that the 'rich are getting richer and the poor are getting poorer' at the core of the fuel poverty problems in high-income countries, particularly when fuel costs increase? A recent review and analysis of income inequalities in 15 rich countries over the past 30 years indicate that the gaps between the affluent and the remainder in some societies are increasing, but in others, they are decreasing [9]. In the majority of the cases considered, the share of the 'poor' in the population has decreased, but with some exceptions, such as in the US and the United Kingdom (UK), while in many other countries the portions of the so-termed 'middle-class' has decreased [9,10]. This could be interpreted both as more middle-class people becoming increasingly poorer or increasingly affluent, i.e., a middle-class 'squeeze' [11]. To some extent,

---

[3] As of 4 September 2023.

8 *Clean energy for low-income communities*

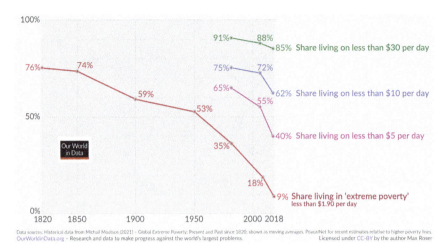

Figure 2.2   Share of population living in poverty by income poverty level [8]

the situation could portray that the affluent can afford to pay increased energy costs while the poor receive substantial help to raise them above threshold poverty, albeit without complete success, but those in the middle, especially at the lower margins, are regularly at risk of slipping into poverty.

It is apparent that, globally, there are major disparities between countries and within countries on what is considered poverty and how it is measured. Moreover, can a household be labelled as living in fuel poverty but not general poverty, i.e., is fuel poverty relative but general poverty absolute, or, within a country, are both comparative measures depending on the income distribution of the whole population and an associated median level? Does having a low income necessarily mean that energy is unaffordable? Is there a financial yardstick that determines whether energy is affordable or not, and if there is, how is the yardstick determined and by whom? Governments do not usually dictate primary energy costs, but they can impact them by implementing particular policies, including some forms of taxation and targeted subsidies. However, it does seem that there has been more political focus on equating income, earned and supplemented, with poverty, presumably in the conviction that eradicating poverty will also ensure the affordability of essential human needs. But are the goals of the elimination of poverty and the simultaneous achievement of affordable energy really compatible? Furthermore, in the case of fuel poverty, if it has not been possible to end it for all after decades of costly efforts with the present energy mix, how can this be accomplished by a transition to clean energy, and, as always, will there be winners and losers in the process [12]?

There have been several questions posed in the preceding opening remarks of this chapter. The search for answers is described in the remaining sections, starting with a necessarily short synopsis of how the terms 'poverty', 'affordable energy', and 'clean energy' are defined by various national and international organisations and how these definitions appear to motivate government policies. The contents of this chapter are aimed, primarily but not exclusively, at a general post-secondary readership. However, those who study and research poverty topics, e.g., political scientists, social scientists, and economists, will be largely familiar with the contents of Section 2.2 and may choose to go directly to Section 2.3.

2.2 Affordable energy, poverty and fuel poverty

Currently, the term 'affordable' is used in all types of political rhetoric by local, national, and international political leaders, as well as many researchers and scholars in academia. However, these are not wholly new occurrences since the term 'affordability' became increasingly common in social policy discussions at least three decades ago, especially in connection with housing [13]. Then, as of now, a persistent challenge has been to define 'affordable' which is universally acceptable and applicable, coupled with agreed measurement methodologies that are globally consistent. The establishment of a one-size-fits-all definition continues to be somewhat elusive, although the adjective is now encountered in many connotations, notably when associated with the United Nations (UN) 'Transforming Our World' sustainable development agenda [14]. These instances include 'affordable housing', 'affordable care', 'affordable food', 'affordable drinking water', and, of course, 'affordable energy', the latter being one of the main focus of the discussions in this chapter. Although a globally acceptable definition of affordability is yet to be formulated, is there a generic and generally understandable meaning of 'affordable'?

Many of the standard dictionaries all have slightly different definitions, such as *'having a cost that is not too high'*, '[affordable housing is] *able to be bought or rented by people who do not earn a lot of money'*, '[something] *that can be afforded; believed to be within one's financial means'* [15–17]. All of which are tantamount to being blinding glimpses of the obvious. Arguably, the statement provided by the US Department of Energy [USDOE] embodies a more practical working description of affordability as it applies to domestic users of energy, i.e., *'Energy affordability is the idea that consumers should be able to pay for their home electricity use—lighting, heating, cooling, powering appliances—while also paying for other basic living expenses, such as food and medication, without having to choose or feel overburdened'* [18]. In the UK, the term 'fuel poverty' is used, which has a somewhat similar description to that of the USDOE, i.e., '[fuel poverty is a] *household's inability to achieve thermal comfort to levels commensurate with a healthy standard of living at a reasonable cost'* [19]. The quoted phrases are sufficiently vague in meaning that they could, and likely do, provide copious opportunities for somewhat self-serving appraisals of achievements. For example, what do the words and phrases 'reasonable', 'basic', 'feel overburdened', and so on really mean in practice? Furthermore, if they can be measured, who sets the benchmark? These queries and observations are not new, and they have been constantly raised by social scientists and economists for at least 40 years, e.g., [20]. Definitive answers continue to remain elusive.

10   *Clean energy for low-income communities*

Nevertheless, such descriptions are useful in helping to understand the concept of affordability, or at least governmental notions of affordability, but more prescriptive or potentially measurable definitions would have greater value. The US and UK definitions, along with the use of similar terms such as 'energy insecurity' and 'energy poverty' by other jurisdictions, while having differences in detail, do have a core commonality in that the affordability of energy or fuel should not be decoupled from the cost of other human essentials, including health and mental well-being and the countless vagaries of living in extreme poverty [21–26]. Moreover, although instinctively, the connection between poverty and affordability is obvious, characterising the precise nature of the linkages between them remains a challenge, especially for policymakers [24,27]. Literature reviews have found that governmental measures taken to tackle affordability seldom seem to take into account the vagaries of how energy poverty is or should be measured [28,29]. These may be somewhat over-critical conclusions since meticulous, if not wholly successful, attempts to define a measurable metric for fuel or energy poverty have been made as a distinct form of poverty in the UK, or as part of the general issue of societal poverty, which is still the case in the US [23,29–31].

For example, the Census Bureau of the US has been publishing various versions of poverty data since 1959 [32]. These data have been used by successive governments and lawmakers to formulate interventions, basically various types of income support, that are frequently appraised to assess their efficacy. One of the outcomes has been to use three different metrics in determining poverty: 'Absolute Poverty', a measure of the minimum needs to afford minimal standards of essentials such as food, water, shelter, etc., but not specifically energy; 'Relative Poverty', an appraisal of income inequality within a societal national framework, and 'Supplementary Poverty', an adjusted rate used by the Census Bureau to account for the effects of government actions in reducing poverty [33,34]. Arguably, the measures taken have had a discernible impact in that poverty rates of over 20% in the early 1950s were reduced to a supplementary poverty rate of 11.7% by 2019, pre-pandemic. However, in terms of actual headcounts, there are now as many impoverished individuals as there were at the start of the 1960s, and the success of the multitude of government support schemes in reducing numbers has been problematic (Figure 2.3), but during the same time period, the official population has doubled and legal immigration has tripled [35–37]. The United States' official 'War on Poverty', consisting mainly of a series of socioeconomic measures to help those considered poor or disadvantaged, could be accurately described as effective, as shown in the diagram, justifying the then President's[4] enthusiasm for the concepts, if not the detailed measures, but their efficacy began to falter in the 1970s [38,39].

The 'oil-crises' of the 1970s, especially the 1973 decision by the Organization of Arab Petroleum Exporting Countries [OAPEC][5] to reduce their oil production

---

[4]Lyndon B. Johnson.

[5]OAPEC is a distinct group from OPEC – the Organization of Petroleum Exporting Countries – albeit some countries are members of both bodies.

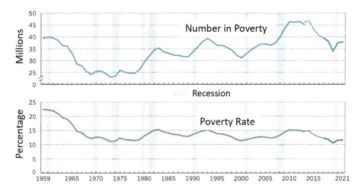

Figure 2.3   Official poverty statistics in the US 1957–2021 [37]

and, in some instances, stop oil shipments to certain countries, led to a quadrupled increase in oil prices, causing severe financial and social trauma for the citizens of many western countries [40]. The governments of the impacted countries brought in measures, some temporary, others permanent, to reduce their dependency on oil imports. In the UK, the Prime Minister of the day urged people to heat only one room in their homes due to the shortages and the escalating prices [41]. The importance of these actions and promptings was the noticeable connection between rapidly rising energy prices and the consequent deterioration of living standards for those whose income levels were below what had been perceived as acceptable pre-crisis, i.e., they were now living in poverty [42,43]. The result was the 1979 encyclopaedic and seminal book on poverty in the UK by Peter Townsen which identified a lengthy litany of the causes of poverty set against a historical background of governmental social policies, and their efficacy, to address what they and other sources perceived as poverty [43].

2.2.1   Fuel poverty benchmarks

In the 1980s, the term 'fuel poverty' appeared in the literature, along with proposed meanings such as *'the inability to afford adequate warmth at home'* [20]. However, this was still a subjective definition – what is meant by 'adequate' warmth? If 'warmth' is taken to correlate with room temperature, does each room in a household have to be at the same temperature continuously, and how do household demographics impact the value of the set temperature to ensure adequate warmth? Moreover, residents, especially those on a fixed low income, i.e., old-age pensioners, may choose not to heat their homes to 'adequate' levels to enable other uses of their disposable income. Conversely, members of similar groups may choose to forgo other essentials in order to enjoy higher room temperatures. These *'Can't Pay, Won't Pay'* situations, as described by Hancock, exasperated governments, and politicians when they attempted to assess affordability measures while acknowledging that fuel poverty was a societal issue not to be ignored [13]. However, Boardman's 1991 seminal work on fuel poverty had already caught the

12   *Clean energy for low-income communities*

attention of politicians of all persuasions in the UK, the European Union (EU), and elsewhere [44,45]. A relatively simple but reasonably measurable parameter was proposed in the works of Boardman from the mid-1980s onwards, which was to define household fuel[6] poverty as being when the cost of energy exceeded 10% of a household's income [44].

By the end of the 1990s, various UK government reports had confirmed Boardman's findings, and, after a great deal of parliamentary discussion, these were instrumental in leading to, and passing, the *'Warm Homes and Energy Conservation Act 2000'* [46,47]. A year later, the government published its policy for eliminating fuel poverty by 2016, defining a household as 'fuel poor' as measured by the Boardman benchmark but with the added codicil of homes reaching specific temperature levels [48]. Boardman's scholarly work continues to the present day, and although rightly celebrated for the 10% rule, she has proposed a number of possible solutions to the challenges of fuel poverty, particularly in terms of the energy efficiency of buildings [22,49]. The use of the 10% income yardstick became standard practice in many countries well into the 21st century, although the '10%' was not consistently defined, with a variety of national interpretations being used [50–52]. Moreover, since its advent, the application of the 10% benchmark, and its use in social policies, have become increasingly problematic. There are many reasons for this concern, including a growing appreciation of the nexus between general poverty, fluctuating fuel and energy prices, human behaviour, and the energy performance of buildings [19,22,53]. Additionally, the societal dimensions of poverty, especially the distinct, but sometimes ephemeral, contributions of fuel or energy poverty, are becoming increasingly understood, such as social inclusion, i.e., participation in society, social justice, and deprivation [43,54–56].

Since the introduction of the Boardman benchmark, successive UK governments, especially since the start of the millennium, have demonstrated an international leadership role in investigating, qualifying, and quantifying fuel poverty [50,51]. In England, the 10% point of reference was used for many years in conjunction with a definition of a living room temperature of 21°C as being an 'adequate standard of warmth'[7] [52]. Eventually, for the reasons outlined in the preceding paragraph, it was understood in the UK that a more all-encompassing approach was needed to determine fuel poverty and the subsequent actions required to eliminate, or at least alleviate the social problem. While this realisation was originally manifest in 'The UK Fuel Poverty Strategy' which acknowledged that fuel poverty was caused by *'a combination of* [3] *factors including, energy efficiency of the home, fuel costs and household income'*, the actions taken since 2001 based on the 10% metric alone were not sufficient to meet the 2016 eradication target as demonstrated by the trends in the largest populated country of the UK, i.e., England[8], as the number of fuel poor households increased from 1.2 million to

---

[6]In the UK it is apparent that the terms 'fuel' and 'energy' are used interchangeably on a regular basis.
[7]A temperature recommended by the World Health Organization.
[8]Fuel Poverty is a devolved policy area in the UK, with England, Scotland, Wales, and Northern Ireland having slightly different policies and measurement methodologies [57].

Providing clean and affordable energy for all 13

4.0 million by 2009, albeit with a drop to 3.5 by 2010 [52,57]. Perhaps this should not have been too much of a surprise, as fuel expenditure had increased by 90% between 2003 and 2010, although average incomes only rose by 24% [57,58].

However, it needs to be noted that the fuel expenditure amounts were determined by the use of the Building Research Establishment Domestic Energy Model [BREDEM], rather than actual spending, for buildings old and new in a so-called Standard Assessment Procedure (SAP) [59–61]. The SAP methodology is used to estimate the annual energy consumption for providing a dwelling's space heating, domestic hot water, lighting, and ventilation at defined levels. The results are then used to quantify the dwelling's 'performance' in terms of energy use per square foot, an environmental impact rating expressed as the level of carbon dioxide [CO_2] emissions, and a fuel-cost-based energy efficiency rating [60]. The latter parameter is the 'SAP' rating. Clearly, the system for assessing fuel poverty was becoming increasingly convoluted, statistically demanding, and very dependent on detailed household and building data.

2.2.1.1 Changing methodologies in the UK

Thus, although in the opening decade of the present millennium, the UK government developed a myriad of measures and surveys to try and quantify and qualify fuel poverty to gauge and guide their policies, the annual data trends remained disappointing, raising many questions [62,63]. Why were the actions taken to eliminate fuel poverty not working? Was the methodology used to determine fuel poverty inadequate? Were the levels of fuel poverty really as high as the data indicated? Likely the political optics and genuine concern for the fuel poverty situation were instrumental in a government-commissioned investigative report in March 2011, referred to informally as the 'Hills[9] reports'. The Hills investigation was to conduct an independent review, *'from first principles'* of fuel poverty and the way it is measured, including *'the extent to which fuel poverty is distinct from poverty, and the detriment it causes'* [64]. The reports were published in two parts, an interim version in 2011 and an extended version in 2012 [64,65]. The extended version also included invited feedback from the earlier report.

Following a comprehensive investigation addressing a substantial compendium of poverty factors and influences, including: income levels, fuel costs and types, dwelling ages and sizes, population demographics, methods of fuel bill payment, building climate change emissions, and their interrelationships, together with observations of the efficacy of poverty definitions and government policies, Hills highlighted that fuel poverty was a distinct and serious problem but that the prevailing methodologies for measuring fuel poverty were oversensitive to fluctuating energy costs [48,65]. Of the six recommendations, two involving new metrics were recommended, the *'Fuel Poverty Gap'*, for determining the depth of fuel poverty issues, and the *'Low-Income High Cost (LIHC)'* methodology for measuring fuel poverty among households and individuals. Using the LIHC methodology, Hills demonstrated (Figure 2.4) that from 2007 onwards, the number of

---

[9]Professor Sir John Hills of the London School of Economics.

14 *Clean energy for low-income communities*

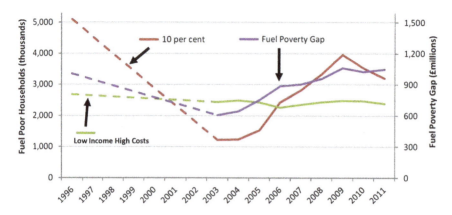

Figure 2.4 Fuel poverty in the UK using LIHC and 10% benchmark 1996–2011 [64]

households suffering from fuel poverty would be determined to be lower than that measured by the 10% benchmark [64]. When the UK government applied the LIHC to 2010 poverty data, the result was a reduction in the percentage of fuel-poverty households to just 12% from 16% using the 10% indicator [58]. This confirmed the Hills LIHC methodology and no doubt it was a welcome outcome for the government.

However, the second metric, 'the fuel poverty gap', a measure of how much a household's spending on fuel would need to reduce so that it would no longer be classified as suffering from fuel poverty, was forecast to keep increasing without the government formulating an *'ambitious strategy for tackling fuel poverty'* [48]. In essence, one indicator would be based on income, while the other, the gap, would be based on relative costs, but the latter would not be simply a measure of fuel prices but how well the fuel was used in a particular household, e.g., a well-insulated home would require less energy to achieve affordable and adequate warmth than one with little or no insulation, and the same situation would be reflected in the use of appliances that were more 'energy efficient' albeit providing comparable services such as refrigeration for food storage. To determine whether a household was 'fuel poor' would then require detailed knowledge of a household's income and its associated fuel costs to maintain a defined adequate level of warmth, along with the other energy requirements such as water heating, lights, appliances, and cooking. The energy costs required for adequate warmth would be modelled using the BREDEM software, as previously mentioned, rather than a household's actual spending. This was meant to ensure that the data from those households where the occupants purposely restricted the room temperature below the prescribed 21°C to lower their actual energy bills was excluded [52,66]. Moreover, as there is no such thing as a standard dwelling, to calculate or measure the household energy requirements, one would need data regarding domicile size, the number of occupants, its 'energy efficiency', and what fuels were used and in what proportions. Eventually, the Hills recommended LIHC, and fuel poverty gap approaches were largely adopted so that household fuel poverty would be determined based on income, energy requirements involving SAP ratings, and fuel costs.

The adoption of LIHC, in England, led not only to a number of continental European nations formulating similar schemes for measuring fuel poverty but also proved to be a stimulus for scholarly research into the many aspects of the subject [50,51,67,68]. The Hills LIHC approach embodied a considerable amount of data but was, perhaps, not as readily understood by the layperson because of its complexities, especially surrounding thresholds for incomes and costs as compared to the 10% income yardstick. Probably, to make the LIHC system more transparent in terms of fuel poverty, a now well-known graphical 'quadrant' depiction was formulated by Hills and used in many subsequent UK government publications, e.g., the 2016 version (Figure 2.5 [66]). As with all attempts to develop policies addressing societal fuel poverty based on statistical formulations, problems were encountered, for example, such as the issue of 'fuel poverty churn' [69]. This phenomenon was caused by households moving in and out of fuel poverty as incomes and costs and their defined thresholds changed. This made it and still makes it difficult to create and implement targeted policies to help those in need since, as the benchmarks can change, so can those who are deemed eligible or not for some form of income support. It appeared to the government from a comparison

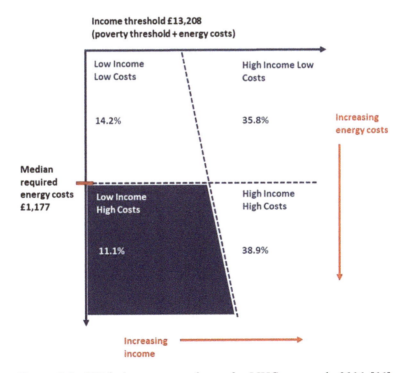

Figure 2.5   UK fuel poverty quadrants for LIHC approach, 2016 [66]

16   *Clean energy for low-income communities*

of their data collections that the number of households annually moving into fuel poverty was very similar to those no longer fuel poor, so while the level of the LIHC indicator remained sensibly stable year-over-year, the identified households were not the same. The government attempted to model this churning to project future household status changes [70].

Nevertheless, it should be appreciated that the descriptions of this and the preceding paragraph represent only a *scratching-the-surface* commentary on the complexities of the Hills recommendations and their applicability to fuel poverty solutions and any subsequent application of the LIHC approach. There is a vast amount of literature on the subject and for those wishing to discover more about the approach, a useful starting point would be the final LIHC methodology handbook published in 2020 prior to the introduction of an updated approach, at least in England, the Low-Income Low Energy Efficiency (LILEE) indicator [66]. The latest indicator has been introduced in England but not in other parts of the UK and it no longer includes the 10% threshold [71,72]. The LILEE approach identifies fuel poverty as a distinct form of poverty and is specifically aimed at providing energy efficiency support for low-income households, separate from any other programmes for reducing income poverty. Households are considered fuel-poor if they have low incomes, and their dwelling is assessed as having low energy efficiency [71]. While these two criteria have their origins in the LIHC scheme, the definitions for income and energy efficiency have been updated and extended.

For instance, 'residual' income is determined by deducting housing, fuel, and other costs from net income and comparing the amount to a prescribed poverty line[10] as illustrated in Figure 2.6, which also shows the different methods used by the individual countries of the UK [71,72]. The energy efficiency determination uses a slightly modified SAP system, mentioned earlier, known as the 'Fuel Poverty Energy Efficiency Rating (FPEER)' methodology, which also takes account of '*the impact of policy interventions that directly affect household energy costs*', to provide an energy efficiency rating, a numerical 0–100 scale divided into eight incremental bands, i.e., A (the highest) to G (the lowest). Those households assessed as being in the lower four lower bands, D to G, are considered to be of low energy efficiency, and if they are also determined to have incomes below the poverty line, see Figure 2.6, only then are they identified as being fuel poor [71–73]. Consequently, households with high energy efficiency ratings, C to A, are no longer considered to be fuel-poor, regardless of income. Arguably, characterising household energy efficiency as the major factor in determining fuel poverty and not income is a thought-provoking change.

Most certainly, the application of the LILEE system in England indicates that there are far fewer households, 3.5 million, suffering from fuel poverty than would be the case using the 10% rule, 8.8 million [57]. The use of different fuel poverty methodologies for Scotland and Northern Ireland, summarised in Figure 2.6, yields fuel poverty rates of 25% and 24%, respectively compared with approximately 13% of households in England [57]. Yet, in terms of poverty measured by

---

[10]In England taken to be 60% of the national median of disposable income.

Providing clean and affordable energy for all 17

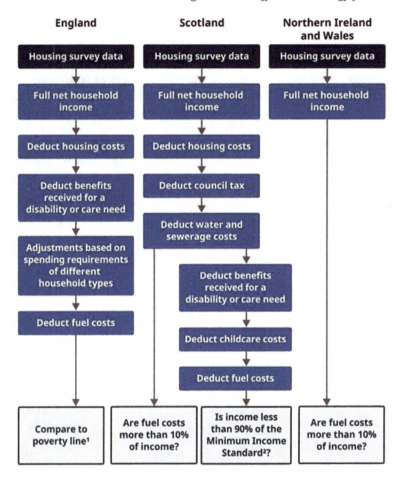

Figure 2.6 How 'Residual Income' is determined in UK Countries [71,72]

disposable incomes, the rates of the different jurisdictions are almost identical [74]. These data suggest that across the UK, poverty rates are very similar, but that fuel poverty in England is slightly more than half of what it is elsewhere. Has the government only found a way to reduce fuel poverty in England? Does the LILEE system markedly underestimate the number of fuel-poor households, or does it provide a way to identify and target only those the government considers to be in the most need? Being able to report that the number of fuel-poor households is decreasing using the new system would be attractive to any government and its policies, as it demonstrates to taxpayers that money is being well spent to reduce fuel poverty. However, the implementation of the LILEE system, which is highly dependent on accurate building and household data, is still too new to reveal its hoped-for value, but concerns have already been expressed about its central rationale [27,75,76].

18 *Clean energy for low-income communities*

2.2.2 Poverty measures and sustainable development

Although the preceding discussion concentrated on the continuing efforts of British governments to address fuel poverty, the same type of concerns, analyses, and policy actions have taken place in other countries and regions such as the EU and the US [77–79]. As in the UK, the EU poverty measures are also not wholly dependent on household income, for example, identifying 13 'deprivation' factors[11], and classifying members of the population who cannot afford at least seven of these components as being at risk of poverty or social exclusion. The status of those people with disposable incomes below the 'at risk' poverty thresholds is similarly grouped. In 2022, over 95 million people in EU countries, 21.6% of the total population, were labelled as being at risk, as shown in Figure 2.7 [79]. However, the at-risk poverty levels are not reflected in the 2022 EU energy poverty data, showing that *'the number of people who were unable to keep their homes adequately warm'* was 9.3%, albeit a concerning rise from the 6.9% a year earlier, which has motivated the EU to take new initiatives to tackle energy poverty [80]. Fuel poverty in high-income natural energy-rich countries like Canada whose population is about a tenth of that of the neighbouring US (according to authoritative international definitions) and has total access to electricity and clean cooking fuels, still uses the definition of a household experiencing fuel poverty when it spends more than 10% of its income on utilities, with the latest published government data estimating that 8% of Canadians are in this predicament, with variations

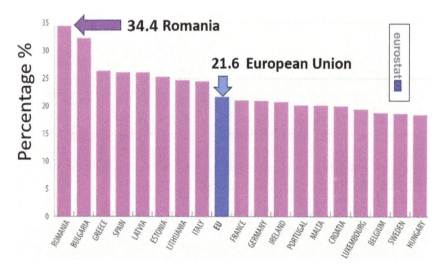

Figure 2.7 Exemplars of European Union populations' 'at risk' of poverty or social exclusions [79]

---

[11]Items "that are considered by most people to be desirable or even necessary to lead an adequate quality of life" [79].

Providing clean and affordable energy for all   19

across its provinces between a low of 6% and a high of 13% [81–83]. Non-government sources have argued that the country's average rate is much higher and that a more thorough approach would be timely [84,85]. In general, until fairly recently, Canada's federal government has not focused on addressing societal poverty to the same extent as its partners in the 'G7' group of rich nations, only in 2018 announcing its first poverty reduction strategy [86].

In their attempts to link the affordability of human necessities – food, water, housing, and related energy needs – with poverty, governments have concentrated on defining and measuring poverty in general and fuel poverty in particular. However, despite all the meticulous analytical and data collection work, especially in this century, on defining, measuring, and quantifying fuel poverty and, to a lesser extent, its relationships to affordability, it appears that the 10% benchmark or similar income-based metrics remain widespread in fuel poverty determinations. Moreover, the establishment of quantifiable, measurable linkages between affordability and poverty and between government actions taken to reduce poverty of all types and affordability remains, at best, a work-in-progress. Nevertheless, in most countries, there is an implicit recognition that the causes and effects of fuel poverty are multifaceted, involving economic, technical, and social issues, such that defining, measuring, and addressing fuel poverty is a complex issue that is still not wholly understood [29,30]. Since the 1960s and 1970s, a number of countries, the US and the UK in particular, along with the World Bank Group[12], have increasingly focused on compiling data on poverty and trying to understand how the problems of poverty can be addressed and *preferably* eliminated [88]. How then can access to affordable energy be gauged if 'affordability' is so difficult to measure, even in rich countries, especially with respect to systemic poverty?

The UN, with the endorsement of its member states, has established sustainable development target indicators involving proxies such as the proportion of a nation's population 'access to electricity' as part of a measure of energy affordability [14,89]. There are 17 SDGs, and for each goal, there are a number of targets, national and international, and indicators associated with them. Each of the 231 unique indicators has a custodian agency or agencies responsible for compiling and verifying national data, developing international standards, and recommending monitoring methodologies. All the gathered information is then submitted to the United Nations Statistics Division (UNSD). The SDG particularly pertinent to the topic of this chapter is goal 7, i.e., *'Ensure access to affordable, reliable, sustainable and modern energy for all'*, and target 7.1 is to achieve this goal by 2030 [14,89]. The progress towards achieving this target is monitored using two indicators, as given in Table 2.1 together with the respective custodians.

Thus, is ensuring energy affordability as straightforward as simply counting the number of people with electricity? However, if a household has some access to electricity but not enough to meet its needs, does this still count as having access? These types of questions had to be answered if access to electricity was to be a universally dependable and acceptable measure of affordable energy. If different

---

[12]Established at the Bretton Woods Conference of '44 Allies' in 1944 [87].

20 *Clean energy for low-income communities*

Table 2.1   Sustainable Development Goal 7, 'Ensure access to affordable, reliable, sustainable and modern energy for all'. Target 7.1 and Indicators [89].

SDG7 Target Number	Description	Target Indicator Number	Description	Custodian Agency
7.1	*'By 2030 ensure universal access to affordable, reliable, and modern energy services'*	7.1.1	*Proportion of population with access to electricity*	*World Bank*
		7.1.2	*Proportion of population with primary reliance on clean fuels and technology [for cooking]*[13]	*World Health Organization*

countries were to use dissimilar measures, would these be needs-based or a case of political expediency in the sense of national bragging rights among international peers? To combat some of these concerns, the International Energy Agency [IEA][14] specified a benchmark level of household electricity delivery, a so-called '*basic bundle*' that would qualify for the designation 'access to electricity' [91]. Initially, the level of the basic bundle was defined as 500 kWh annually for urban households, reduced to 250 kWh for rural dwellings, at best over 20 times less than the average household usage in the US [92]. The level was increased to 1,250 kWh in the 2020 IEA definition, sufficient to power:

- four lightbulbs for 5 h per day,
- a cell phone charger,
- a television for 4 h per day,
- a fan for 6 h per day, and
- a refrigerator

which it is claimed could be reduced to 420 kWh if a household uses 'energy efficient' appliances [91]. If this is the benchmark, then everyone in high-income countries has access to electricity and therefore affordable energy, but the earlier discussions on general and fuel poverty unquestionably show that this is not the case. If the UN's SDG indicator 7.1.1 is to be used as one of the crucial determinants for affordable energy, a better definition of a basic bundle of household electricity will be necessary. To address this need, as discussed in more detail in Section 2.4.4, the custodian agencies of SDG 7 together with their cooperating partners have formulated a comprehensive 'Multi-Tier Framework' [MTF] approach to electricity access. The implementation of the MTF system has encountered some data obstacles; hence, at least in terms of reported data,

---

[13]The originally agreed 7.1.2 indicator statements and the 2023 indicator framework update did not contain the phrase 'for cooking' but this has been added in the 2023 tracking report [89].

[14]Formed in 1974 as an autonomous agency to be an energy think-tank by several member nations [90].

the basic-bundle approach appears to remain the yardstick, although this will change as the MTF system matures [89]. Regardless of the system used, in 2023, the custodian agencies of SDG 7 reported that 675 million people, or 8.5% of the world's population, still lacked access to electricity. Can it then be inferred that achieving universal access to electricity, and thus energy afford-ability for all by 2030, though a commendable goal for humanity, is an example of impractical political idealism [89]? These issues are further discussed in Section 2.4.

2.2.2.1 Impact of global population growth

If the provision of clean and affordable energy and the eradication of poverty are to be 'for all', what does the 'for all' really mean? It is safe to presume that it means the entire global population. This would be the most obvious interpreta-tion. However, as the growth of the global population has significantly increased since the start of the 20th century, and is predicted to continue well into the latter quarter of the 21st century, can the rate of progress towards achieving clean and affordable energy *for all* match the rate of population growth as well as addressing the existing challenges? Even if the targets of the complete set of sustainable development goals are met by 2030 what will the global situation be if the population continues to increase beyond 2030? Projecting the size of the global population into the future is a precarious activity because of the many factors involved. These include changes in existing demographics, fertility rates, life expectancies, trends in migration, economic development, government policies, environmental vulnerabilities, and so on, as pertinent to individual countries or regions. For all of these reasons, in the UN's periodic updates of world population prospects, it is emphasised that all their projections are prob-abilistic, and their analyses are qualified by stated levels of confidence as to their likelihood of happening [93,94]. In publicly released UN population data, it is commonplace to categorise the probabilistic projections as being 'high', 'medium', or 'low', not merely to simplify the results of their analyses, but more likely as part of a vital communications strategy [95].

The latest UN population prospects report, together with detailed supporting data, was published in 2022, 5 years after the previous report [93,96]. Projections for each year from 2023 up to 2100 are provided, and they are presented and analysed by different categories, namely:

- Sustainable Development Goal (SDG) regions, e.g., sub-Saharan Africa, Europe, and North America,
- World Bank Income Groups, e.g., high, medium, and low-income countries,
- UN Development Groups, e.g., more developed, less developed, least devel-oped countries, and
- Geographical by country and region.

These are very useful formats for academic scholars, policymakers, and media commentators. Overall, the 2022 projections are lower than those of the 2017 revisions, but nevertheless, there could be a further 2.3 to 2.4 billion people added

22 *Clean energy for low-income communities*

to the global population by the end of this century, with about 1.8 billion over the next three decades. This is an increase equivalent to another 5–7 countries the size of the US being added to the present needs to provide access to clean and affordable energy. However, the scale of the projected population growth will not be globally consistent across categories, regions, or country groups. For example, although the global population will peak in 2086, many countries will reach individual peaks earlier, indeed much earlier in some instances, such as China, Japan, Germany, and Italy, which have already reached their maximum populations, while countries such as France, the UK, and India will reach their peaks in 2041, 2056, and 2063 respectively. In terms of sustainable development goals, net-zero commitments, and climate change mitigation desires, the years 2030, 2050, and 2100 are particularly relevant. Population changes of note for these years as well as the peak year of 2086 are given in Table 2.2.

As can be seen in the table, by 2050, the target date for the net-zero scenario, the number of low-income people will have increased by a staggering factor of 3.44 above 2022 levels. Does this imply that energy poverty will become more prevalent because the cost of renewable energy has not met the expected reduced level or that the capital cost of transition has proved to be overwhelming in some countries and regions? Rapidly decreasing populations in China and Japan should provide opportunities for less-stressful transitions to clean renewables to be possible in the second half of the century. The population of the US will steadily increase throughout the 21st century, and it is not clear when it will reach its peak. The measures taken to keep poverty in check, as illustrated in Figures 2.1 and 2.2, may maintain present rates, but there is a need to further increase 'welfare' expenditure. Maybe if the pace of the transition from fossil fuel sources is lessened, or the use of nuclear, hydropower, and electricity storage is significantly

Table 2.2 Exemplars of population changes 2022–2100 with highlighted key years [94,253]

Group/ Region/ Country	Population in Thousands					
	2022	2030	2050	2086	2100	Ratio 2100/ 2022
		Paris Agreement/ SDGs	**Net-Zero**	**Peak Global**		
World	7,975,105	8,546,141	9,709,492	10,430,926	10,349,323	1.29
Low-Income	737,605	906,731	2,536,467	2,108,260	2,265,048	3.07
Sub-Saharan Africa	1,166,766	1,417,346	2,111,548	3,190,631	3,443,347	2.95
India	1,417,173	1,514,994	1,670,491	1,622,428	1,519,850	1.07
China	1,425,887	1,417,975	1,312,636	1,064,759	911,277	0.64
Nigeria	218,541	262,580	377,460	525,078	546,092	2.49
United States	338,290	352,162	375,392	392,364	394,041	1.16
Japan	123,952	118,515	103,077	79,818	73,664	0.59

Providing clean and affordable energy for all 23

accelerated, the transitions can be achieved in the desired timelines. However, both strategies will require firm, continuous, and perspicacious political leadership. Although the US has the economic capacity to attempt to offset the impact of population growth, the possibility of inadvertently increasing energy prices or poverty rates should not be overlooked, and this could be detrimental to the present transition schedules. In general, countries that know these projections may change their energy, health care, and welfare policies, as well as their planned scales of immigration.

The biggest, and perhaps more profound, challenges from population growth will occur on the African continent, especially in the region known as sub-Saharan Africa, where the population will increase by a factor of almost 3 to a third of the total global population by the end of the century, up from 14.6% currently. This means that the population of sub-Saharan Africa could be 1 billion more than that of India and China, the world's two most populous countries combined. Although some African countries have achieved poverty reduction in recent years, regrettably, many of the 49+[15] countries of sub-Saharan Africa are among the World Bank's and UN's presently categorised *poorest* and so-called *least developed countries* (LDCs) [97–99]. Consequently, global poverty is concentrated in sub-Saharan Africa, which has a poverty rate about four times higher than the world's next poorest region [99]. Given the close connection between general poverty, affordability, and access to energy, the challenges associated with providing clean and affordable energy for the LDCs could be overwhelming, as discussed in Section, 2.2.3.

2.2.3 *Poverty and energy challenges – Least Developed Countries (LDC)*

In 1971, the UN began to categorise certain countries as 'Least Developed'. By October 2023, 46 countries distributed across four continents [33 in Africa, eight in Asia, three in the Pacific, and one in the Caribbean[16]] were designated as LDCs [100]. Their combined population is about 1.1 billion people, of which three-quarters live in poverty [101]. But what is an LDC? There is a wealth of literature on global poverty, most of which refers to LDCs as low-income countries confronting severe structural impediments to sustainable development [98]. However, the precise UN definitions have been changing, including the term low-income being replaced by 'lower income' and now, just gross national income (GNI) per capita [102]. Therefore, it is no longer a prerequisite for a country to be classified as low-income by the World Bank, for the UN to categorise it as an LDC. The UN's system is based on three criteria, GNI per capita, a human assets index (HAI), and an economic and environmental vulnerability index [EVI] [102]. The latter two indexes are the accepted measures of the 'structural impediments'. Consequently, not all low-income countries are LDCs, and not all LDCs are low-income. In fact,

---

[15]Depending on the source the number of acknowledged countries in the region varies between 49 and 53.
[16]By convention the Caribbean is considered to be part of the continent of North America

24   *Clean energy for low-income communities*

Table 2.3   World Bank income classification thresholds 2023–2024 [97]

Thresholds US Dollars (USD) per capita			
Low	**Lower Middle**	**Upper Middle**	**High**
$\leq 1,135$	1,136–4,465	4,466–13,845	>13,845

only 26 of the current 46 LDCs are low-income[17], the rest being lower-middle-income countries, apart from Tuvalu[18] which is classified as an upper-middle-income country. The World Bank updates its income classifications on 1 July annually, depending on a country's financial performance in terms of GNI per capita. For the fiscal year 1 July 2023 to 30 June 2024, the levels are given in Table 2.3, all thresholds being higher than for 2022–2023 [97].

In 2022, the average GNI per capita for low-income countries was 721 USD, while for LDCs it was 1,225 USD, and for high-income countries it was 51,087 USD. It was the lowest for Afghanistan at 390 USD and the highest for Bermuda at 125,240 USD [103]. Nevertheless, for sub-Saharan Africa, which includes both the majority of global LDCs and low-income countries, the region's GNI per capita was in the lower-middle range at 1,665 USD. Thus, while the difference between the low-income countries in regions like sub-Saharan Africa and the region's average income is less than 1,000 USD, just about sufficient to buy a mid-range smartphone, the average to low ratio is considerable, i.e., over 230%, about the same as comparing the incomes of the US to Greenland, or Canada to Lithuania. Arguably, while all countries in sub-Saharan Africa are poor, some are extremely poor. In the Democratic Republic of Congo (DRC), for example, the GNI per capita of 590 USD is measurably lower than the average low-income country, but its population is projected to increase even faster than most of the sub-Saharan region by the end of the century, quadrupling to 423 million people. The DRC's population will then be larger than that of the US in 2100. At the moment, the US has a per capita income over 130 times that of the DRC, along with access to electricity five times higher, but still, a significant share of its population experiences both general poverty and energy poverty. With such a lack of income and energy and a hugely growing population, it is extremely difficult to envisage that the DRC will be able to achieve the global sustainable development and climate change goals by 2100, let alone by 2030 or 2050. Maybe if the DRC, as an LDC, could remove, or significantly remove, its structural impediments to development, as manifest in its HAI and EVI levels, then there could be hope that substantial socioeconomic improvements can be experienced by its population?

The HAI index, a version of the UN's Human Development Index, is a calculated measure based on six defined indicators that provide an aggregated score of

---

[17]The two low-income countries not presently categorised as LDCs are North Korea and Syria.
[18]A group of islands in the Pacific formerly known as the Ellice Islands.

Providing clean and affordable energy for all 25

between 0 and 100. The higher the score, the greater the country's development of its human capital. In 2021, the range of scores for LDCs was 18.3 to 89.4, with a threshold of 66 or more being required as one of the precursors to moving out of the LDC category [102]. The DRC's HAI score was 47.9, indicating that its development of human capital was still low; indeed, of the 46 LDCs, it was the 12th lowest. The EVI index uses eight indicators shared between economic and environmental measures, like the HAI, on a 0 to 100 scale, but in this case, lower scores indicate less vulnerability to economic and environmental shocks, including situations caused by conflicts, natural disasters, instability of food production, and so on. In this case, the 2021 UN analysis gave EVIs scores of between 23.3 and 57.1, with the threshold for category change cut-off being 32. Of the 34 LCDs above the threshold, 28 countries were located in the sub-Saharan region, although, interestingly, that number did not include the DRC, which has an EVI score below the threshold of 28.3.

In the 2022 World Bank report on 'Poverty and Shared Prosperity', their investigations into recent pre-COVID and post-COVID economics, including the immediate effects of current inflation and the Ukraine-Russia conflict, found that food inflation globally was more than 5% above core inflation, but that in 36 of the 50+ sub-Saharan countries it has been twice as high as non-food inflation. Energy prices have also risen by 11% and may rise even further to 50% by 2023 [99]. Yet these increases are overshadowed by the rising prices of 'agricultural inputs', such as fertiliser. In the 2 years before March 2022, these increased by 217%. No doubt, it will be interesting to know what the impact of these inflationary trends will be on the EVIs of LDCs, especially those in sub-Saharan Africa, in the coming years. Yet, ironically, the incomes of some food production and agricultural workers in the region have increased because of increasing inflation, especially since food exports have become more valuable. This has resulted in the lowering of poverty rates in some instances [99].

Globally, by 2050, just over one in five people will live in sub-Saharan Africa, and by 2100, this will increase to one in three, whereas in 2022 it was about one in seven [14.6%]. The 2022 sub-Saharan global population share was approximately the same as in Europe and North America combined, but by 2050, this share will be closer to twice that of the latter regions and by 2100, over three times their population contributions. The relative change in population share between China and sub-Saharan Africa is even more noteworthy. By 2100, the population of China will be just over a quarter that of the sub-Saharan region, yet it will be almost a quarter more in 2022 [94]. The importance of this comparison to Europe and North America and to China is that the former provides significant overseas development aid to African countries, especially sub-Saharan Africa and its LCDs, while China is known for its level of investments in the same continent [104–107]. For a variety of reasons, the amount of such aid to and investments in Africa has been faltering recently [104]. As the population in sub-Saharan Africa, throughout the remainder of the 21st century, rapidly increases and those in Europe and China decline, albeit with modest rises in North America, it seems highly likely that the relative levels of aid and investment will not keep pace with these changes. But such overseas help may not be necessary.

26 Clean energy for low-income communities

This possible lack of need is because of the plans outlined in a report by the Organization of African Unity (OAU) in 2013, i.e., 'Agenda 2063: The Africa We Want', with a vision to establish 'An integrated, prosperous and peaceful Africa, driven by its own citizens, representing a dynamic force in the international arena' [108]. The implementation of the vision for Africa's 'inclusive and sustainable socioeconomic development' over the 50-year period involves a series of five 10-year plans encompassing a set of 20 goals [109]. The target for Goal #20 is that 'Africa Takes Full Responsibility for Financing Her Development' [110]. Their latest performance progress report involving data from 70% of the OAU members acknowledges that the pandemic, coupled with structural difficulties including low technical and financial capabilities, akin to HAI and EVI perhaps, have impacted the levels of achievement of the implementation targets, especially Goal 20 [111]. In terms of Goal 7, which includes the establishment of an environmentally sustainable climate for Africa, the performance has also not met expectations. Nevertheless, overall progress has been made, resulting in the report being presented in a positive, but perhaps overly optimistic, manner. The implementation of Agenda 2063 is a very ambitious undertaking, arguably more so than either achieving the goals of the UN's 2030 sustainable development agenda or the Paris Agreement.

However, it is not only energy prices and the lack of financial and technical capabilities that will impact the growing populations in low-income countries and LDCs, but the infrastructure needs to greatly improve access to energy. For example, as measured by the access to electricity indicator, the access of extremely poor households is only 38% compared with the global average of 78%, and in terms of communication, only 55% of such households had access to a cell phone whereas globally, 84% of households had access to at least one such device [99]. Given all the vagaries associated with considerable population growth in poorer countries, it is difficult to share the sanguinity of organisations like the OAU.

2.3 Clean energy

When used as a signifier with the term energy, 'clean' may appear easier to define than affordable, but its use can frequently lead to misinterpretations because of its somewhat eclectic application. Indeed, 'clean', 'green', and 'renewable' are frequently used interchangeably, and their definitions are topics of debate, all of which are responsible for misunderstandings and confusion [112]. For instance, clean energy and green energy are said to be different since the latter is considered to be generated from natural resources that could emit greenhouse gases and other air pollutants, e.g., biomass combustion, whereas the former is energy produced from sources that do not release certain designated atmospheric pollutants as defined by national clean air standards, e.g., the US Environmental Protection Agency's (USEPA) legislated list of 'criteria air pollutants' or globally by the World Health Organization (WHO) [113–115]. Nevertheless, the absence of these criteria pollutants, or contaminants, as they are referred to in Canada, in the emissions from clean energy generation does not necessarily mean that the emissions will not contain traces of health hazardous materials known as 'air toxics', which are mainly organic chemicals [116].

If clean energy, by definition, is not to pollute, then certain sources of renewable energies, e.g., biogas and biodiesel, which do pollute the atmosphere, should not be considered to be clean. Yet governments such as Canada consider 'clean fuels' to include certain biofuels, biogas, sustainable aviation fuels, and synthetic fuels, whereas others may describe them as green or renewable [83]. In an attempt to resolve this somewhat muddled situation, use is made throughout this chapter of the often-repeated statements stemming from the Sustainable Development Goal 7, its targets, and indicators of the UN's sustainable agenda as explained in the next paragraph [14]. However, it should be noted that in the definitions extracted from these sources and the supporting literature, the term 'green' is rarely encountered and 'clean' originally used in the context of describing desirable fuels has become increasingly popular in labelling certain types of energy sources. For example, the declared aim quoted, SDG Goal 7 has been modified in a more recent statement by the Secretary-General of the United Nations to be, '*affordable, renewable, and sustainable energy for all*', or '*clean renewable energy*', albeit the wording of indicator definitions has remained the same [117,118]. The achievement of this 'imperative', as it was referred to in the statement, was claimed not only to meet the sustainable development goal for energy but also to prevent a climate catastrophe. But how does all this help specify what is clean energy?

Can it be inferred from the Secretary-General's statements that renewable energy can also be unclean or that reliable and modern energy are not as important attributes as being renewable and sustainable? These questions may seem to be nitpicking, '*giving too much attention to details that are not important*', or is it just a matter that the use of the term 'clean' when applied to energy sources and conversions is somewhat equivocal in the same way as when it is applied to air and drinking water [119]? For air and water to be described as clean, they must meet certain defined national or international standards with respect to the upper limits of concentration levels of prescribed pollutants and contaminants above which there are or could be, adverse health implications [120,121]. Consequentially, in the absence of a universally acceptable definition of what exactly clean energy is, similar approaches to specifying what constitutes clean air and water are used for clean energy, and this mainly involves the terminology of the targets and indicators of SDG 7. As shown in Table 2.1, the original 'clean fuel' indicator, 7.1.2, now has an added codicil 'for cooking'. This is likely because, as of 2023, the global data collected for monitoring this indicator focuses on the primary fuel used for cooking with non-solid fuels considered clean when used in combination with technologies that meet the WHO's indoor air quality emission standards [122,123].

The clean fuels identified in the WHO standards are biogas, ethanol, LPG (Liquefied petroleum gas), natural gas, electricity, and under some conditions solar, but solid fuels, wood, animal dung, crop wastes, charcoal, and coal are considered as polluting and *non-modern* cooking fuels [89,122,124]. Biogas, sometimes

28 Clean energy for low-income communities

known as renewable natural gas or biomethane, biomass, and solar are considered to be renewable fuels, and electricity can be counted as clean if produced from renewable and sustainable sources. Arguably then there are tensions between the use of the respective contributions of indicator 7.1.1 and the modified 7.1.2, in measuring access to clean and affordable energy, which are compounded by the obvious UN's emphasis on the use of renewable energy sources and the linkage of such sources with the concept of clean energy. As a result, another SDG 7 target, 7.2, i.e., *'By 2030 increase substantially the share of renewable energy in the global energy mix'* and accompanying indicator, 7.2.1, *'Renewable energy share in the total final energy consumption'*, is also included in national assessments of the overall goal of SDG 7, and in this case, there are three custodian agencies, the IEA, the International Renewable Energy Agency (IRENA), and the UNSD [89,124]. However, further methodological tensions can arise, as in addition to the now familiar 'renewables' solar and wind sources, but the 7.2.1 indicator also identifies many other sources as renewable, such as hydro, solid biofuels that include fuel-wood, animal and vegetable waste, black liquor[19], bagasse[20], and charcoal, liquid biofuels like biodiesel and bio-gasoline, geothermal, marine, and renewable municipal waste. This is probably because most of these other 'renewables' are used to generate electricity, but if *polluting and non-modern fuels* are considered unclean for cooking, how can they be identified as being capable of producing clean electrical power?

Nevertheless, the SDG 7.1 and 7.2 indicators provide the parameters that nations have agreed to use to determine the progress towards universal access to clean energy, although they do not actually unequivocally specify what clean energy is, and it appears that exactly what constitutes a clean fuel depends on what use will be made of a particular fuel. Moreover, as discussed, different nations and organisations do not interpret the notion of clean energy or clean fuel in a consistent manner, e.g., [83,114]. The hoped-for existence of precise definitions of 'clean' and 'affordable' as stated in the opening sentence of the introductory remarks still remains unfulfilled. This absence necessitates any available clean and affordable data to be treated with a degree of caution. This circumstance is recognised within the implementation of Agenda 2030 by a 'Tier' classification system for determining the quality and quantity of indicator data developed by the 'Inter-agency and Expert Group on SDG Indicators' [IAEG-SDGs] [125,126]. The three original Tiers are described in Table 2.4 but as of 31 March 2023 there are no longer any indicators classified as Tier III [127]. The sustainable development indicators relevant to the considerations of clean and affordable energy, i.e., 7.1.1,7.1.2, and 7.2.1 are all categorised as Tier 1. This suggests that while the precise definition scenarios have idiosyncrasies, there is an accepted consistency, at least by the majority of the member states of the United Nations, in the approaches being used to measure the progress in achieving the national goals of SDG7. Can it then be concluded that 'access to electricity' is an effective way of determining

---

[19]A by-product of the process for converting pulpwood into paper pulp.
[20]For example, sugar-cane or sorghum pulp.

Table 2.4 Tier classification for global SDG indicators [127]

Tier I	'Indicator is conceptually clear, has an internationally established methodology and standards are available, and data are regularly produced by countries for at least 50 per cent of countries and of the population in every region where the indicator is relevant'.
Tier II	'Indicator is conceptually clear, has an internationally established methodology and standards are available, but data are not regularly produced by countries'.
Tier III	'No internationally established methodology or standards are yet available for the indicator, but methodology/standards are being (or will be) developed or tested'.

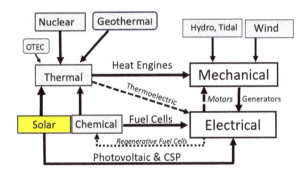

Figure 2.8 Energy sources and convertors

energy affordability and that an energy supply is only clean if it is renewable, in accordance with the *'clean renewable energy'* statement [118]? Moreover, how does the satisfaction of the chosen indicators ensure that the energy supply is reliable and modern?

2.4 Access to electricity

Why, apart from the difficulties encountered in defining affordability in terms of income, poverty, and household energy efficiency, as discussed in Section 2.2, should access to electricity be regarded as one of the main measures of energy affordability? Is it a superior criterion? Is the total amount of household energy supplied and consumed synonymous with a household's entire utilisation of and expenditure on electricity? What is the present state of electricity access? Will a 100% access to electricity ensure that energy becomes universally affordable?

2.4.1 Why electricity?

Electricity is an incredibly useful form of power and energy that can be generated from a multitude of different energy sources and used in so many ways, to the extent that it could be described as a veritable *Lingua Franca* of modern energy services (Figure 2.8). However, a decade ago, a review of the definitions of modern

30   *Clean energy for low-income communities*

energy found that '*there is not always complete agreement or clarity within the literature about how modern energy services are defined*', and it is apparent that this situation has barely changed as there is still no universally accepted definition [91,128]. As defined by indicators 7.1.1 and the modified 7.1.2, access to modern energy services involves access to electricity and clean cooking fuels. Although the WHO indoor air quality guidelines included electricity as a clean cooking fuel, some carbonaceous fuels were also listed, but following the advent of the Global Methane Pledge (GMP) launched at the 26th UN Climate Change Conference of the Parties (COP26) in 2021, there is to be a major global effort to significantly reduce the use of carbonaceous cooking fuels by 2030 and beyond [129,130]. Moreover, several studies have associated childhood asthma with the use of gas cooking stoves, and although some other researchers do not agree, the asthma claims have led to some municipal authorities prohibiting gas installations in new homes, in some cases requiring all cooking and heating systems to be electric [131,132]. Along with the growing electric vehicle market, it does seem that the decision has been made for an almost all-electric future, apart from the occasional exception such as the use of geothermal district heating [133].

If the all-electric scenario is indeed the future, then it makes sense to determine energy access by measuring the share of the populace that has access to electricity, but is 2030 the future? If not, when will the electric future arrive? Perhaps 2050, if driven by the more recent determination by world leaders to limit the global warming increase to 1.5°C by the end of the century rather than just the original commitment to '*pursue efforts*' towards this benchmark while '*holding the increase in the global average temperature to well below 2°C*' [134,135]. To limit the temperature increase to no more than 1.5°C, the UN states that greenhouse gas emissions (GHG) need to be reduced by 45% by 2030 and almost 0% by 2050, but are still referred to as 'Net-Zero' [136]. Although formal universal agreements on the net-zero scenario do not yet exist in the same manner as the Paris Agreement or Agenda 2030, at COP27, in 2022, the US launched the 'Net-Zero Government Initiative' along with 18 other countries, but this was a follow-on from the 'Race to Zero' campaign launched by the UN in June 2020 to involve so-called 'non-state actors' in a 'Global Ambition Alliance' to achieve carbon neutrality by 2050 [136–139]. As of September 2022, over 11,000 of these 'actors' including businesses, financial institutions, cities, states and regions, and educational and healthcare institutions, have joined the Race-to-Zero [138,140].

However, to join the race, participants have to apply and must be deemed able to meet the '5P[21]' criteria, i.e., '*Pledge, Plan, Proceed, Publish, Persuade*', including, as part of the pledge, '*phasing down and out all unabated fossil fuels as part of a global, just transition*', and as a member, '*accelerate the deployment of market-ready, existing technologies such as energy efficiency and renewable energy*' [141]. In addition, the 'High-Level Expert Group' [HLEG] appointed by the UN Secretary-General to establish the criteria for assessing commitments to

---

[21]Not to be confused with the 5Ps of the Agenda 2023 for Sustainable Development, i.e., People, Planet, Prosperity, Peace, Partnership.

net-zero also recommended that, '*Non-state actors cannot claim to be net zero while continuing to build or invest in new fossil fuel supply*' [142]. The world's largest coalition of financial institutions, the Glasgow Financial Alliance for Net Zero (GFANZ), that claims over 500 members in 50 countries have embraced this particular HLEG recommendation, and their members shall, i.e., must, end financial support for fossil fuel expansion, should align themselves with the IPCC[22] and IEA scenarios for achieving net-zero and also not make '*unrealistic assumptions on deployment and development of future technologies*' [143]. The IEA has published a roadmap in a special report for the global energy sector for achieving net-zero by 2050, involving over more than 400 milestones of what needs to be done and when, '*to transform the global economy from one dominated by fossil fuels into one powered predominantly by renewable energy like solar and wind*', and which will need a huge level of investment [144].

From these net-zero 'club' requirements, and its rapidly increasing membership, it could be concluded that if the all-electric future is to be achieved by 2050, then all electricity will need to be generated from renewable energy sources with no fossil fuels involved in the global energy mix. Moreover, it appears that to ensure this scenario becomes a self-fulfilling outcome, financial assistance from governments and financial services from 'non-state-actors', i.e., essentially the banking sector, will be phased out and eventually eliminated sooner rather than later. The primary aim of the net-zero initiative is to limit the extent of global warming as a priority rather than meet the requirements of SDG 7.1, but if all generated energy is to be electrical and used for all purposes, including cooking, then there will be some beneficial effects of the climate action initiative on sustainable development, hopefully. Consequently, as an increasing number of world leaders become convinced about the merits of the zero GHG emissions approach, it makes sense to use 'access to electricity' and 'clean cooking fuels' as effective measures of universal access to energy. This does not mean, necessarily, that an all-electric future must only involve 'clean' renewable energy sources because, as shown in Figure 2.8, electricity can be generated from a number of sources.

Indeed, does the future have to be all-electric? Unless the WHO household indoor air quality guidelines are to be wholly revamped with many sources of presently defined clean cooking fuels, e.g., biogas, ethanol, LPG, and natural gas, being reclassified as unclean, the complete electrification of *all* energy delivery may not be necessary. This will depend on the global scale of adoption of the Net-Zero plan, which requires a complete transformation away from fossil fuels. To a large extent, the IEA roadmap proposes that access to electricity and access to clean cooking fuels be treated as distinct issues. For example, by 2030, the energy roadmap for clean cooking will only include 20% of electricity, with the remaining 80% being LPG, natural gas, and 'modern' bioenergy. Then, in the next 20 years, the use of natural gas as a cooking fuel will be almost completely replaced by increasing amounts of electricity and modern bioenergy. The IEA's plan also highlights that technically, sufficient amounts of bio-LPG could be produced from

---

[22]Intergovernmental Panel on Climate Change – A United Nations body.

32   *Clean energy for low-income communities*

municipal solid waste to meet the clean cooking fuel needs of over 750 million people by 2050 [144]. However, as the goal of the net-zero approach is to eliminate all human-generated GHG emissions, especially carbon dioxide (CO_2), how can it be reconciled with pathways that promote the use of GHG-generating cooking fuels? The simple answer is that presumably, in the same way as solid biomass can be classified as carbon-neutral or net-zero using the IPCC definition of net-zero as, '[w]*hen anthropogenic emissions of greenhouse gases to the atmosphere are balanced by anthropogenic removals over a specified period*', so can liquid and gaseous biomass [145].

About three-quarters of all anthropogenic GHG emissions come from the global energy sector, so it is not surprising that this sector is one of the foremost concerns in the mitigation strategies aimed at controlling global warming and climate change. However, of this sector's emissions, only about 14% come from energy use in households, and of this share, it is not clear how much is from cooking alone [146]. In terms of timelines, it is apparent that the electrification of industry, transport, and energy production is more pressing than clean cooking in achieving the desired climate control, as typified by the Paris Agreement and the Net-Zero commitments. This will require some acknowledgement that it is unlikely that many of the goals of the sustainable development agenda, particularly with regard to energy, can be wholly achieved by 2030. It could be argued that success with climate control, i.e., global warming, will impact everyone on Earth, whereas ensuring universal access to clean cooking fuel will impact only about 40% of the population, and, in any case, the fall-back position of the increasing use of bio-energies for cooking will lessen the effects of delayed electrification. Nevertheless, measuring access to electricity should provide a suitable metric for assessing progress towards clean energy and arresting climate change. It may also present some insights into the reliability of energy supply, affordability, and poverty, especially in particular geographic regions [147,148].

2.4.2   *Energy measurement and data gathering*

The history of human civilisation is inextricably linked with the history of energy [149]. The ancient scholars attempted to determine how much 'work' could be done from their machines, real and imaginary, but it was in the 18th century CE that the English engineer, Thomas Savery, and, particularly, the Scottish engineer James Watt who, by observation and rudimentary calculation, quantified, in readily understandable terms, the amount of power that their steam engines could produce[23] [149,150]. This provided the basis for the modern measurement systems dealing with power and energy. However, the terms power and energy are often used interchangeably, but this is incorrect as energy is the capacity to do work, whereas power is the rate at which work is done or produced, i.e., power is energy per unit time. So, for example, a device that is producing, say, p units of power will deliver $p \times$ time units of energy. Moreover, the measurement units of power and energy are also different, e.g., kilowatts (kW) and kilowatt-hours (kWh),

---

[23]For marketing rather than scientific purposes [150].

respectively. Similar confusions are frequent occurrences in the media surrounding the use of the terms electricity and energy [151].

Electricity is a form of energy, but banner headlines of the type 'all energy is being produced by renewables' are invariably misrepresentative, usually due to a lack of scientific knowledge or maybe an attempt to influence opinions, i.e., propaganda. What such headlines should say is 'all electricity *etc.*' These examples of the misuse of terminologies are not academia's pedantic nuances because the interchangeable use of the terms energy and electricity can lead to many false assessments being made regarding progress in universal access to energy and any consequent political policies and actions based on these evaluations. Clearly, it is essential that, on a global scale and per capita basis, there is a need to know how much energy is produced by source and by what means, and how much energy is consumed and for what purposes. Additionally, in harmony with climate change mitigation strategies, there is a need to know the GHG emission profiles of the various energy extraction, transformation, and conversion techniques, along with any other harmful environmental impacts associated with the practices.

A number of international agencies, governmental, public, and private, collect colossal amounts of energy data, such as the IEA, IRENA, the United States Energy Information Administration (USEIA), World Bank, Eurostat, and BP[24]. Since 1952, an annual statistical review of world energy has been published by BP, but in 2023, this review was transferred and became the responsibility of the Energy Institute (EI) with the Centre for Energy Economics Research and Policy at Heriot-Watt University in Scotland undertaking the data compilations [152–154]. These open-access statistical reviews always attempted to present objective, non-political information by providing, *'the fullest, most reliable account of energy production, consumption, trade and emissions'* [153]. Most certainly, the reviews have proved a trustworthy and valuable source of energy data, used by many organisations including the media and the scholastic and academic communities. Since 2013 in particular, the Global Change Data Lab, which operates out of the Oxford Martin School at Oxford University, has made extensive use of the data produced by BP/EI along with data from other key agencies, including the United Nations, to produce their massive open-access website, 'Our World in Data' (OWID) [155]. OWID also makes use of energy data, especially in connection with electrical energy, which is collated by the not-for-profit energy think-tank 'Ember-Climate' [156] This is mentioned here because the work of the OWID teams is broadly used and acknowledged in the discussions in the remaining parts of this chapter.

Other organisations also produce detailed energy data, but on a commercial basis, such as the IEA's very detailed 'World Energy Statistics and Balances data service', which is updated bi-annually and requires an individual subscription of €890[25] per annum [157]. However, the IEA also produces a multitude of open-access reports and web documents, as well as 'highlights' of the World Energy data service information. They also produce what is described as a *'handy, pocket-sized*

---

[24]Formerly known as British Petroleum.
[25]As of August 2023.

34 *Clean energy for low-income communities*

annual publication showing key worldwide energy statistics', which includes not only energy consumption but analyses of energy prices, emissions data, forecasts of future energy supplies, and so on [158]. There is then a wealth of useful and authoritative data on which careful observations and assessments of the viability and timelines of universal access to clean and affordable energy can be undertaken. However, the different organisations involved in the collation and analyses of energy data, to some extent, use different methodologies in their analytical investigations and presentations, which OWID converts to a more consistent format when using and interpreting the data from various sources.

Energy and electricity can be measured using the same units, but, unfortunately, the data monitoring and collection agencies, both national and international, use different representations. In reviewing the literature, especially dealing with statistical information, a variety of units are likely to be encountered, such as British Thermal Units (BTU), Quads (an amount of energy equal to 10^{15} BTU), Joules, including multiples such as Petajoules (PJ)[26] and Exajoules, Watt-hours, including familiar multiples such as kilowatt-hours, Megawatt-hours and perhaps less familiar Terawatt-hours, Million or Mega tonnes of oil equivalent (Mtoe – energy released when burning 1 mega tonne of crude oil), calories, and so on. These variants are caused largely by nationally adopted measurement systems such as the International System of Units (SI), the former British Imperial System, the US variant of the imperial system, or traditional industry standards. This can cause some confusion when analysing published data. Consequently, where possible, the SI units are used here. A further quirk is that primary energy, i.e., raw energy before conversion, production data can be presented using the 'direct' method, the IPCC's preferred 'substitution method', or the USEIA approaches, which include elements of both methodologies, as discussed later [159]. In general, the latter method normally takes account of the energy losses when fossil fuels are converted to usable energy by using an assumed 'efficiency factor' and the vagaries of the reporting of electrical energy generation from renewables. Let us say the efficiency factor is 40%, or 0.4, the value used in 2006[27] reports, to generate 50 TWh using only fossil fuels, it would require 50/0.4, i.e., 125 TWh of fossil energy input. However, in reporting the electrical energy generated from renewables and nuclear, only the direct gross output is reported [159].

The argument presumably is that electricity generated by non-fossil devices is raw energy. To some, this situation has the appearance of a numerical sleight-of-hand, but that is not the case as long as those responsible for reporting energy data make the basis for the actual statistics very clear, which is not always the case. Both BP and now EI explain the methodologies used in their annual global surveys, with the substitution method being described as the 'input-equivalent' but the detailed descriptions are presented at the end of their reports, so they could be missed in a perfunctory reading. The definition of 'input-equivalent' energy is associated with attempts to compare the number of energy sources used in fossil-fuelled thermal

---

[26]1 PJ = 10^{15} Joules; 1 Exajoule = 10^{18} Joules.
[27]Chosen for numerical simplicity in this illustrative example.

Providing clean and affordable energy for all 35

power stations and the electricity output of non-combustion generation devices and is so defined as '*the amount of fuel that would be required by thermal power stations to generate the reported electricity output*' [153,160]. The USEIA, which also publishes vast amounts of global and federal energy data and analyses, uses a similar approach to 'input-equivalent' called 'fossil fuel equivalency' which has the same impact, i.e., the adjusted data represents the amount of energy that would have been consumed if any form of electricity had been generated by fossil fuels [161]. An alternative to the fossil fuel equivalency approach, also used by the USEIA, is the 'Captured Energy Approach' defined as the net energy available for consumption after a non-combustible renewable resource is transformed into electricity but not including conversion losses [161,162]. The main difference between the USEIA and the BP/EI data is that the USEIA presents their information in terms of BTUs, and any energy data collected from agencies using kilowatthours are converted to BTUs or Quads[28]. Additionally, the USEIA also has what they refer to as an 'Incident Energy Approach' which is a measure of the energy *input* to *non-combustible* renewable devices before any conversion takes place (Table 2.5) [161].

If a substantial database of sufficiently consistent measurements could be established for determining the gross energy input to the converter devices shown in Table 2.5, then the incident energy approach would be an improvement on the present substitution methods, which use a single thermal efficiency factor of 40.7% for such devices [154]. For example, in a 2016 research study, the USEIA found that the efficiency factor for solar was less than 20% but up to 90% for utility-scale hydropower [161]. Regrettably, the incident energy approach was found to be difficult to universally apply and an onerous challenge for data providers, so a single factor efficiency remains in use for the Table 2.3 devices. Moreover, as with non-combustible renewable energy devices, the input-equivalent, or fossil-fuel equivalence, substitution approaches are also used for solid biomass energy data, but, after 2021, a separate thermal equivalent efficiency factor to the one used with

Table 2.5   Measurable gross input energy to non-combustible renewable energy converters [161]

Non-combustible renewable energy convertor device	Measurable energy input to converter device, i.e., the '*energy contained in …*'
Hydrodynamic	'*the water passing through the penstock*'.[29]
Geothermal	'*the hot fluid at the surface of the well bore*'.
Wind	'*the wind that passes through the rotor disc*'.
Solar	'*the sunlight that strikes the panel or collector mirror*'.

[28]The conversion factor of 1 kWh = 3,412 BTU is internationally accepted.
[29]*The penstock is a closed conduit for carrying water to the turbines*" [161].

36 *Clean energy for low-income communities*

the non-combustible devices has been used to convert electrical generation from biomass to a primary energy equivalent [160]. In 2022, the biomass conversion factor was defined as 32% [154]. Modern biomass, which has been declared renewable by many governments and a near-zero-emission fuel, accounts for over half the global supply of renewable energy and five times more than solar and wind in terms of final energy demand [163,164].

However, so-called 'Traditional' biomass, when used for cooking and heating, is not included in modern biomass data, especially if combusted in simple stoves and open fires, and therefore not in compliance with the WHO guidelines but, arguably, could conform to the IPCC's net-zero definition [123,145]. These types of situations can prove problematic in the gathering of data regarding the usage of clean cooking fuels, as discussed in Section 2.5. The foregoing discussions have highlighted that defining, acquiring, and analysing generated and consumed energy data can be problematic. Moreover, depending on how primary energy, the most commonly employed and published energy metric, is defined can distort analyses, at least in terms of the comparative quantities of the various energy forms used in global mixes.

However, for energy consumers, perhaps the most important factors are the final delivered amounts and how much of the delivered amounts are usefully utilised for their intended purpose since energy losses or wastes are inevitable along the pathways between the original energy sources and the end-user. Ritchie has defined four meaningful energy measurement locations along the routes of these '*energy-chains*' from primary to useful energy and, using three simple examples for carbonaceous fuels, has summarised the impact of losses between them as illustrated in Figure 2.9 [165]. Thus, to access useful energy, whatever the form, transformation, transmission, distribution, and equipment losses have to be accounted for in determining the amount of primary energy that is required at the start of the energy chain. Consequently, affordability is not only a function of the expense of the primary energy source but also of the costs associated with lost and

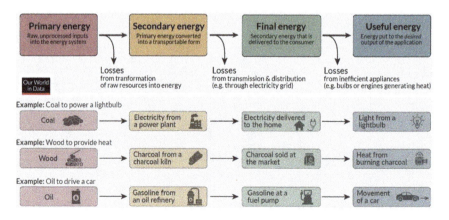

Figure 2.9 Ritchie's four ways of measuring energy [165]

wasted energy along the energy chain. While scientific laws and the existence of certain material properties restrain these losses from ever being wholly eliminated, making the various elements of the 'energy-chain' more efficient can either reduce the amount of primary energy needed or increase the availability of useful energy. The importance of energy efficiency in reducing energy poverty, and hence affordability, and achieving the aims of the net-zero scenario has been advocated by many researchers, agencies, and governments [22,49,141,166,167]. However, the acquisition costs of energy-efficient devices are usually more expensive than existing technology devices, e.g., incandescent light bulbs versus LED, and while future operating expenditure may be cheaper, this is of little comfort to low-income people who cannot afford to replace their 'energy inefficient' devices without significant government intervention. Regrettably, the Ritchie approach and terminology have not yet been universally adopted, but that does not detract from their usefulness.

2.4.3 Electricity versus energy

While the choice of using the metric 'access to electricity' as the indicator for determining 'access to energy' in the sustainable development agenda is defensible, as discussed in Section 2.4.1, as of 2022, only about 20% of global final energy consumption is in the form of electricity, only some 3% more than a decade ago [154,157]. At that rate of progress, it could be the next century and beyond, not 2030 or 2050, before full electrification, i.e., all final energy consumption being electricity, is achieved. If only 20% is electricity, then obviously 80% of consumed energy comes from other forms of energy. In countries with low rates of electrification, such as Nigeria [1.9%] and Indonesia [14%], biomass is a dominant energy form, although they export traditional fossil fuels, whereas another oil and gas exporting country, Norway, with a high rate of electrification [47%], has an energy mix heavily reliant on hydropower [168–170]. The overall energy mix can also be reflected in the energy consumed in dwellings, for example, in the UK, the average household uses between 18%–20% electricity and 80%–82% gas [171]. With such a wide range of global energy mixes and the dominance of other energy types over electricity, perhaps the SDG 7.1.1 indicator is more a measure of electrification than access to energy? Moreover, the actual usage of electricity and access to it is not the same metric. Indeed, the energy threshold for determining electricity access is inordinately low, as discussed in Section 2.2.2, so, as illustrated in Figure 2.10, for the exemplar access trends from 2000 to 2020, a deceptive impression can be perceived of the scale of the challenges of electrification [172]. Maybe this is because it is only since the global acceptance of the Paris Agreement and Agenda 2030 that access to electricity has become far more than a mere uninteresting statistic?

Regardless of access shares and benchmarks, it is not necessary to have all useable energy in the form of electricity, especially for residential heating and cooking, where acceptable, clean alternatives can be used. An obvious alternative to electricity for certain purposes in the residential sector would be natural gas, and

38   Clean energy for low-income communities

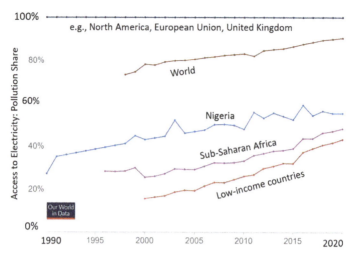

Figure 2.10   Access to electricity for 4 h per day for lighting, phone charging, or radio power, exemplars [172]

its use is in harmony with the WHO's indoor air quality guidelines. Awkwardly, if the recent Global Methane Pledge is to be fully implemented, then national gas as an alternative to electricity will be prohibited as its use will adversely impact the likelihood of success of climate change mitigation measures. Consequently, while the full electrification of final energy may not be necessary from an access point of view, it may be inevitable. In the meantime, a less vague term than 'proportion' may need to be used for the 7.1.1 indicator, e.g., a specific percentage of people having access to electricity, along with the adoption of more representative access benchmarks, see Section 2.4.4. If full, or even wide-ranging, electrification is to become a reality, then having access to clean energy also means having access to clean electricity. This is also more of a challenge than is generally appreciated since not all electricity generation, at present, is produced from primary sources that are considered to be clean or at least low-carbon (Figure 2.11) [173]. Particularly noteworthy items from the data presented, in the author's opinion, are given in Table 2.6, hopefully reinforcing that the absence of universally accepted definitions of what is 'clean' and 'renewable' can result in varied and different interpretations of the energy data by both protagonists and antagonists alike who may be advocating particular energy mixes.

However, yet again, great caution needs to be exercised in likening the mix of primary energy consumption for producing electricity with that of total primary energy consumption. For instance, while oil only contributes 3.1% of the primary energy used to generate electricity, its share of the overall energy mix is 33.1%, over ten times higher. Indeed, almost 89% of the total primary energy consumed by the source is from fossil fuels and nuclear, rising to 95% if hydropower is included [174]. Even if only fossil fuels are to be replaced, can it really be possible to substitute their almost 85% contribution to global primary energy sources and

Providing clean and affordable energy for all 39

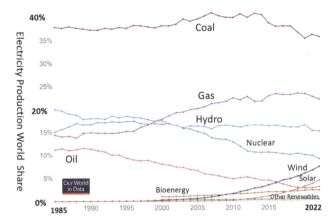

Figure 2.11 Energy source share of global electricity production [153]

Table 2.6 Observations of electricity mix energy data

Observation Number	Details
1	The combined percentage share of traditional fossil fuels in the global total electricity production has been reducing since the Paris Agreement
2	In 2022, oil only accounted for 3.1% of electricity production
3	Wind, solar, and 'other' clean renewables accounted for 15.1%
4	Considering hydropower and bioenergy to be wholly renewable, but not necessarily clean, then the total renewable share rises to 32.7%
5	If nuclear is considered as at least a low-carbon source, then the total share of non-fossil generated electricity increases to over 41%

approximately 60% to global electricity production, both by 2030? Even the most optimistic research projections have indicated that the transition to 100% renewables by 2030 is highly unlikely, but not because of technological obstacles or high costs [175–177][30]. However, for the US, it was concluded that the full transition by 2035 is possible and could be achieved without the need for fossil fuels, nuclear energy, or bioenergy, the main impediments being the lack of political will and institutional inertia [175] – sentiments echoed by the UN, the UNFCCC, and others [136,178,179].

Notwithstanding when, or if, the attainment of the full renewable transition occurs, the upward trend in the renewable share of global primary energy consumption (Figure 2.12) indicates that the requirements of SDG indicator 7.2.1 may be satisfied, albeit with no particular percentage threshold being defined [174].

---

[30]References [175,176] have also published accompanying supplementary materials and, in the case of reference [176], some 474 citations are listed.

40 *Clean energy for low-income communities*

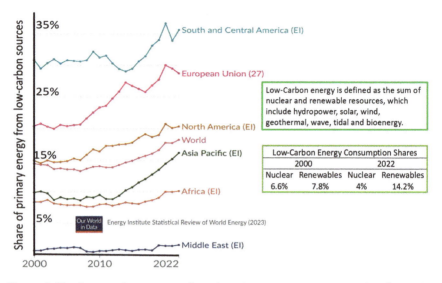

Figure 2.12 Low-carbon source share in primary energy consumption by region [174]

As previously mentioned, a major issue in the data gathering system is the misleading perception of the extent of electricity access when using the original IEA/World Bank 'basic bundle' benchmark. This benchmark considers a household to have sufficient electricity if it can provide basic lighting and the ability to charge a phone or power a radio for 4 h a day throughout the year. This amount of electricity is approximately equivalent to powering a modern electric oven for 6 min or a 100-watt light bulb for 5 h [171]. Obviously, an enhanced benchmark of 1,250 kWh per annum as suggested by the IEA, Section 2.2.2, would be more realistic, but not, for example, for households in the US, which on average use 10,632 kWh and as high as over 14,400 kWh in the state of Louisiana, whereas in the UK, depending upon the region, the average household is between 3,000 and 4,000 kWh per year [91,180,181]. Nevertheless, there are regions of the world, especially in sub-Saharan Africa, where 500 kWh per annum of total electrical consumption would be most welcomed [182].

Apart from the physical size, i.e., floor area, of an individual residence, the number of occupants and their demographics, their income, the average time per day spent 'at-home', and the indoor air temperature, there are many other factors that impact the amount of energy they consume. These features include the form of construction and the quality of the materials used, the geographical location of the dwelling, its orientation as described by the cardinal direction that the front of the residence faces, the Köppen–Geiger climate classification zone which it is sited, and so on. In other words, there are a multitude of parameters that have to be accounted for in determining the energy requirements of an existing or to-be-built home. This is why, in the UK, see Section 2.2.1, the SAP approach using the BREDEM software is so popular, at least with their government and building industry [60,61]. Thus, a great deal of effort has to be expended in determining, whether by modelling or direct metered measurement, how much electricity is consumed by households. However, these endeavours alone are not sufficient to fully grasp the scale of global electricity use, as the share of electricity usage attributed to the residential sector is only about one-quarter of the total consumption. Industrial use dominates electricity consumption, with transport accounting for less than 2% (Figure 2.13) [158].

Using the measurement of household electricity generation and usage alone does not provide an inclusive portrayal of either the scale of global access to electricity or the quantity of electricity consumed. Moreover, the type of data depicted in Figure 2.10, based on a low-access benchmark, can mask the challenges faced by particular countries, regions, and groups. For example, the average person in the US consumes three times the amount of electricity they use as a household member in other activities of their normal daily life. Another 'masking' example is the fact that the links between population size, population growth, economic growth, and electricity consumption can be ambiguous. This can be illustrated using the period 2014–2022[31] as an exemplar era during which global increases in electrical generation have outpaced global population growth, but the increases in global electricity consumption have only been marginal. Moreover, such trends in particular countries have not been consistent with the global scenario (Table 2.7) [158,183–187].

The countries listed in Table 2.7 were chosen to illustrate the change in electricity consumption and generation rates across the various global regions based on population size and growth rates for the period from 2014 (i.e., just before the member states agreed to the UN's triumvirate of universal accords on climate change, sustainable development, and disaster reduction), to the latest collected data in 2022. Those chosen can be placed into two categories: (a) countries that have

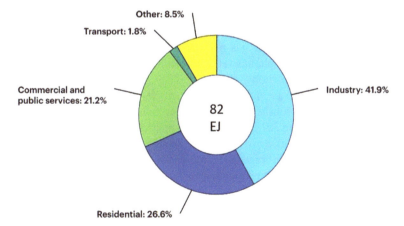

Figure 2.13   Sector share of global electricity consumption in 2019 [158]

---

[31]Chosen because of the availability of data in the Paris Agreement and Agenda 2030 eras.

42 Clean energy for low-income communities

Table 2.7   Exemplars of electricity consumption and generation[32] 2014–2022 [158,183–187]

Country/ Region/ Group	Consumption 2014 per capita kWh	Generation	Consumption 2022 per capita kWh	Generation	Population Millions 2014	2022
Niger	51	22	50.7	18	19.4	26.2
Eritrea	96.3	96	111.8	124 (2021)	3.3	3.7
Nigeria	142	171	124.6	147	179	219
Senegal	232.7	261	314	333 (2021)	14.0	17.3
Pakistan	420	515	595	645	208	236
Indonesia	1,084	927	1,034	1,211	256	276
India	797.3	965	1,025	1,297	1,307	1,417
Brazil	2,611	2,860	2,694	3,162	203	215
WORLD	**3,105**	**3,192**	**3,210**	**3,577**	**7,238**	**7,905**
China	3,905	4,164	5,526	6,199	1,385	1,426
European Union	6,021	6,394	6,070	6,286	507	448
France	6,904	8,904	6,543	7,264	66.5	64.6
Saudi Arabia	9,048	9,708	9,645	11,030	31.1	36.4
Australia	10,071	10,295	9,217	9,614	23.5	26.2
South Korea	10,497	10,376	10,971	11,705	50.6	51.8
United States	12,994	12,719	11,975	12,702	322	338
Norway	23,000	27,431	24,117	28,095	5.1	5.4
Iceland	53,832	54,326	49,725	53,924	0.33	0.37

consumption rates still below the 'enhanced' IEA benchmark of 1,250 kWh, or the average world consumption amount of 3,120 kWh per capita, and (b) those whose consumption rates are above the global average, considerably so in some cases. For the first category of exemplars, although there has been population growth, per capita consumption rates have increased, apart from Nigeria. Why is Nigeria an exception since it holds the largest natural gas reserves on the continent as well as being its leading oil producer? A simple answer is that conflict, wanton crude oil theft, and rising national rates of inflation – now over 22% – have adversely impacted the country's capacity to produce and distribute electricity, as well as people's ability to pay, despite government subsidies [188,189]. Yet, household access to electricity in Nigeria has increased since 2010, reaching 55% in 2020, and at a price per kWh prior to 1 July 2023, that was 40% lower than the heavily subsidised costs in Indonesia, 65% lower than the global average, and almost 90% lower than in the UK [187,190]. However, the July date is pertinent as a 40% tariff is being added to the wholesale price of electricity as part of Nigeria's effort to accelerate its transition to zero-carbon energy by 2060 [191,192].

---

[32]Electricity is generated as a primary or secondary product in power plants whereas consumption is the electricity delivered and purchased by the end-use sector consumers including industry, transport, households, and others.

Providing clean and affordable energy for all 43

For the remaining exemplars in Nigeria's category, countries of varying sizes and populations have been able to increase both the generation of electricity and its consumption, with a notable exception being Niger. In this sub-Saharan African country, the levels of consumption in 2014 and 2022 have barely changed, but the amount of generated electricity has decreased, probably indicating the importation of more electrical power from neighbouring countries than in 2014. Nevertheless, the consumption rate improvements in this category likely indicate that access to electricity has been enhanced since 2014, or maybe the share of the population experiencing energy poverty levels has been reduced, or, more likely, some combination of both. However, the COVID pandemic, the Ukraine-Russia war, and the ongoing transitions to renewable energy sources have somewhat skewed the state of access to electricity along with the electricity pricing structures in many countries over the past 2–3 years, although the general trend in recent years has been for prices to increase. While consumption and generation rates will have a significant influence on the energy transitions embedded in the sustainable development and climate change agendas, the rates of transition will be impacted as much by the costs of energy, and specifically electricity, as by national policies. It would be remiss not to discuss, although briefly, the effects that the increasing changes, or not, to clean or low-carbon energy sources are having on electricity prices in the exemplar and similar countries.

As discussed earlier in Section 2.2 of this chapter, to counter the possible negative effects of price rises on levels of energy poverty, countries have introduced a number of schemes to help low-income households, in particular, to adjust to changing energy mixes and, in some instances, to encourage all consumers to support the changes with new incentives and subsidies for producers and consumers alike. In this regard, the economic and political well-being of a country largely dictates how 'helpful' governments can be or want to be. Examples include Pakistan, which has recently announced electricity price increases of between 100% and 200% because the terms of a large International Monetary Fund financial bailout preclude measures such as subsidies and incentives to ease the 'real' costs of electricity. Whereas countries like Iran, where almost 90% of electricity is generated using fossil fuel sources, have one of the lowest, if not the lowest, household electricity prices in the world despite having inflation rates considerably higher than Nigeria's [190,193].

But how have population growth and changing energy mixes affected countries that both consume and generate electricity well above the IEA benchmark and the global average? Of the chosen exemplars and similar countries, apart from France, they have all experienced population growth, but while some have experienced decreases in their per capita electricity generation, others have increased their outputs. Saudi Arabia and, to a lesser extent, Norway are typical examples of countries that have increased their production while maintaining high per capita consumption rates; they also export large quantities of electricity. Norway, which generates nearly all its electricity using hydropower, has now become Europe's largest grid electricity exporter, and, since 2014, Saudi Arabia has also become a significant net exporter [194,195]. In some countries, not only have generation rates fallen but also consumption rates, which could imply that energy use

44　*Clean energy for low-income communities*

efficiency measures are having a measurable effect. The ongoing transitions to renewable energy-sourced electricity, especially in some European countries, are also having noticeable but somewhat paradoxical impacts. In countries such as Denmark, Germany, and the UK that have enthusiastically embraced the transition to renewable energy-fuelled electricity, especially wind and solar power, households have experienced some of the highest consumer prices globally, while others, such as Iceland and Norway, which use geothermal and hydropower sources, respectively, have prices that mirror the global average.

Nevertheless, in countries like Iran and Saudi Arabia that, at the moment, generate most of their electricity from non-renewable sources, the household price of electricity is extremely low, although this may be because of government subsidies rather than lower actual costs of generation. Moreover, the least expensive country for household electricity in the EU is Bulgaria, which generates almost 90% of its electricity also from non-renewables, namely nuclear and fossil fuels, in equal measure [196]. However, Belgium, which also generates about half of its electricity from nuclear sources, has household prices among the highest in the EU, probably because, in the absence of its own natural resources, it has to import fossil fuels, especially natural gas, whose supplies have been adversely impacted recently by the Russia-Ukraine war [190,197]. Thus, while the share of electricity generation by nuclear sources is very similar in Belgium and Bulgaria the household costs are four times higher in Belgium, perhaps suggesting that a transition involving increasing amounts of nuclear power does not necessarily result in lower household electricity prices. Conversely, France, long cited as 'emblematic' by nuclear advocates in terms of stable and lower electricity prices, has been replacing nuclear-sourced electricity with renewables, and this has led, so far, to rising costs for households [198]. In the past, about 75% of France's energy and 70% of its electricity have been generated from nuclear sources, including recycled nuclear fuel, but at the start of the 'exemplar' era in 2014, the government announced an energy plan that would reduce nuclear generation to 50% by 2025, a date that was later modified to 2035 but finally abandoned in 2023 [199,200].

While Figures 2.11 and 2.12 and Tables 2.6 and 2.7 provide authoritative and useful data regarding electricity generation and consumption, together with the associated energy sources and sector uses, the data illustrated in Figure 2.10, as discussed, are far less meaningful; not only because of the low 'access' threshold but also because, if below the threshold, it was deemed that there was no access. These benchmark problems were recognised several years ago by measurement and monitoring agencies, and there have been great efforts to develop a more realistic approach to determining electricity access. This has led to the establishment, as discussed in more detail in the next section, of the 'Multi-Tier Framework' especially pertinent to defining access to electricity for SDG 7 monitoring purposes and the global tracking of the progress of the associated indicators [201]. Specific versions of the framework have also been developed for the measurement of access to clean cooking fuels and access to household space heating [201].

Overall, while the exemplars provide only a thumbnail sketch of the available data on electricity consumption and generation, they do represent the circumstances for 60% of the global population. Although road maps for complete global electrification have been attempted, the discussion regarding the chosen exemplars purposefully suggests there will be many somewhat perplexing challenges to overcome to achieve the sustainable development goals and simultaneously successfully combat climate change. Not the least of these problems will be convincing the general public that electrification will be less expensive and more accessible if the use of fossil fuel energy is abandoned.

2.4.4 Household electricity access benchmarks redefined

The history of electricity generation and distribution goes back almost 150 years. Initially aimed at the electrification of industry, its uses have become commonplace, especially in households, in what the UN describes as 'developed countries' or 'developed economies'. Following the seminal 1987 Brundtland report to the UN, *'Our Common Future, From One Earth to One World'*, the links between poverty, human development, economic growth, and energy access became topics of much debate and investigation, especially by the United Nations Development Programme (UNDP) with regard to 'developing countries' and energy poverty [202,203]. Presumably with a view to describing the situation in developing countries, the UNDP suggested that energy poverty could be described as the *'inability to cook with modern cooking fuels and the lack of a bare minimum of electric lighting to read or for other household and productive activities at sunset'* [203,204]. This background is mentioned because it likely provided the rationale for SDG indicators 7.1.1 and 7.1.2 and the low threshold of the basic-bundle approach to the measurement of access to electricity.

When the basic-bundle approach is applied to developed country data, and others who are close to the transition from developing to developed, the results are meaningless as all will be either at 100% access or within a few percent. The data collection for the basic-bundle approach also involved determining access only from nationally designated grid-distributed electricity [205]. However, with the many technical developments that have taken place relatively recently, especially with respect to the use of renewable energy sources, it has been recognised that not all consumers of electricity need access to nationwide integrated grid systems. Consumer access could also be provided by mini-grids, local, municipal, regional, and off-grid stand-alone systems. Moreover, as most countries, rich or poor, now have all these different ways of providing electricity, the simple grid/no-grid can provide a deceptive picture of access and be too influential in guiding government policies. For these and other reasons associated with sustainable development, the World Bank, in partnership with UN Energy[33] and the international organisation 'Sustainable Energy for All' [SE4ALL][34] developed the Multi-Tier Framework [201,205–207].

[33]Established in 2004 as a mechanism *'for inter-agency collaboration in the field of energy'*.
[34]SE4ALL established in 2011 by the then UN Secretary-General is now an independent agency.

46 Clean energy for low-income communities

Although specifically developed for the measurement of SDG 7 indicators as they relate to households, the format of the associated information-gathering processes is expected to provide crucial data pertinent to all aspects of an envisaged sustainable future and the mitigation of climate change [208]. In this framework, six tiers of access are defined, from Tier 0 to Tier 5, with Tier 0 being no access at all and the other tiers being defined, in the case of households, by different levels of electricity consumption and the availability of supply, together with examples[35] of the devices that could be powered by the different tiers of electrical consumption, as given in Table 2.8.

Apart from capacity and availability, the MTF system also gathers information on 'reliability, quality[36], affordability, formality[37], and health and safety' [209]. All the information is gathered through an extensive and evolving questionnaire, which national statistics organisations are encouraged to use [205]. Global implementation of the MTF system is still in its infancy, but an illustrative example of its usefulness to identify not only the scale of a nation's electrification but also where government action is needed to lift a household to a higher tier of access was provided in a 2018 report on energy access in the sub-Saharan country of Rwanda [205,209]. Using the grid/non-grid and basic-bundle criteria for access to electricity for households, Rwanda was assessed at 23% [205]. However, when the MTF system was applied to the available information, the results indicated that the access rates were higher, at almost 27% (Figure 2.14), the largest share being Tier 3 access [209].

Table 2.8 The MTF tier system for household electrical energy consumption [201,205,206]

Level Of Access ->	TIER					
	0	1	2	3	4	5
Annual Household Consumption per person, kWh		22	224	696	1,800	2,195
Minimum Daily Availability, hours		4	4	8	16	23
Indicative Appliance Use		Task lighting, phone charging or radio	Tier 1 plus (general lighting, air circulation and television)	Tier 2 plus (refrigerator, washing machine)	Tier 3 plus (microwave, space heating, ironing)	Tier 4 plus (energy-intensive device, e.g., air conditioning)

---

[35]Referred to as 'indicative' services.
[36]As measured by voltage fluctuations.
[37]Is the supply from formal or informal connections.

Providing clean and affordable energy for all 47

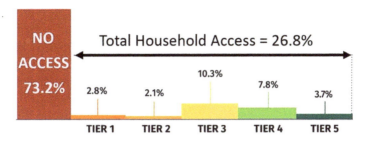

Figure 2.14 Rwanda exemplar of MTF assessment of tiered access to electricity [171]

The Rwanda MTF result may prove unique, or indicative of the usefulness of the approach. As the full MTF system partly relies on national surveys, which are, at best, only conducted every 2–3 years and often less frequently, the system may not be fully operational much before the 2030 deadline. Nevertheless, the example results suggest that the MTF-type approach should convey deeper insights, especially for governments, into the policies and actions needed to address access to electricity issues. Even so, there should be no expectations regarding the development of a better measure of energy affordability using the MTF and household survey approaches, as the benchmarks suggested by the protagonists of the systems are tantamount to being the same as the Boardman and Hills measures of energy poverty [22,44,64,201,207]. Arguably, 'affordability' remains a synonym for poverty, as does poverty for income level. Moreover, as shown in Table 2.8, it is really only in the higher Tiers that cooking and heating become factors in electricity consumption. Does this mean, for instance, that if only electricity can be used as a clean cooking fuel, then a far higher proportion of the global population than presently assessed will be deemed to have no access to such fuels? For many people in lower-income nations, especially in sub-Saharan Africa, an increase from their present Tier classification to a Tier 4 or 5 will prove way beyond their financial means without considerable external assistance. These issues are discussed in the next section.

The MTF access data will likely become an increasingly important addition to the data from many different household surveys undertaken by international agencies, which are compiled by the World Bank in conjunction with national censuses and surveys to produce the meta-database for the SDG 7.1.1 indicator monitoring [124,201,205]. A similar overall multi-sourced system is used for compiling the metadata for the SDG 7.1.2 indicator dealing with the types of household fuel and associated technologies used, including those for cooking.

2.5 Clean cooking fuels

Since early humans first discovered how to make and control fire – likely by accident – they found it useful for cooking certain foods to make them more

48   *Clean energy for low-income communities*

digestible, flavourful, nutritious, energy-rich, and safe. As a result, biomass has been used as a cooking fuel and continues to be widely used today [210]. However, over the past two or three centuries, an energy transition took place from biomass to traditional fossil fuels, and, in many countries, these became the cooking fuels of choice; particularly gaseous fuels manufactured from coal and, since the latter part of the 20th century, natural gas. The use of electricity as a cooking fuel, although dating back to the late 19th century, is a relatively recent development, with electric ovens only starting to become competitive with gas stoves during the second half of the 20th century, albeit a little earlier in the US [211]. Until the 1990s, an individual's choice of cooking fuel was invariably based on cost and availability and not on whether it was 'clean' or 'unclean' [212,213]. However, the growing concerns with air quality, levels of air pollution, and the associated adverse health impacts from the 1960s onwards eventually led to clean air legislation in many countries and the identification of key air pollutants, such as particulate matter (PM) and carbon monoxide (CO), whose levels had to be kept below a defined threshold level [214,215]. Eventually, the WHO produced guidelines on what constitutes ambient (outdoor) air pollution and indoor air pollution [120,122]. With the advent of these guidelines and various national versions, the majority of traditionally used cooking fuels were now considered to be 'unclean' because of their emissions, especially PM and CO levels, which were above the recommended guidelines. Indeed, in the year 2000, over 3 billion people were determined to be using such fuels [216].

One obviously alarming parameter associated with the use of PM-producing fuels is the number of human deaths they cause, with annual estimates[38] varying between 2.3 million and 3.8 million from indoor pollution alone, and between 5.5 million and 8.7 million from all anthropogenic sources, such as the burning of fossil fuels, although the largest global source of PM is the airborne dust in deserts [217]. To place these statistics in a broader context, it is worth mentioning that in 2019, annual deaths globally were 55 million, with obesity responsible for 5 million, transport accidents for 1.3 million, homicides for over 415,000, and occurrences like the European summer heat wave of 2022 resulting in 63,000 deaths [218,219]. Clearly, deaths from the emissions of unclean cooking fuels cannot be ignored, but some global progress has been made in providing access to clean fuels since the start of this century. India and China have made considerable progress in increasing access to clean cooking fuels, although some other countries have struggled, and there is evidence of worsening situations as indicated by the 'access' exemplars given in Table 2.9 [216,220].

Nevertheless, despite the progress made in India and China, the latest World Bank data shows that the number of deaths attributable to household air pollution in India and China is over 811,000 and 728,000, respectively, compared to zero deaths in countries and regions like the US, the UK, and the EU [221]. It seems highly unlikely that there are no attributable deaths to indoor air pollution in some countries, but, as with all the data related to air pollution, indoor and outdoor,

---

[38]Pre-Covid pandemic

Providing clean and affordable energy for all   49

Table 2.9   Exemplars of clean cooking fuel access 2000–2021 [216,220]

Country	Year		
	2000	**2016**	**2023**
	Millions of people		
World	3,090	3,020	~2,035
India	823	781	404
China	672	562	240
Nigeria	121	177	176
Pakistan	110	115	105

inconstancies in methodologies and, in some instances, a paucity of relevant information mean that, at best, the published numbers are estimates. However, with some certainty, it can be stated that hundreds of millions do not enjoy access to clean cooking fuels and that over the past two decades, the number of people without access to clean cooking fuels has been three to four times greater than the number without access to electricity. Moreover, while the foremost concern is associated with the adverse health impacts of polluting cooking fuels, in-depth studies of such effects on childhood mortality, estimated at close to 240,000 deaths annually, and natural development are still largely lacking [222].

As previously discussed, the WHO guidelines on air quality and what constitutes a clean cooking fuel are somewhat at odds with climate change mitigation and net-zero strategies given that not only do non-electric cooking fuels produce PM but also GHGs. This situation will have to be rectified in the future, but how? As with access to electricity efforts, far more information and data are required, especially at national levels, to identify and make clear to governments and consumers what specific actions can be taken in an effective manner to meet climate and sustainable development targets. Arguably, this process is underway already, as the MTF tiered approach adopted for gathering and analysing 'access to electricity' data has also been developed for determining access to '*modern energy cooking services*' [201,205]. The cooking fuel framework uses six tiers in a similar manner to the MTF access to electricity methodology, but, in this case, the primary attribute of the tier is not energy consumption but the level of an individual's exposure to PM and CO. Of the other five attributes, *availability* and *affordability* are included as with MTF electricity, but with different definitions. For example, with Tier 3 modern cooking services, primary energy must be '*readily available when needed*' for at least 80% of the year, whereas Tier 3 electricity requires availability for 8 h per day [205]. Tier 3 access to electricity is considered to be affordable if it costs less than 5% of household income, whereas for cooking it is less than 10% of such income, which could suggest that energy for cooking is more expensive than general household energy or that cooking takes a larger share of the household's consumed energy [205].

50 *Clean energy for low-income communities*

The MTFs for electricity access and clean cooking are intended to contribute directly to the assessment of progress towards SDG indicators 7.1.1 and 7.1.2 but detailed analyses using the approaches are yet to be published. Metadata for access to cooking fuels and cooking devices is collected and compiled in much the same way as that for access to electricity analyses except that more use is made of targeted national-level surveys and censuses along with specialised household surveys conducted by such agencies as the WHO, the United Nations Children's Fund (UNICEF) and the United States Agency for International Development (USAID) [123,124,205,220]. The WHO, the SDG custodian for 7.1.2, is working with an increasing number of countries to entrench the MTF questionnaire for clean cooking fuels into their national surveys and official statistics [124]. So, it is likely that some of the findings published in the annual SDG 7 progress reports may have used some MTF information, where it exists, along with that garnered from national household surveys [89,223,224]. The WHO is also working with UN member states to expand the MTF-type questionnaire to assess how households use fuels and technologies for applications other than cooking.

2.6 Space heating and warm households

Although not part of the SDG 7.1 target indicators 7.1.1 and 7.1.2, household access to energy and the use that is made of the consumed energy and for what means are now becoming topics of increased interest. Recently, the EU published an analysis of how, specifically, household energy is used [225]. A summary of the results for 2021 is given in Figure 2.15, where it can be seen that the household energy expended on space heating is an order of magnitude greater than that used for cooking and almost five times as much as that used for lighting and powering

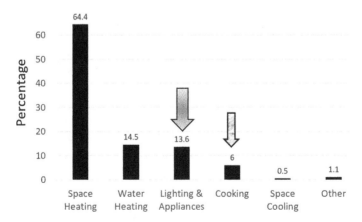

Figure 2.15   EU household energy use by purpose in 2021 [187]

Providing clean and affordable energy for all 51

appliances. For those experiencing energy poverty and affordability, reduced heating, i.e., lowering room temperatures below recommended thresholds, maybe a cost-coping measure. However, it has to be appreciated that the pattern of household energy use shown in Figure 2.14 is unlikely to be representative of all global populations. In countries, or parts of countries, located closer to the warmer equatorial regions, people will not need to heat their homes as much as those living in temperate or cold climates, but, conversely, homes in warmer countries, if the occupants can afford to, will likely consume more energy on space cooling. Such scenarios define the utilisation boundaries of household energy use, as there will be many instances where those who experience hot summers and cold winters will use as much energy for cooling as they do for heating, depending upon the length and depth of the seasonal changes.

Thus, keeping homes at comfortable temperatures is a significant factor in household energy consumption, but the ability to cook food is essential for human well-being and survival, and providing lighting after sunset is an important factor in energy poverty mitigation [21,46,76,203]. The added codicil 'for cooking' associated with the updated definition of indicator 7.1.2 would seem to exclude a specific focus and measurement dealing with the use of modern energy services, clean fuels, and technologies involved in household space heating [89]. Yet, as shown in Figure 2.14, in terms of overall household energy use, space heating can dominate consumption. To address this issue, an MTF for space heating has been developed, and the WHO has made significant progress in including heating data in their global database, covering the period 1977–2020 for some 71 countries [124,201,205]. Similarly, the WHO has also included data in their huge energy database with respect to lighting, but for more countries than for heating [126], and information for a longer period of time, 1963–2020 [124,125].

2.7 Affordable energy for other end-use sectors

The discussions so far regarding clean and affordable energy for sustainable development have focused largely on households' needs, as these impact almost everyone[39]. However, most people have to travel to work, shop, attend school, visit family and friends, for recreation, and so on, which means that energy is required for transportation along with that needed for the operation of the associated infrastructure. Travel may be by on-road vehicles, private and public, rail, marine, or air, and involve not only people but also local, national, regional, and international import and export trade. Many trade activities involve the manufacture of a wide range of goods produced from chemicals, petroleum and coal products, and paper to food items, all of which use machinery that requires

---

[39]There are also a large number of homeless, both permanent and temporary, with the possible exception of Japan, in many countries, regardless of income. However, not all countries gather data in a consistent manner, if at all, so a global figure is not available.

52 *Clean energy for low-income communities*

Table 2.10 World total energy consumption by sector and source 2020 [228]

End-Use Sector	Energy consumption TWh	Sector % of Total	Source % of Sector Energy Consumption					
			Coal	Oil	Natural Gas	Biofuels and waste	Electricity	Heat
Total Energy Consumption[a]	99,278	100	10.2	35.0	16.7	11.6	22.7	3.8
Industry	33,361	33.6	26.6	9.4	21.5	8.7	28.6	5.2
Transport	27,917	28.1	0.1	90.0	4.7	3.8	1.5	0.04
Households	23,556	23.7	2.7	13.6	23.3	27.8	26.7	5.9
Other[b]	14,444	14.6	4.2	23.3	17.9	7.1	43.5	4.0

[a]In 2020, a further 11,472 TWh of non-combusted fossil fuel sources was consumed for non-energy purposes giving a world total consumption figure, according to the UN, of 110,750 TWh [228].
[b]Includes Commercial and Public Services and other energy consumption not covered in the main sector categories as per UN energy statistics methodologies.

power derived from a variety of energy sources. The needs of the manufacturing sub-sector dominate the industrial uses of energy, but non-fuel mining, construction, and agricultural activities also consume energy [226]. Energy is also required to generate electricity. Furthermore, not all fossil fuel sources are combusted but are consumed directly as raw materials in the generation of a multitude of non-energy products that include lubricants, solvents, fertilisers, plastics, synthetic rubber, fabrics and dyes, skin-care treatments, construction materials for roads and roofing, and many more [227]. Indeed, the 'non-energy' use of fossil fuels accounts for about 10% of the world's total final energy consumption [228]. This may seem to be a contradictory situation, but non-energy consumption is included as being essential for the determination of GHG emissions, albeit as a separate category from the emissions calculations for combusted fuels [229].

As can be appreciated from the UN's global energy consumption data, shown in Table 2.10, industry and transport consume the most energy, but while industry uses more electricity than households, transport hardly uses any electricity, relying almost completely on oil. However, many governments have committed to all forms of transport, especially on-road vehicles, being progressively electric over the next 15–30 years. To ensure that their commitments are achieved, several strategies have been adopted, including providing financial incentives to purchasers of electric vehicles (EVs) and banning the manufacture of new cars that are not electric. In France, the government recently brought in legislation to ban domestic short-haul flights. These 'carrot and stick' approaches may work as new on-road EV sales have increased dramatically over the last decade, for example, over 6% of all new vehicle sales in the US in 2022 were electric, and in China, some 3.5 million EVs were sold in 2021 [230]. Nevertheless, despite the efforts to move to an all-electric transportation scenario, only 2.2% of on-road vehicles worldwide are electric [230]. The electrification of transportation is the largest challenge to sustainable development, climate change, and net-zero emissions, not only because of vehicular costs and consumer purchasing reluctance but also due to the lack of infrastructure, i.e., battery charging facilities, and the need to generate more electricity. Overall, over 4.5 times more electricity will be required to meet the demands of an all-electric world, and, in the case of transportation, the use of electricity will have to increase by a factor in the region of 70, based on the latest UN data [228]. If electricity is to be clean, then some 60 +% of the global primary energy sources presently used will have to be replaced. Can this really be achieved in a relatively short time span with no incremental energy costs?

2.8 Discussion possible, practical, propaganda

There is no doubt that, for the purposes of measuring SDG 7 indicators, especially 7.1.1 and 7.1.2, an ever-increasing amount and breadth of insightful data is being collected, correlated, and analysed. But have these undertakings resulted in government actions, both nationally and globally, to ensure that sufficient progress is being made to achieve the desired level of access to clean and affordable energy *for all*? The links between affordability, poverty, and incomes are inescapable but complicated to qualify and quantify on either a generalised global basis or nationally. If global poverty is to be eradicated, the primary objective of Agenda 2030, then surely energy/fuel poverty must first be eliminated? However, as discussed in the earlier parts of this chapter, energy poverty is far from being eliminated or even reduced, despite, in some cases, government methodologies being changed to enable them to present a more positive outcome of their poverty reduction strategies. Such an observation maybe unfair. Governments could argue that such changes have allowed a more accurate insight into the causes of household energy poverty and, hence, have been instrumental in formulating more effective policies to combat such poverty. Whatever the rationale for fighting energy poverty over the past half-century or more, the problem has remained remarkably persistent. Providing affordable energy for all by 2030 is not possible, whatever the energy mix.

Some of the problems encountered in the attempts to reduce energy poverty have been self-inflicted and not always justified. This issue centres around the conflicting interpretations of what is 'clean' energy and what is 'renewable energy'. If clean means, there are to be no GHG emissions, then nuclear energy would qualify for the 'clean' category, but nuclear is not deemed renewable, and the thrust of the leadership of the United Nations and the Net-Zero coalition is for only renewables to be considered clean. Yet, many of the fuels considered by the WHO to be clean do produce GHG emissions, including some renewables, such as biomass. Without nuclear, biomass, and all fossil fuels, even natural gas, due to the 'Global Methane Pledges', can the world produce sufficient energy from solar, wind, water, i.e., *clean* hydro and tidal, and geothermal sources to meet the global

54 *Clean energy for low-income communities*

energy consumption needs and at affordable prices? Groups of researchers have published models and reviews indicating that this can be achieved, if not by 2030, then soon after, and at a low cost, but to do so also requires the implementation of many recommendations by governments worldwide [144,175,176]. From the findings and analyses of the most recent SDG7 progress report, it appears that global governments have not fulfilled, or been able to fulfil, all the pledges and commitments made with regard to their contributions to the Paris Agreement or the Sustainable Development Agenda [118,136,208,223].

If the average rate of progress, made between 2015 and 2021, towards SDG 7 indicators 7.1.1 and 7.1.2 continues, then it is projected that by 2030 at least 660 million and maybe as high as 673 million will still be without access to electricity and that as many as 1.9 billion will still be using unclean cooking fuels [144,224,231]. The promise of lower costs of electricity with the transition to the increasing use of clean renewables has also not materialised, as yet, as evidenced by the cost of 'wholesale' electricity in the EU tripling in the first half of 2022 and rising globally by an average of 30% in 2022, although the IEA suggests these increases are due to the effects of conflicts and higher fossil fuel prices [232]. Presumably, the implication is that the sooner clean renewable energy replaces all other forms of energy, the cheaper electricity will become. To date, this argument does not seem to have gained significant traction with global policymakers, as the world percentage share of low-carbon sources, nuclear plus renewables, has only started to increase over the last 10 years to above 18%[40] in 2022, consisting of 14% from renewables and 4% from nuclear [223,233]. At the start of the century, the energy share of renewables was about 1% more than that of nuclear, but the recent gains in the renewable share have coincided with a declining share of nuclear. Even so, it is projected that the renewable share of global energy will only reach about 23% by 2030 and 35% by 2050 [144,224].

If the SDG targets are highly unlikely to be met, what are the likely energy mixes in 2030 and 2050? Countries have already announced and enacted energy and emission policies to meet the Paris Agreement targets, mainly through the so-called 'National Determined Contributions' (NDC). What impact will these commitments have on the energy mix? As the most ambitious, perhaps radical, energy transition strategy is the net-zero approach, to which an increasing number of countries have pledged their support, what further impact will such pledges have on future energy mixes? The IEA, in their global net-zero roadmap, has attempted to project the mixes using these two scenarios, which they have labelled as the '*stated policies scenario*' (STEPS) and the '*Announced Pledges Case*' (APC) that assume the net-zero pledges will be fully implemented in a timely manner [144]. The results are summarised in Figure 2.16. As can be seen, the same energy sources used between 2000 and 2020 will still be in the energy mix in 2030 and 2050, whatever the scenario, although the respective usage shares will change [144]. As the IEA's remit is to work with governments and industries globally '*to shape a secure and sustainable energy future for all*', it is

---

[40]Using the substitution method

Providing clean and affordable energy for all 55

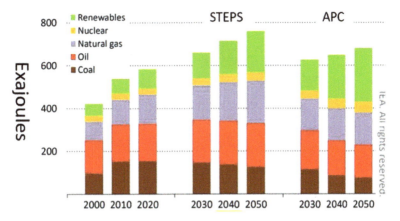

Figure 2.16   IEA's global energy supply forecasts to 2050 using STEPS and APC [119]

reasonable to assume that their reports are thorough and rigorous, although there have been some past criticisms about the IEA's undervaluing the role of renewables and their declining costs [234]. There may be some validity to these reproaches, but they may have more to do with the IEA's interpretations and presentation of facts than the actual facts and forecasts themselves. The IEA is not the agency to produce such forecasts. Indeed, there are many organisations that have attempted to model the energy transition trends to 2050 using a variety of assumptions [235].

For example, the USEIA model predicts future energy trends not only for the US but for the world as a whole. In their 2021[41] international report, the constituents of 2030 and 2050 energy mixes are the same as the IEA's, but the scale of the individual components has some differences in magnitude. In terms of electricity generation energy sources, fossil fuels will still dominate in 2030 but renewables will take the lead by 2045, and in 2050, are forecast to have a share of over 56% [236]. The USEIA's analysis also projects that electricity will account for 50% of residential energy consumption in 2050, with natural gas providing just less than a third of the overall consumption. Moreover, world crude oil production in 2050 will increase by 30% compared to 2020 [236]. A *'Resources for the Future'* report considered 14 different modelling scenarios, eight of which forecast that the demand for natural gas would be higher by 2050 than it is today, and all had forecasted energy mixes that included coal, gas, and oil [235]. Taking all these forecasts into account, it could be concluded that 'access to electricity' and 'access to clean energy' *for all* will not take place they are far from it. Consequently, the SDG7 energy targets will not be achieved by 2030, and their fulfilment is central to the achievement of all the other goals of Agenda 2030, see Figure 2.17 [237]. The overall implication is that many people will still be impoverished, a situation

---

[41] A new report is due in October 2023

56 *Clean energy for low-income communities*

Figure 2.17   Central role of SDG7 in sustainable development agenda 2030 [194]

supported by a recent World Bank analysis projecting that approximately 574 million people will be struggling with extreme poverty in 2030 [238].

However, is the purpose of all these forecasts to inform everyone of the apparent hopelessness of meeting the sustainable development and climate change goals within the ambitiously defined deadlines, including the 2050 net-zero pledges, or is it part of an orchestrated campaign to increase global funding to ensure the original deadlines and aims are met? Although there have been some improvements in meeting the various SDG targets, albeit even with the adverse impact of the COVID pandemic, the various progress reports have all shown that the rate of progress is insufficient to meet the deadlines, which has led the UN leadership, with the support of many UN member states, to call for a 'Rescue Plan' [239]. In essence, this, and other reports over the past few years, have indicated that global financial contribution commitments and pledges have not been kept, so there is a shortfall of US$135 trillion in investments needed to achieve the SDG goals by 2030 out of the estimated total cost of US$176 trillion, equivalent to close to twice the world's annual GDP; five times that of the US, 33 times that of Europe's richest country, Germany, 40 times that of India, and over 250 times that of the United Arab Emirates [240–243]. Exactly how much of these investment funds will need to be used specifically for the transition to renewables, a key indicator of SDG 7, is a matter of debate, although it was estimated, pre-COVID, that at least US$50 trillion would be needed [244].

It should be noted that the published amount of estimated funding required to accelerate progress in the sustainable development and climate change agendas to meet the 2030 deadline does seem to vary in the many UN reports and press releases, which is perhaps not wholly surprising given the difficulties encountered in forecasting future rates of economic growth and financial markets. Notwithstanding the somewhat puzzling situation, it can be construed that the amount will involve tens or even hundreds of trillions of US dollars. But where is this money going to come from over the next 6–7 years? The literature reveals many suggested possibilities, such as nations increasing their annual contributions to achieving the desired goals, eliminating all fossil fuel subsidies, not just the 'inefficient ones', SDG 12.c.1, transferring large shares of the 'liquid' assets of rich countries to poor countries, and providing low or no-interest loans to the least developed countries to aid them in addressing sustainable development goals [14,245]. It appears that the uncertainties in the cost of the energy transitions needed to fully implement the Paris Agreement and Sustainable Agenda commitments are matched by the elusiveness of how they are to be funded. However, a recent paper modelling future growth using the IPCC's 'Shared Socioeconomic Pathways' scenarios, along with the data from such organisations as the World Bank, has concluded that by 2050 the SDG goal for the elimination of extreme poverty will be achieved along the disappearance of the low-income countries group and an increase to twice the number of members of the high-income group, but even these somewhat optimistic projections only include the doubling of global electricity consumption [246,247].

Based on the presented discussions, in the author's opinion, technical, economic, and political impediments will not only prevent universal access to clean and affordable energy by 2030 but also well into the second half of this century, *if at all*. The infrastructure needs alone will provide overwhelming engineering challenges. Claims that it will *only* require political will and financial transfers from rich to poor to overcome, in a timely manner, all the stumbling blocks likely to be encountered in a universal energy transition from the present mix to an all-clean, affordable, renewable, and sustainable mix are tantamount to opinions bordering on propaganda. But is propaganda a fair depiction? It depends. If propaganda is taken to be a form of '*biased communication, a way of altering the views of a population towards a certain cause, or engaging psychological mechanisms of influence*', then the answer is yes [248]. Others consider propaganda as '*a narrowly selfish attempt to get people to accept ideas and beliefs, always in the interest of a particular person or group and with little or no advantage to the public*', in which case the portrayal is unreasonable [249]. However, especially among academics and scholars, there is no consensus about a precise definition or meaning of the word 'propaganda' or its use. Nonetheless, all forms of propaganda are powerful tools for the achievement of personal or political objectives, particularly in a connected world. Perhaps, the most useful definition of propaganda, as suggested almost 80 years ago, is that it has to do with, '*any ideas or beliefs that are intentionally propagated*', in attempts to reach a stated goal [250].

58   *Clean energy for low-income communities*

Consequently, there appears to be a general, if not universal, belief of sufficient magnitude that the mitigation of climate change and attainment of sustainable development are seen as essential for human and earth's survival. This has resulted in the development of an overall ideology around these goals. Yet, these beliefs are based more on the pronouncements of many world leaders, who rely mostly on numerous computer software models rather than on indisputable measurements. Rarely are these circumstances mentioned in the media aimed at the general public, or so it seems. If not propaganda, then what description should be used to describe the unlikely achievement of the SDGs, specifically SDG 7, by 2030?

2.9   Concluding remarks

In this chapter, the intricacies of energy generation and consumption were discussed against the background of providing clean and affordable energy for households. Arguably, 'clean' is a scientific axiom, but it is nevertheless contentious. Does it ensure the absence of unhealthy atmospheric pollution or the utilisation of energy sources that mitigate climate change? Can these two viewpoints be reconciled by eliminating GHG and particulate matter emissions, while simultaneously encouraging the increased use of bioenergy, and discouraging nuclear power? What yardsticks can be used to determine whether an energy source is clean or unclean; the WHO's guidelines, or the many actions, e.g., the net-zero roadmap, the methane pledge, being proposed and implemented to eliminate any human generation of carbon emissions, apart from exhaling? It appears that the only way to achieve the desired 'clean' outcome is to wholly switch to renewable sources that generate electricity. However, as discussed earlier, evidence indicates that when such switches are made, energy costs, especially electricity prices, rise rapidly, although cheaper energy in the future remains an often-repeated mantra. Even so, for some consumers, electricity prices have been abated by the subsidised installation of solar panels and operational incentives such as 'feed-in-tariffs' [251,252]. Only those households with the financial means have been able to take advantage of such schemes. Those who could not afford to do so, or were not able to, for a variety of reasons (demographics, location, property ownership, etc.), have had to cope with increasingly unaffordable grid electricity prices, partly due to having to underwrite the costs of the incentives for others. This has likely resulted in more people experiencing energy poverty and an even greater need for government welfare-type financial interventions. Affordability is then a political measure rather than a scientific or mathematical artefact and it is in the eye of the beholder based on their income and poverty circumstances.

Despite more than five decades of considerable effort, not only to eliminate poverty but also to provide a universally accepted definition of poverty and an appropriate measurement system, the problems of global poverty continue. No country is immune. A similar situation exists with the definition and measurement of affordability. It seems that the term 'affordability' is encountered on a daily basis when societal issues are discussed in the mainstream media, by politicians of all ideological persuasions, activists, and lobbyists. But in the same way as a simple climate change question – what should be the ideal surface temperature of the planet – has yet to be answered, so too have questions on how much, in the Germane currency, an affordable home or affordable energy cost the consumer. The underlying philosophy seems to be that if poverty is eliminated, then everything will be affordable. So, utterances such as the government is committed to building a specific number of homes of which the majority will be affordable, without stating the price of such dwellings, do seem to be meaningless or have more to do with political propaganda, i.e., influencing public opinion in terms of their policies, or maybe motivational 'feel-good' attempts, than actuality. When the quest began to eradicate poverty, and later to provide affordability, there was little or no focus on or acknowledgement of anthropogenic climate change and sustainable development. This situation has radically changed, and the general public is becoming increasingly persuaded that action must be taken on these issues, but, in most countries, it is the economy and an individual's financial well-being that are the foremost concerns of the majority.

Sufficient food, clean water, clean air, and adequate shelter are necessary for human survival. These can only be provided for all by defeating poverty. Moreover, ensuring access to clean and affordable energy for everyone is one of the key strategies for eliminating energy poverty and general poverty. Such access is then a laudable goal, but there are no indications that it can be achieved by the Agenda 2030 or Net-Zero timelines or indeed in the foreseeable future. Nevertheless, on all socioeconomic fronts, apart from conflict resolution, progress has been made in this millennium and, in the absence of a series of global catastrophes, will likely continue. It is the speed and depth of progress, especially regarding energy transitions, that is the main concern of political leaders, and the cost of the progress is the concern of the general population. As Smil has repeatedly highlighted, past energy transitions have taken much longer than perceived, a great deal of effort had to be expended, and availability and cost were dominant factors [149,212,213]. It appears not much has changed since Smil's insightful investigations.

The use of the 'access to electricity' indicator as a measure of achieving clean and affordable energy for all is a harbinger of the all-electric future desired by world political leaders, yet the monumental task to achieve this, as reflected by the data given in Table 2.10, is likely, not doable in any reasonable timeframe associated with the present century. Access to energy that is both affordable *for all* and simultaneously meets the clean zero-emissions requirements seems to be more of a *forlorn hope* than a truly practical endeavour, but the eradication of global poverty and the exclusion of anthropogenic environmental harm are attractive propositions. Regardless of recent global setbacks involving pandemics and armed conflicts, and notwithstanding continuing population growth, there have been some global improvements in human development areas over the last several decades, such as income, life

60   *Clean energy for low-income communities*

expectancy, access to education, and household living standards, as well as extreme poverty reduction [8]. There are reasons for optimism for further, if not wholly satisfactory, improvements to be achieved, rather than the pessimism apparently being espoused by some current political and scientific leaders who could do well to take the advice of the American Businessman and former CEO of the Disney company, Bob Iger, i.e.,

'What I have really learned over time is that optimism is a very, very important part of leadership. However, you need a dose of realism'.

References

[1] United Nations. *United Nations Framework on Climate Change*. New York: United Nations; 1992.

[2] Litman T. *Transportation Affordability: Evaluation and Improvement Strategies*. Victoria, B.C.: Victoria Transport Policy Institute; 2021.

[3] Anacker KB. Introduction: housing affordability and affordable housing. *International Journal of Housing Policy*. 2019; 19(1): 1–16.

[4] Sodsriwiboon P, Srour G. *Does a Minimum Wage Help Workers?* [Online]; 2019 [cited 3 September 2023]. Available from: https://www.imf.org/en/Publications/fandd/issues/2019/03/does-a-minimum-wage-help-workers-basics.

[5] Tanner MD. *Poverty and Welfare: 2022 . CATO Handbook for Policymakers. Handbook*. Washington, DC: CATO Institute; 2022.

[6] Hanna R, Olken BA. Universal basic incomes versus targeted transfers: anti-poverty programs in developing countries. *Journal of Economic Perspectives*. 2018; 32(4): 201–226.

[7] Worldometer. *GDP by Country*. [Online]; 2022 [cited 3 September 2023]. Available from: https://www.worldometers.info/gdp/gdp-by-country/.

[8] Roser M. *The History of the End of Poverty Has Just Begun*. [Online]; 2022 [cited 4 September 2023]. Available from: https://ourworldindata.org/history-of-poverty-has-just-begun' [Online Resource].

[9] Gornick J, Johnson N. *Income Inequality in Rich Countries: Examining Changes in Economic Disparities*. Insight Briefing. New York: Social Science Research Council; 2022.

[10] Foster JE, Wolfson MC. Polarization and the decline of the middle class: Canada and the U.S. *The Journal of Economic Inequality*. 2010; 8(2): 247–273.

[11] Chauvel L, Haim EB, Hartung A, and Murphy E. Rewealthization in twenty-first century Western countries: the defining trend of the socio-economic squeeze of the middle class. *The Journal of Chinese Sociology*. 2021; 8(4): 17.

[12] Carley S, Konisky DM. The justice and equity implications of the clean energy transition. *Nature Energy*. 2020; 5: 569–577.

[13] Hancock KE. 'Can pay? won't pay?' or economic principles of 'afford-ability'. *Urban Studies*. 1993; 30(1): 127–145.

[14] United Nations. *Transforming Our World: The 2030 Agenda for Sustainable Development*. Washington, DC: United Nations, Economic and Social Affairs; 2015.

[15] Merriam-Webster. *Affordable*. [Online]. [cited 10 July 2023]. Available from: https://www.merriam-webster.com/dictionary/affordable.

[16] Cambridge Dictionary. *Affordable*. [Online]; 2023 [cited 10 July 2023]. Available from: https://dictionary.cambridge.org/us/dictionary/english/affordable.

[17] Dictionary.com. *Affordable*. [Online]; 2018 [cited 10 July 2023]. Available from: https://www.dictionary.com/browse/unaffordability.

[18] USDOE. *Office of Energy Efficiency & Renewable Energy (eere)*. [Online]; 2023 [cited 10 July 2023]. Available from: https://www.energy.gov/eere/energy-accessibility-and-affordability.

[19] Burlinson A, Giulietti M, Law C, and Liu HH. Fuel poverty and financial distress. *Energy Economics*. 2021; 102(105464): 18.

[20] Bradshaw J, Hutton S. Social poverty options and fuel poverty. *Journal of Economic Psychology*. 1983; 3(3–4): 249–266.

[21] Fefferman N, Chien-Fei C, Bonilla G, Nelson H, and Kuo CP. How limitations in energy access, poverty, and socioeconomic disparities compromise health interventions for outbreaks in urban settings. *iScience*. 2021; 24(12): 1–17.

[22] Boardman B. Fuel poverty synthesis: lessons learnt, actions needed. *Energy Policy*. 2012; 49: 143–148.

[23] Bednar DJ, Reames TG. Recognition of and response to energy poverty in the United States. *Nature Energy*. 2020; 5(6): 432–439.

[24] Moore R. Definitions of fuel poverty: implications for policy. *Energy Policy*. 2012; 49: 19–26.

[25] Li K, Lloyd B, Liang XJ, and Wei YM. Energy poor or fuel poor: what are the differences? *Energy Policy*. 2014; 68: 476–481.

[26] González-Eguino M. Energy poverty: an overview. *Sustainable and Renewable Energy Reviews*. 2015; 47: 377–385.

[27] Lawson A, Davies R. *Pressure Rises on Hunt as 2m More Households Fall into Fuel Poverty*. [Online]; 2023 [cited 4 August 2023]. Available from: https://www.theguardian.com/society/2023/feb/28/pressure-jeremy-hunt-as-2m-more-households-fall-into-fuel-poverty.

[28] Heindel P, Schuessler R. Dynamic properties of energy affordability measures. *Energy Policy*. 2015; 86: 123–132.

[29] Cong S, Nock D, Qiu YL, and Xing B. Unveiling hidden energy poverty using the energy equity gap. *Nature Communications*. 2022; 13(2456): 12.

[30] Baker KJ, Ronald Mould R, and Restrick S. Rethink fuel poverty as a complex problem. *Nature Energy*. 2018; 3: 610–612.

[31] UK Department of Business, Energy and Industrial Strategy. *Sustainable Warmth Protecting Vulnerable Households in England. Presented to*

62 *Clean energy for low-income communities*

Parliament. London: UK Government, Department of Business, Energy and Industrial Strategy; 2021. Report No.: ISBN 978-1-5286-2412-1.

[32] United States Census Bureau. Poverty Thresholds. [Online]; 2023 [cited 24 August 2023]. Available from: https://www.census.gov/data/tables/time-series/demo/income-poverty/historical-poverty-thresholds.html.

[33] Fay B. *Poverty in the United States*. [Online]; 2021 [cited 23 August 2023]. Available from: https://www.debt.org/faqs/americans-in-debt/poverty-united-states/.

[34] DeSilver D. *Pew Research CenterWho's Poor in America? 50 Years into the 'War on Poverty,' a Data Portrait*. [Online]; 2014 [cited 24 August 2023]. Available from: https://www.pewresearch.org/short-reads/2014/01/13/whos-poor-in-america-50-years-into-the-war-on-poverty-a-data-portrait/.

[35] Creamer J, Shrider EA, Burns K, and Chen F. *Poverty in the United States: 2021; Current Population Reports P60-277*. Washington, DC: U.S. Department of Commerce, U.S. Census Bureau; 2022.

[36] The United States Government. *Historical Population Change Data (1910-2020)United States Census Bureau*. [Online]; 2021 [cited 25 August 2023]. Available from: https://www.census.gov/data/tables/time-series/dec/pop-change-data-text.html.

[37] Wikipedia contributors. *Immigration to the United States*. [Online]; 2023 [cited 25 August 2023]. Available from: https://en.wikipedia.org/w/index.php?title=Immigration_to_the_United_States&oldid=1172019096.

[38] McKee GA. and Lyndon B. *Johnson and the War on Poverty*. [Online]; 2014 [cited 27 August 2023]. Available from: https://prde.upress.virginia.edu/content/WarOnPoverty2.

[39] Haskins R. *The War on Poverty: What Went Wrong?* [Online]; 2013 [cited 27 August 2023]. Available from: https://www.brookings.edu/articles/the-war-on-poverty-what-went-wrong/.

[40] Hayes A. *1973 Energy Crisis: Causes and Effects*. [Online]; 2023 [cited 11 August 2023]. Available from: https://www.investopedia.com/1973-energy-crisis-definition-5222090.

[41] HISTORY.com Editors. *Energy Crisis (1970s)*. [Online]; 2022 [cited 11 August 2023]. Available from: https://www.history.com/topics/1970s/energy-crisis.

[42] PSE. *Poverty and Social Exclusion; Defining, Measuring and Tackling Poverty*. [Online]; 2013 [cited 11 August 2023]. Available from: https://www.poverty.ac.uk/.

[43] Townsend P. *Poverty in the United Kingdom*. London: Allen Lane and Penguin Books; 1979.

[44] Boardman B. *Fuel Poverty: From Cold Homes to Affordable Warmth London*. New York: Belhaven Press; 1991.

[45] Green KP, Jackson T, Herzog I, and Palacious M. *Energy Costs and Canadian Households: How Much Are We Spending?* Vancouver; 2016. Report No.: ISBN 978-0-88975-384-6.

[46] UK Government. *Warm Homes and Energy Conservation Act 2000*. [Online]; 2000 [cited 23 August 2023]. Available from: https://www.legislation.gov.uk/ukpga/2000/31/data.pdf.

[47] Gore D. *The Warm Homes and Energy Conservation Bill*. Research Paper 00/26. London: House of Commons; 2000.

[48] Hills J. *Getting the Measure of Ffuel Poverty: Executive Summary. Summary*. London: London School of Economics and Political Science, Great Britain. Dept. of Energy and Climate Change; 2012. Report No.: ISSN 1460-9770.

[49] Boardman B. *Fixing Fuel Poverty: Challenges and Solutions London. Sterling, VA*: Earthscan; 2010.

[50] Fabbri K. Building and fuel poverty, an index to measure fuel poverty: an Italian case study. *Energy*. 2015; 89: 244–258.

[51] Imbert I, Nogues P, and Sevenet M. Same but different: on the applicability of fuel poverty indicators across countries—insights from France. *Energy Research & Social Science*. 2016; 15: 75–85.

[52] Inter-Ministerial Group on Fuel Poverty. *The UK Fuel Poverty Strategy*. London: UK Government, Department of the Environment Food and Rural Affairs; Department of Trade and Industry; 2001.

[53] MDPI/Energies. *Energy Perfomance in Buildings and Quality of Life*. Fabbri K, editor. Basel: MDPI; 2020.

[54] Barron P, Cord L, Cuesta J, Espinoza SA, Larson G, and Woolcock M. Social sustainability in development: meeting the challenges of the 21st Century. *New Frontiers of Social Policy.*Washington, DC: World Bank; 2023.

[55] Eurostat. *Living Conditions in Europe – Poverty and Social Exclusion*. [Online]; 2023 [cited 17 August 2023]. Available from: https://ec.europa.eu/eurostat/statistics-explained/index.php?title=Living_conditions_in_Europe_-_poverty_and_social_exclusion&oldid=584082#Poverty_and_social_exclusion.

[56] United Nations Department of Economic and Social Affairs. *Leaving No One Behind: The Imperative of Inclusive Development; Report on the World Social Situation 2016*. New York: United Nations, Department of Economic and Social Affairs; 2016. Report No.: ISBN 978-92-1-130336-0; eISBN 978-92-1-057710-6.

[57] Hinson S, Bolton P, and Kennedy S. *Fuel Poverty. Research Publication and Briefing*. London, UK: House of Commons Library, Commons Library; 2023.

[58] Department of Energy and Climate Change (DECC). *Annual Report on Fuel Poverty Statistics 2012*. London: GOV.uk, Department of Energy & Climate Change; 2012.

[59] Kelly S, Pollitt MG, and Crawford-Brown D. Building performance evaluation and certification in the UK: is SAP fit for purpose? *Renewable and Sustainable Energy Reviews*. 2012; 16(9): 6861–6878.

64 *Clean energy for low-income communities*

[60] Department for Energy Security and Net Zero and Department for Business,Energy and Industrial Strategy. *Guidance: Standard Assessment Procedure.* [Online]; 2022 [cited 28 August 2023]. Available from: https://www.gov.uk/guidance/standard-assessment-procedure#full-publication-update-history.

[61] Henderson J and Hart J. *BREDEM 2012.* Watford: BRE; 2015.

[62] Kidson M and Norris E. *Implementing the Fuel Poverty Strategy. Case Study.* London: Institute for Government; 2014.

[63] Stockton H and Campbell R. *Time to Reconsider UK Energy and Fuel Poverty Policies?* [Online]; 2011 [cited 28 August 2023]. Available from: https://www.jrf.org.uk/sites/default/files/jrf/migrated/files/fuel-poverty-policy-summary.pdf.

[64] Hills J. *Fuel Poverty: The Problem and Its Measurement: Interim Report of the Fuel Poverty Review.* London: UK Department of Energy and Climate Change, Centre for the Analysis of Social Exclusion; 2011. Report No.: ISSN 1465-3001.

[65] Hills J. *Getting the Measure of Fuel Poverty; Final Report of the Fuel Poverty Review.* The London School of Economics and Political Science, Centre for Analysis of Social Exclusion; 2012. Report No.: CASE Report 72, ISSN 1465-3001.

[66] Oxley S. *Fuel Poverty Methodology Handbook (Low Income High Costs). Handbook.* London: UK Government, Department for Business, Energy and Industrial Strategy; 2020.

[67] Romero JC, Linares P, and López X. The policy implications of energy poverty indicators. *Energy Policy.* 2018; 115: 98–108.

[68] Burlinson A, Giulietti M, and Battisti G. The elephant in the energy room: establishing the nexus between housing poverty and fuel poverty. *Energy Economics.* 2018; 72: 135–144.

[69] Department for Energy Security and Net Zero and Department for Business, Energy and Industrial Strategy. Energy Trends: December 2018, Special Feature Article – Do Households Move In and Out of Fuel Poverty? London: UK Government, Department for Energy Security and Net Zero and Department for Business, Energy & Industrial Strategy; 2018.

[70] Department for Business, Energy & Industrial Strategy. *Annual Fuel Poverty Statistics Report, 2018 (2016 Data) England.* Annual Report. London: UK Government, National Statistics; 2018.

[71] Massey C and Waters M. How fuel poverty is measured in the UK: March 2023. Article. Newport,Wales: Office for National Statistics (UK); 2023.

[72] Department for Energy Security & Net Zero and BRE. *Fuel Poverty Methodology Handbook (Low Income Low Energy Efficiency). Handbook.* London: UK Government, Department of Energy Security and Net Zero; 2023.

[73] Department of Energy and Climate Change. *Fuel Poverty Energy Efficiency Rating Methodology.* URN 14D/273. London: Gov.UK, Department of Energy and Climate Change; 2014.

[74] Francis-Devine B. *Poverty in the UK: Statistics*. #SN07096. London: Gov. UK, Commons Library Research Briefing, 6 April 2023; 2023.

[75] Croon TM, Hoekstra JSCM, Elsinga MG, Dalla Longa F, and Mulder P. Beyond headcount statistics: Exploring the utility of energy poverty gap indices in policy design. *Energy Policy*. 2023 June; 177(113579): 19.

[76] Deller D, Turner G, and Waddams Price CM. Energy poverty indicators: inconsistencies, implications and where next? *Energy Economics*. 2021; 103(105551): 15.

[77] Widuto A. Energy poverty in the EU. Briefing. EPRS | European Parliamentary ResearchService, Members' Research Service; 2022. Report No.: PE 733.583 – July 2022.

[78] Office of State and Community Energy Programs (SCEP). *Low-Income Community Energy Solutions*. [Online]; 2023 [cited 7 September 2023]. Available from: https://www.energy.gov/scep/slsc/low-income-community-energy-solutions.

[79] Eurostat. *Living Conditions in Europe – Poverty and Social Exclusion*. [Online]; 2023 [cited 7 September 2023]. Available from: https://ec.europa.eu/eurostat/statistics-explained/index.php?title=Living_conditions_in_Europe_-_poverty_and_social_exclusion#Poverty_and_social_exclusion.

[80] EU Directorate-General for Energy. *Energy Poverty in the EU*. [Online]; 2023 [cited 7 September 2023]. Available from: https://energy.ec.europa.eu/topics/markets-and-consumers/energy-consumer-rights/energy-poverty-eu_en.

[81] Ritchie H, Roser M, and Rosado P. *Energy*. [Online]; 2022 [cited 8 September 2023]. Available from: https://ourworldindata.org/energy.

[82] Canada Energy Regulator. *Market Snapshot: Fuel Poverty Across Canada – Lower Energy Efficiency in Lower Income Households*. [Online]; 2023 [cited 8 September 2023]. Available from: https://www.cer-rec.gc.ca/en/data-analysis/energy-markets/market-snapshots/2017/market-snapshot-fuel-poverty-across-canada-lower-energy-efficiency-in-lower-income-households.html.

[83] International Energy Agency (IEA). *Canada 2022 Energy Policy Review*. Paris, France; 2022.

[84] Riva M, Makasi SK, Dufresne P, O'Sullivan K, and Toth M. Energy poverty in Canada: prevalence, social and spatial distribution, and implications for research and policy. *Energy Research & Social Science*. 2021; 81 (102237): 12.

[85] Canadian Urban Sustainability Practitioners. *Energy Poverty in Canada: A CUSP Backgrounder*. Canadian Urban Sustainability Practitioners; 2019. Available from: https://energypovert.ca.

[86] Employment and Social Development Canada. *Opportunity for All – Canada's First Poverty Reduction Strategy. Report to Parliament*. Ottawa: Government of Canada, Employment and Social Development Canada; 2018/2022. Report No.: ISSN: 978-0-660-26905-4.

66 *Clean energy for low-income communities*

[87] World Bank Group. *World Bank Group Archivists' Chronology 1944-2013*. Washington, DC: World Bank Group, Library and Archives of Development; 2014.

[88] World Bank. *World Development Report 1978 (English). World Development Indicators*. Washington, DC: World Bank; 1978. Report No.: Library of Congress Catalog Number: 78-67086.

[89] IEA, IRENA, UNSD, World Bank, WHO. *Tracking SDG 7. The Energy Progress Report Chapter 7 - Tracking Progress Toward sdg 7 Across Targets: Indicators and Data*. Washington, DC: World Bank; 2023.

[90] Scott R. *The History of IEA: The First 20 years; Origins and Structure Paris*. France: OECD; 1994

[91] IEA. *Defining Energy Access: 2020 Methodology*. [Online]; 2020 [cited 9 September 2023]. Available from: https://www.iea.org/articles/defining-energy-access-2020-methodology.

[92] U.S. Energy Information Administration. *Electricity Consumption in U.S. Homes Varies by Region and Type of Home*. [Online]; 2019 [cited 10 September 2023]. Available from: https://www.eia.gov/energyex-plained/use-of-energy/electricity-use-in-homes.php.

[93] United Nations Department of Economic and Social Affairs. *World Population Prospects 2022: Summary of Results*. New York: United Nations, Population Division ; 2022. Report No.: ISBN: 978-92-1-148373-4.

[94] United Nations Department of Economic and Social Affairs. *Probabilistic Population Projections based on the World Population Prospects 2022*. [Online]. New York: United Nations; 2022 [cited 7 November 2023]. Available from: https://population.un.org/wpp/Download/.

[95] Ritchie H, Rodés-Guirao L, Mathieu E, *et al. Population Growth*. [Online]. Oxford: Our World in Data; 2023 [cited 15 November 2023]. Available from: https://ourworldindata.org/population-growth' [Online Resource].

[96] United Nations Population Fund. *World Population Dashboard*. [Online]; 2023 [cited 15 November 2023]. Available from: https://www.unfpa.org/data/world-population-dashboard.

[97] Hamadeh N, Van Rompaey C, and Metreau E. *World Bank Group Country Classifications by Income Level for FY24* (July 1, 2023–June 30, 2024). [Online]; 2023 [cited 12 November 2023]. Available from: https://blogs.worldbank.org/opendata/new-world-bank-group-country-classifications-income-level-fy24.

[98] United Nations Department of Economic and Social Affairs. *Least Developed Countries (LDCs)*. [Online]; 2023 [cited 13 November 2023]. Available from: https://www.un.org/development/desa/dpad/least-developed-country-category.html.

[99] World Bank. *Poverty and Shared Prosperity 2022: Correcting Course*. Washington, DC: World Bank; 2022. Report No.: ISBN (paper): 978-1-4648-1893-6.

Providing clean and affordable energy for all   67

[100]   UNCTAD. *UN List of Least Developed Countries.* [Online]; 2023 [cited 18 November 2023]. Available from: https://unctad.org/topic/least-developed-countries/list.

[101]   UN News. *5 Things You Need to Know About the World's Least Developed Countries.* [Online]; 2023 [cited 17 November 2023]. Available from: https://news.un.org/en/story/2023/03/1134087.

[102]   United Nations Department of Economic and Social Affairs. *Handbook on the Least Developed Country Category: Inclusion, Graduation and Special Support Measures*: 4th edition. Handbook. New York: United Nations, Committee for Development Policy; 2021. Report No.: ISBN: 9789211046984.

[103]   The World Bank. *The World Bank Data.* [Online]; 2023 [cited 19 November 2023]. Available from: https://data.worldbank.org/indicator/NY.GNP.PCAP.CD.

[104]   Harcourt S. *Official Development Assistance (ODA).* [Online]; 2023 [cited 18 November 2023]. Available from: https://data.one.org/topics/official-development-assistance/#where-does-aid-go.

[105]   OECD. *Development Aid at a Glance: Statistics by Region: 2.* Africa. Paris, France: OECD; 2018.

[106]   Husted TF, Blanchard LP, Arieff A, and Cook N.*U.S. Assistance for Sub-Saharan Africa: An Overview.* Washington, DC: Congressional Research Services; 2023.

[107]   House Foreign Affairs Committee GOP. *China Regional Snapshot: Sub-Saharan Africa.* [Online]; 2022 [cited 19 November 2023]. Available from: https://foreignaffairs.house.gov/china-regional-snapshot-sub-saharan-africa/.

[108]   African Union Commission. *Agenda 2063: The Africa We Want (Popular Version).* Addis Ababa, Ethiopia: African Union; 2015. Report No.: ISBN: 978-92-95104-23-5.

[109]   African Union. *The First-Ten Year Implementation Plan.* [Online]. [cited 19 November 2023]. Available from: https://au.int/agenda2063/ftyip.

[110]   African Union. *Goals and Priority Areas of Agenda 2063.* [Online]. [cited 19 November 2023]. Available from: https://au.int/agenda2063/goals.

[111]   African Union Development Agency – NEPAD. *Second Continental Report: On the Implementation of Agenda 2063.* Johannesburg, South Africa: African Union; 2022. Report No.: ISBN: 978-1-77634-908-1.

[112]   Wex. *Clean Energy.* [Online]; 2021 [cited 12 July 2023]. Available from: https://www.law.cornell.edu/wex/clean_energy.

[113]   TWI Ltd. *What is Clean Energy? How Does It Work? Why is It So Important?* [Online]; 2023 [cited 12 July 2023]. Available from: https://www.twi-global.com/technical-knowledge/faqs/clean-energy.

[114]   USEPA. *Criteria Air Pollutants.* [Online]; 2022 [cited 11 July 2023]. Available from: https://www.epa.gov/criteria-air-pollutants.

[115]   WHO. *Air Pollutants.* [Online]; 2023 [cited 11 July 2023]. Available from: https://www.who.int/teams/environment-climate-change-and-health/air-quality-and-health/health-impacts/types-of-pollutants.

68   *Clean energy for low-income communities*

[116]   Nathanson JA. *Air Pollution*. [Online]; 2023 [cited 11 July 2023]. Available from: https://www.britannica.com/science/air-pollution.

[117]   UNSTATS. *Revised List of Global Sustainable Development Goal indicators*. New York; 2017.

[118]   United Nations. *Clean, Renewable Energy Is 'the Difference between Life and Death', Will Prevent Climate Catastrophe, Secretary-General Tells High-Level Dialogue*. [Online]; 2021 [cited 21 July 2023]. Available from: https://press.un.org/en/2021/sgsm20932.doc.htm.

[119]   Cambridge Dictionary. *nitpicking*. [Online]. [cited 22 July 2023]. Available from: https://dictionary.cambridge.org/us/dictionary/english/nitpicking.

[120]   World Health Organization. *WHO Global Air Quality Guidelines. Particulate Matter (PM2.5 and PM10), Ozone, Nitrogen dioxide, Sulfur dioxide and Carbon monoxide*. Geneva; 2021. Report No.: ISBN 978-92-4-003421-1.

[121]   World Health Organization. *Guidelines for Drinking-Water Quality*: 4th edition Incorporating the First and Second Addenda. Geneva; 2022. Report No.: ISBN 978-92-4-004507-1 (print version); ISBN 978-92-4-004506-4 (electronic version).

[122]   World Health Organization. *WHO Guidelines for Indoor Air Quality: Household Fuel Combustion. Guidelines*. Geneva: WHO; 2014. Report No.: ISBN 9789241548878.

[123]   World Health Organization (WHO). *WHO Publishes New Global Data on the Use of Clean and Polluting Fuels for Cooking by Fuel Type*. [Online]. Geneva; 2022 [cited 5 October 2023]. Available from: https://www.who.int/news/item/20-01-2022-who-publishes-new-global-data-on-the-use-of-clean-and-polluting-fuels-for-cooking-by-fuel-type.

[124]   United Nations Statistics Division. *SDG Indicators Metadata Repository*. [Online]; 2023 [cited 10 September 2023]. Available from: https://unstats.un.org/sdgs/metadata/.

[125]   United Nations Statistics Division. *Welcome to the Sustainable Development Goal Indicators*. [Online]; 2023 [cited 12 September 2023]. Available from: https://unstats.un.org/sdgs/iaeg-sdgs/tier-classification/.

[126]   United Nations Statistics Division. *UNSD Sustainable Development Goals*. [Online]; 2023 [cited 12 September 2023]. Available from: https://unstats.un.org/sdgs/iaeg-sdgs/.

[127]   United Nations Statistics Division. *Tier Classification for Global SDG Indicators as of 31 March 2023*. Periodic Update. New York: United Nations, Economic and Social Affairs; 2023.

[128]   Watson J, Byrne R, Morgan Jones M, *et al. What are the Major Barriers to Increased Use of Modern Energy Services Among the World's Poorest People and are Interventions to Overcome These effective? Systematic Review*. Collaboration for Environmenatl Evidence Library; 2012. Report No.: CEE Review 11-004.

[129]   U.S. Department of State. *Global Methane Pledge: From Moment to Momentum*. [Online]; 2022 [cited 13 September 2023]. Available from: https://www.state.gov/global-methane-pledge-from-moment-to-momentum/.

[130] Mountford H, Waskow D, Gonzalez L, *et al.* *COP26: Key Outcomes from the UN Climate Talks in Glasgow.* [Online]; 2021 [cited 13 September 2023]. Available from: https://www.wri.org/insights/cop26-key-outcomes-un-climate-talks-glasgow.

[131] Jeffrey D. *Gas Stoves Are in Homes All over Australia. And the Evidence Says They're Making Kids Sick.* [Online]; 2023 [cited 14 September 2023]. Available from: https://www.9news.com.au/national/gas-stove-health-risks-indoor-air-pollution-childhood-asthma-explainer/b07a7c8c-0e00-4c8d-a414-b460326fde7a.

[132] Willers SM, Brunekreef B, Oldenwening M, *et al.* Gas cooking, kitchen ventilation, and asthma, allergic symptoms and sensitization in young children–the PIAMA study. *Allergy.* 2006; 61(5): 563–568.

[133] Bochove D, *Bloomberg. While Europe Shivers, Iceland Basks in Clean and Abundant Geothermal Power from Its Bounty of Steaming Water.* [Online]; 2023 [cited 14 September 2023]. Available from: https://fortune.com/2023/02/28/iceland-clean-geothermal-power-energy/.

[134] United Nations. *Paris Agreement.* New York and Paris: United Nations Framework Convention on Climate Change (UNFCCC); 2015.

[135] United Nations. *Climate Change. The Paris Agreement; What is the Paris Agreement?.* [Online]; 2023 [cited 15 September 2023]. Available from: https://unfccc.int/process-and-meetings/the-paris-agreement.

[136] United Nations. *For a Livable Climate: Net-Zero Commitments Must be Backed by Credible Action.* [Online]; 2023 [cited 15 September 2023]. Available from: https://www.un.org/en/climatechange/net-zero-coalition.

[137] United Nations. *Framework Convention on Climate Change. Cities, Regions and Businesses ramp up Ambition on Climate Change to Deliver Healthier Economies in the Wake of the Pandemic: Race to Zero COP25 Chile.* [Online]; 2020 [cited 16 September 2023]. Available from: https://cop25.mma.gob.cl/en/cities-regions-and-businesses-ramp-up-ambition-on-climate-change-to-deliver-healthier-economies-in-the-wake-of-the-pandemic/.

[138] Hitchings-Hales J. *What Is the 'Race to Zero'? Everything to Know About the Mission to Cut Emissions.* [Online]; 2021 [cited 16 September 2023]. Available from: https://www.globalcitizen.org/en/content/race-to-zero-net-zero-emissions-climate/.

[139] The White House. *CCEQ Launches Global Net-Zero Government Initiative, Announces 18 Countries Joining U.S. to Slash Emissions from Government Operations.* [Online]; 2022 [cited 16 September 2023]. Available from: https://www.whitehouse.gov/ceq/news-updates/2022/11/17/ceq-launches-global-net-zero-government-initiative-announces-18-countries-joining-u-s-to-slash-emissions-from-government-operations/.

[140] United Nations Climate Change. *Race To Zero Campaign.* [Online]; 2023 [cited 16 September 2023]. Available from: https://unfccc.int/climate-action/race-to-zero-campaign.

[141] Climate Champions. *UNFCCC. Interpretation Guide: Race to Zero Expert Peer Review Group.* Washington, DC; 2022.

70 *Clean energy for low-income communities*

[142] The High-Level Expert Group on the Net Zero Emissions Commitments of Non-State Entities. *Integrity Matters: Net Zero Commitments by Businesses, Financial Institutions, Cities and Regions.* New York: United Nations; 2022.

[143] Reclaim Finance. *Strengthened Race to Zero Criteria Require GFANZ to Support Fossil Fuels Phase-out Reclaim Finance.* [Online]; 2022 [cited 17 September 2023]. Available from: https://reclaimfinance.org/site/en/2022/09/16/strengthened-race-to-zero-criteria-require-gfanz-to-support-fossil-fuels-phase-out/.

[144] IEA. *Net Zero By 2050 A Roadmap for Global Energy Sector.* Paris, France; 2021.

[145] UNFCCC. *Race to Zero Lexicon.* New York: United Nations; 2021.

[146] Ritchie H, Roser M, and Rosado P. *CO$_2$ and Greenhouse Gas Emissions.* [Online]; 2020 [cited 18 September 2023]. Available from: https://ourworldindata.org/co2-and-greenhouse-gas-emissions.

[147] Fujimori S, Hasegawa T, Oshiro K, *et al.* Potential side effects of climate change mitigation on poverty and countermeasures. *Sustainability Science* 2023; 18: 2245–2257.

[148] Dimitrov R, Hovi J, Sprinz DF, Sælen H, and Underdal A. Institutional and environmental effectiveness: will the Paris Agreement work? *WIREs Climate Change.* 2019 February 27; 10(4): 12.

[149] Smil V. *Energy and Civilization: A History.* Cambridge, MA: MIT Press; 2017.

[150] Longley R. *The Origins of the Term, 'Horsepower'.* [Online]; 2021 [cited 19 September 2023]. Available from: https://www.thoughtco.com/where-did-the-term-horsepower-come-from-4153171.

[151] Ritchie H, Roser M, and Rosado P. *Electricity Mix.* [Online];2022 [cited 20 September 2023]. Available from: https://ourworldindata.org/electricity-mix.

[152] Bousso R. https://www.reuters.com/business/energy. [Online]; 2023 [cited 20 September 2023]. Available from: https://www.reuters.com/business/energy/bp-ends-70-years-publishing-statistical-review-world-energy-2023-02-28/.

[153] Energy Institute. *About the Statistical Review.* Energy Institute. [Online]; 2023 [cited 20 September 2023]. Available from: https://www.energyinst.org/statistical-review/about.

[154] Energy Institute. *Statistical Review of World Energy.* 72nd edition. Annual Review. London, England: Energy Institute; 2023.

[155] Roser M. *About Our World in Data.* [Online];2023 [cited 20 September 2023]. Available from: https://ourworldindata.org/about.

[156] Wiatros-Motyka M. *Global Electricity Review 2023.* [Online];2023 [cited 28 September 2023]. Available from: https://ember-climate.org/insights/research/global-electricity-review-2023/.

[157] IEA. *World Energy Balances; Energy Balances for 156 Countries and 35 Regional Aggregates World Energy Statistics and Balances.* [Online];2023

[cited 26 September 2023]. Available from: https://www.iea.org/data-and-statistics/data-product/world-energy-balances.

[158] IEA. *Key World Energy Statistics*. Paris, France: IEA; 2021.

[159] Ritchie H. *Primary Energy Production is Not Final Energy Use: What Are the Different Ways of Measuring Energy?* [Online]; 2021 [cited 20 September 2023]. Available from: https://ourworldindata.org/energy-substitution-method.

[160] BP. *bp Statistical Review of World Energy 2022*. 71st edition. Annual Review. London, UK: BP; 2022.

[161] USEIA. *Monthly Energy Review August 2023*. DOE/EIA-0035(2023/8). Washington, DC: U.S. Department of Energy, Office of Energy Statistics; 2023.

[162] USEIA. *EIA Offers Two Approaches to Compare Renewable Electricity Generation with Other Sources*. [Online]; 2019 [cited 23 September 2023]. Available from: https://www.eia.gov/todayinenergy/detail.php?id=41013.

[163] Bains P, Moorhouse J, and Hodgson D. *Bioenergy*. [Online];2023 [cited 24 September 2023]. Available from: https://www.iea.org/energy-system/renewables/bioenergy.

[164] Turgeon A, Morse E, and Crooks M. *Biomass Energy*. [Online];2023 [cited 24 September 2023]. Available from: https://education.nationalgeographic.org/resource/biomass-energy/.

[165] Ritchie H. *The Four Ways of Measuring Energy*. [Online];2022 [cited 25 September 2023]. Available from: https://ourworldindata.org/energy-definitions.

[166] IEA. *Energy Efficiency Policy Toolkit 2023; From Sønderborg to Versailles*. Paris: IEA; 2023.

[167] IEA. *Energy Efficiency – The Decade for Action. Ministerial Briefing*. Paris: IEA; 2023.

[168] Enerdata. *Global Energy Trends 2023*. [Online];2023 [cited 27 September 2023]. Available from: https://www.enerdata.net/publications/reports-presentations/world-energy-trends.html.

[169] Enerdata. *Electricity: Electrification: Share of Electricity in Total Final Energy Consumption*. [Online]; 2023 [cited 27 September 2023]. Available from: https://yearbook.enerdata.net/electricity/share-electricity-final-consumption.html.

[170] Lawrence Livermore National Laboratory. *Energy Flow Charts*. [Online]; 2023 [cited 27 September 2023]. Available from: https://flowcharts.llnl.gov/commodities/energy.

[171] Ofgem. *Average Gas and Electricity Usage*. [Online]; 2020 [cited 27 September 2023]. Available from: https://www.ofgem.gov.uk/information-consumers/energy-advice-households/average-gas-and-electricity-use-explained.

[172] Ritchie H, Rosado P, and Roser M. *Access to Energy*. [Online]; 2019 [cited 27 September 2023]. Available from: https://ourworldindata.org/energy-access.

72   *Clean energy for low-income communities*

[173] Ritchie H and Rosado P. *Electricity Mix*. [Online];2020–2023 [cited 28 September 2023]. Available from: https://ourworldindata.org/electricity-mix.

[174] Ritchie H and Rosado P. *Energy Mix*. [Online]; 2023 Update [cited 28 September 2023]. Available from: https://ourworldindata.org/energy-mix.

[175] Jacobson MZ, von Krauland AK, Coughlin SJ, Palmer FC, and Smith MM. Zero air pollution and zero carbon from all energy at low cost and without blackouts in variable weather throughout the U.S. with 100% wind-water-solar and storage. *Renewable Energy*. 2022; 184: 430–442.

[176] Breyer C, Khalili S, Bogdanov D, *et al.* On the history and future of 100% renewable energy systems research. *IEEE Access*. 2022; 10: 78176–78218.

[177] Levin K, Fransen T, Schumer C, Davis C, and Boehm S. *What Does "Net-Zero Emissions" Mean? 8 Common Questions, Answered. World Resources Institute*. [Online];2023 [cited 29 September 2023]. Available from: https://www.wri.org/insights/net-zero-ghg-emissions-questions-answered.

[178] UNFCCC. *Technical Dialogue of the First Global Stocktake*. [Online]; 2023 [cited 29 September 2023]. Available from: https://unfccc.int/sites/default/files/resource/sb2023_09_adv.pdf.

[179] Black R, Cullen K, Fay B, *et al. Taking Stock: A global Assessment of Net Zero Targets*. Oxford: University of Oxford, The Energy & Climate Intelligence Unit and Oxford Net Zero; 2021.

[180] USEIA. *Frequently Asked Questions (FAQS) How Much Electricity Does an American Home Use?* [Online]; 2023 [cited 29 September 2023]. Available from: https://www.eia.gov/tools/faqs/faq.php?id=97&t=3.

[181] Statista. *Mean Domestic Electricity Consumption per Household in Great Britain in 2021, by Region (in Kilowatt-Hours)*. [Online]; 2023 [cited 29 September 2023]. Available from: https://www.statista.com/statistics/517845/average-electricity-consumption-uk/.

[182] Blimpo MP and Cosgrove-Davies M. *Electricity Access in Sub-Saharan Africa: Uptake, Reliability, and Complementary Factors for Economic Impact*. Washington, DC: World Bank, Africa Development Forum; 2019. Report No.: ISBN (paper): 978-1-4648-1361-0.

[183] Ritchie H, Max R, and Rosado P. *Energy Our World in Data*. [Online]; 2022 [cited 1 October 2023]. Available from: https://ourworldindata.org/energy.

[184] Statista. *Electricity Consumption per Capita Worldwide in 2022, by Selected Country*. [Online]; 2023 [cited 1 October 2023]. Available from: https://www.statista.com/statistics/383633/worldwide-consumption-of-electricity-by-country/.

[185] CountryEconomy. *Electricity Consumption*. [Online]; 2023 [cited 1 October 2023]. Available from: https://countryeconomy.com/energy-and-environment/electricity-consumption.

[186] World Bank. *World Development Indicators: Electric Power Consumption (kWh per capita)*. [Online]; 2023 [cited 1 October 2023]. Available from:

https://databank.worldbank.org/source/world-development-indicators/Series/ EG.USE.ELEC.KH.PC.

[187] World Bank. *Metadata Glossary: Access to Electricity (% of Population)*. [Online]; 2023 [cited 2 October 2023]. Available from: https://databank. worldbank.org/metadataglossary/world-development-indicators/series/EG. ELC.ACCS.ZS.

[188] The World Bank. *Nigeria Development Update: Seizing the Opportunity*. Washington, DC: *World Bank*; 2023.

[189] USEIA. *Table 1. Nigeria's Energy Overview*, 2021. [Online]; 2023 [cited 19 November 2023]. Available from: https://www.eia.gov/international/ content/analysis/countries_long/Nigeria.

[190] GlobalPetrolPrices.com. *Electricity Prices*. [Online]; 2023 [cited 19 November 2023]. Available from: https://www.globalpetrolprices.com/ electricity_prices/.

[191] Jeremiah K. *Households Under Pressure as New Electricity Tariff Due July 1*. [Online];2023 [cited 7 November 2023]. Available from: https://guardian.ng/news/households-under-pressure-as-new-electricity-tariff-due-july-1/.

[192] Igbinadolor N. *Tariff Hike: Nigerians in Panic Bulk Energy Buying*. [Online];2023 [cited 19 November 2023]. Available from: https://businessday.ng/business-economy/article/tariff-hike-nigerians-in-panic-bulk-energy-buying/.

[193] Trading Ecconomics. *Iran Inflation Rate*. [Online]; 2023 [cited 19 November 2023]. Available from: https://tradingeconomics.com/iran/inflation-cpi.

[194] Farmer M. *How Norway Became Europe's Biggest Power Exporter*. [Online];2021 [cited 20 November 2023]. Available from: https://www. power-technology.com/features/how-norway-became-europes-biggest-power-exporter/?cf-view&cf-closed.

[195] GlobalData. *The Power Export in Saudi Arabia (2017–2020, GWh)*. [Online];2023 [cited 20 November 2023]. Available from: https://www. globaldata.com/data-insights/power-and-utilities/the-power-export-in-saudi-arabia-1083787/.

[196] U.S. Department of Commerce International Trade Administration. *Bulgaria – Country Commercial Guide: Energy*. [Online];2022 [cited 20 November 2023]. Available from: https://www.trade.gov/country-commercial-guides/bulgaria-energy.

[197] U.S. Department of Commerce International Trade Administration. *Belgium – Country Commercial Guide Energy*. [Online];2023 [cited 20 November 2023]. Available from: https://www.trade.gov/country-commercial-guides/belgium-energy.

[198] Countryeconomy. *countryeconomy.com/energy-and-environment; France – Household Electricity Prices*. [Online]; 2023 [cited 20 November 2023]. Available from: https://countryeconomy.com/energy-and-environment/electricity-price-household/france.

74 *Clean energy for low-income communities*

[199] World Nuclear Association. *Nuclear Power in France.* [Online]; 2023 [cited 20 November 2023]. Available from: https://world-nuclear.org/information-library/country-profiles/countries-a-f/france.aspx.

[200] U.S. Department of Commerce International Trade Administration. *France – Country Commercial Guide Energy (ENG).* [Online]; 2023 [cited 19 November 2023]. Available from: https://www.trade.gov/country-commercial-guides/france-energy-eng.

[201] Energypedia. *Global Tracking Framework for Measuring Energy Access.* [Online]; 2021 [cited 1 October 2023]. Available from: https://energypedia.info/wiki/Global_Tracking_Framework_for_Measuring_Energy_Access.

[202] Brundtland GH. *Our Common Future: The World Commission on Environment and Development.* 1st edition. Oxford: Oxford University Press; 1987.

[203] Gaye A. *Human Development Report 2007/2008: Access to Energy and Human Development. Human Development Report Office Occasional Paper.* New York: United Nations Development Program; 2008.

[204] Takada M, Charles NA. *Energizing Poverty Reduction; A Review of the Energy-Poverty Nexus in Poverty Reduction Strategy Papers.* New York: UNDP; 2007.

[205] World Bank and the World Health Organization. *Measuring Energy Access: A Guide to Collecting Data Using 'the Core Questions on Household Energy Use. Living Standards Measurement Study (LSMS) Guidebook.* Washington, DC: World Bank and the World Health Organization; 2021. Report No.: WHO/HEP/ECH/AQH/2021.9.

[206] Ritchie H. *Definition: Access to Electricity.* [Online]; 2021 [cited 3 October 2023]. Available from: https://ourworldindata.org/definition-electricity-access.

[207] Bhatia M, Angelou N. *Beyond Connections: Energy Access Redefined.* ESMAP Technical Report; 008/15. Washington, DC: World Bank, The International Bank for Reconstruction and Development; 2015.

[208] United Nations and UN Energy. *Energy Compacts Annual Progress Report 2023.* Annual Progress Report. New York: United Nations, UNDP and UN Energy; 2023.

[209] Koo BB, Rysankova D, Portale E, Angelou N, Keller S, and Padam G. Rwanda – beyond connections: energy access diagnostic report based on the multi-tier framework. ESMAP Papers. Washington, DC: World Bank; 2018.

[210] Wrangham RW. *Catching Fire: How Cooking Made Us Human.* 1st edition. New York: Hachette Book Group/Basic Books; 2009.

[211] Bellis M. *History of the Oven from Cast Iron to Electric.* [Online]; 2018 [cited 5 October 2023]. Available from: https://www.thoughtco.com/history-of-the-oven-from-cast-iron-to-electric-1992212.

[212] Smil V. *Energy Transitions: History, Requirements, Prospects.* Santa Barbara: Praeger; 2010.

[213] Smil V. Examining energy transitions: a dozen insights based on performance. *Energy Research & Social Science.* 2016; 22: 194–197.

[214] USEPA. *NAAQS Table.* [Online]; 2023 [cited 5 October 2023]. Available from: https://www.epa.gov/criteria-air-pollutants/naaqs-table.

[215] Reader GT. Clean air, clean water, clear conscience. In Ting DSK, Stagner JA. *Sustainable Engineering Technologies and Architectures*. Melville, New York: AIP; 2021. p. 34.

[216] Ritchie H. *How Many People Do Not Have Access to Clean Fuels for Cooking?*. [Online]; 2021 [cited 5 October 2023]. Available from: https://ourworldindata.org/no-clean-cooking-fuels.

[217] Roser M. *Data Review: How Many People Die from Air Pollution?* [Online]; 2021 [cited 5 October 2023]. Available from: https://ourworldindata.org/data-review-air-pollution-deaths.

[218] Dattani S, Spooner F, Ritchie H, and Roser M. *Causes of Death*. [Online]; 2023 [cited 6 October 2023]. Available from: https://ourworldindata.org/causes-of-death.

[219] Dattani S. *How Do Researchers Estimate the Death Toll Caused by Each Risk Factor, Whether It's Smoking, Obesity, or Air Pollution?* [Online]; 2023 [cited 6 October 2023]. Available from: https://ourworldindata.org/how-do-researchers-estimate-the-death-toll-caused-by-each-risk-factor-whether-its-smoking-obesity-or-air-pollution.

[220] WHO. *Population with Primary Reliance on Polluting Fuels and Technologies for Cooking (in Millions)*. [Online]; 2023 [cited 6 October 2023]. Available from: https://www.who.int/data/gho/data/indicators/indicator-details/GHO/population-with-primary-reliance-on-polluting-fuels-and-technologies-for-cooking-%28in-millions%29.

[221] WHO. *Household Air Pollution Attributable Deaths*. [Online];2022 [cited 6 October 2023]. Available from: https://www.who.int/data/gho/data/indicators/indicator-details/GHO/household-air-pollution-attributable-deaths.

[222] Grippo A, Zhu K, Yeung EH, *et al.* Indoor air pollution exposure and early childhood development in the Upstate KIDS Study. *Environmental Research*. 2023; 234: 116528.

[223] Sustainable Energy for All. *SEforALL Analysis of SDG7 Progress – 2023*. [Online];2023 [cited 6 October 2023]. Available from: https://www.seforall.org/data-stories/seforall-analysis-of-sdg7-progress.

[224] Sustainable Energy for All. *Deep Dive Analysis: Tracking SDG7: The Energy Progress Report 2023*. Annual Progress Report. Vienna, New York, Washington, DC: Sustainable Energy for All; 2023.

[225] Eurostat. *Energy Consumption in Households*. [Online]; 2023 [cited 7 October 2023]. Available from: https://ec.europa.eu/eurostat/statistics-explained/index.php?title=Energy_consumption_in_households.

[226] United Nations Department of Economic and Social Affairs. *2020 Energy Balances: Concepts and Definition*. New York: United Nations, Statistics Division; 2022. Report No.: ISBN 978-92-1-259222-0.

[227] USEIA. *Monthly Energy Review: October 2023. Monthly Review*. Washington, DC: U.S. Department of Energy , Office of Energy Statistics; 2023. Report No.: DOE/EIA-0035(2023/10).

76 Clean energy for low-income communities

[228] United Nations Department of Economic and Social Affairs. *2023 Energy Statistics Pocketbook*. New York: United Nations, Statistics Division; 2023. Report No.: ISBN 9789210024389.

[229] Krtková E, Danielik V, Szemesová J, Tarczay K, Kis-Kovács G, and Neuzil V. Non-Energy use of fuels in the greenhouse gas emission reporting. *Atmosphere*. 2019; 10(7): 12.

[230] Kilgore G. *How Many Electric Cars Are There in the United States? We Found Out*. [Online]; 2023 [cited 24 November 2023]. Available from: https://8billiontrees.com/carbon-offsets-credits/cars/how-many-electric-cars-in-the-us/.

[231] IEA. *A Vision for Clean Cooking Access for All*. World Energy Outlook Special Report. Paris: IEA; 2023.

[232] IEA. *World Energy Outlook*. Revised version. Paris: IEA; 2022.

[233] Ritchie H. *Is the World Making Progress in Decarbonizing Energy?* [Online]; 2021 [cited 8 October 2023]. Available from: https://ourworldindata.org/decarbonizing-energy-progress.

[234] Wikipedia Contributors. International Energy Agency. [Online]; 2023 [cited 9 October 2023]. Available from: https://en.wikipedia.org/w/index.php?title=International_Energy_Agency&oldid=1178385423.

[235] Raimi D, Zhu Y, Newell RG, Prest BC, and Bergman A. *Global Energy Outlook 2023: Sowing the Seeds of an Energy Transition*. Washinton, DC: Resources for the Future; 2023.

[236] USEIA. *International Energy Outlook 2021*. [Online];2021 [cited 9 October 2023]. Available from: https://www.eia.gov/outlooks/ieo/.

[237] World Bank. *State of Electricity Access Report (SEAR)*. Washington, DC: Energy Sector Management Assistance Program (ESMAP); 2017.

[238] World Bank. *Poverty and Shared Prosperity 2020: Reversals of Fortune*. Washington, DC: World Bank; 2020. Report No.: ISBN (paper): 978-1-4648-1602-4.

[239] Secretary-General U. *Progress towards the Sustainable Development Goals. UN A/78/XX-E/2023/XX*. New York:, General Assembly Economic and Social Council; 2023.

[240] Jessop S. *Reuters: Sustainable Business*. [Online];2022 [cited 9 October 2023]. Available from: https://www.reuters.com/business/sustainable-business/cost-hit-un-sustainability-goals-rises-176-trillion-report-2022-09-08/.

[241] A Force for Good. *Capital as a Force for Good: Capitalism for a Secure and Sustainable Future*. London: Force for Good; 2022. Report No.: ISBN 978-1-7399529-4-5; ISBN (pdf) 978-1-7399529-6-9.

[242] Battisti F. SDGs and ESG criteria in housing: defining local evaluation criteria and indicators for verifying project sustainability using Florence Metropolitan Area as a case study. *Sustainability*. 2023; 15(2-9372): 37.

[243] Worldometer. *GDP by Country*. [Online];2023 [cited 9 October 2023]. Available from: https://www.worldometers.info/gdp/gdp-by-country/.

[244] Morgan Stanley. *Decarbonization: The Race to Zero Emissions*. [Online]; 2019 [cited 9 October 2023]. Available from: https://www.morganstanley.com/ideas/investing-in-decarbonization.

[245] United Nations. *Financing for Sustainable Development Report 2023: Financing Sustainable Transformations*. New York: United Nations, Inter-agency Task Force on Financing for Development; 2023. Report No.: ISBN: 978-92-1-101465-5.

[246] Kenny C and Gehan Z. *Scenarios for Future Global Growth to 2050*. Working Paper 634. New York; London: Center for Global Development; 2023.

[247] Hausfather Z. *Explainer: How 'Shared Socioeconomic Pathways' Explore Future Climate Change*. [Online];2018 [cited 10 October 2023]. Available from: https://www.carbonbrief.org/explainer-how-shared-socioeconomic-pathways-explore-future-climate-change/.

[248] Norwich University. *History of American Propaganda Posters: American Social Issues Through Propaganda*. [Online]; 2023 [cited 24 November 2023]. Available from: https://online.norwich.edu/history-american-propa-ganda-posters-american-social-issues-through-propaganda.

[249] American Historical Association. *Defining Propaganda II*. [Online]; 2023 [cited 24 November 2023]. Available from: https://www.historians.org/about-aha-and-membership/aha-history-and-archives/gi-roundtable-series/pamphlets/em-2-what-is-propaganda-%281944%29/defining-propaganda-ii.

[250] Casey RD. *EM2: What Is Propaganda? (1944)*. [Online]; 2023 [cited 24 November 2023]. Available from: www.historians.org/about-aha-and-membership/aha-history-and-archives/gi-roundtable-series/pamphlets/em-2-what-is-propaganda-%281944%29.

[251] Fields S. *Feed-In Tariffs: What You Need to Know*. [Online]; 2023 [cited 24 November 2023]. Available from: https://www.energysage.com/solar/feed-in-tariffs-a-primer-on-feed-in-tariffs-for-solar/.

[252] Kenton W. *Feed-In Tariff (FIT): Explanation, History and Uses*. [Online]; 2021 [cited 24 November 2023]. Available from: https://www.investopedia.com/terms/f/feed-in-tariff.asp.

[253] World Bank. *Population 2022*. [Online] Washington, DC; 2023 [cited 14 November 2023]. Available from: https://databankfiles.worldbank.org/public/ddpext_download/POP.pdf.

Chapter 3

Low-cost and energy-efficient housing design: a review on research trends

Abbas Shadmand[1] and Semra Arslan Selçuk[1]

Energy, one of the basic needs of modern life, has begun to be consumed excessively due to the side effects of improper construction. The increase in primary energy consumption brings with it the emergence of serious problems. Considering the share of the built environment in energy consumption, it reveals the potential for reducing the issues arising from the energy consumption of the sector in question. One way to mitigate these problems is to make buildings more energy-efficient. However, in today's world, many people are deprived of the right to shelter. Moreover, it is not only sufficient to have shelter, but the necessity of comfortable housing is also essential. In this context, low-cost housing projects have started to be implemented, and their importance is increasing day by day. In this respect, this study focused on how comfortable, energy-efficient housing can be provided for low-income groups. For this purpose, a comprehensive literature review was conducted within the scope of the study, and a bibliometric analysis was carried out with 195 articles drawn from the Web of Science database to identify research trends on the subject. Nineteen articles with high-impact values were selected from the literature, and an attempt was made to identify the potential study areas needed through content analysis.

Keywords: Energy consumption; Low-cost Housing; Energy efficiency; Energy justice; Energy poverty

3.1 Introduction

Energy stands as one of the most fundamental requisites for the modern world, finding indispensable application in nearly every facet of daily life, encompassing domains like transportation, agriculture, telecommunications, and the energy industry, as well as the heating, cooling, lighting, and energizing of various electrical devices within buildings. This unremittingly escalating demand has prompted forecasts by the International Energy Agency, indicating a projected 53% surge in global energy consumption by 2040 [1], as depicted in Figure 3.1. This trend has

[1]Department of Architecture, Gazi University, Türkiye

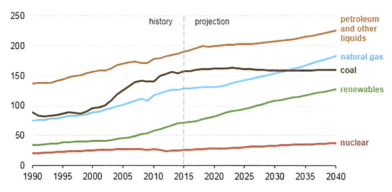

Figure 3.1   World energy consumption chart according to energy sources [1]

amplified the utilization of unsustainable energy resources and heightened reliance on fossil fuels, resulting in many issues, including resource scarcity, climate alteration, environmental degradation, and societal disparities. Within this framework, it is well established that the construction sector assumes a significant role in the overall global consumption of energy and resources.

Built environment is responsible for approximately 40% of global energy consumption, 40% of waste, and 16% of water consumption. Buildings, which have a share of 10% in global employment, affect the natural environment and ecology both locally and globally due to their long lifespan. In addition, the construction sector has a 7.9% share in global CO_2 emissions [1,33].

Research shows that more than 80% of greenhouse gas emissions from buildings arise from factors such as heating, ventilation, and air conditioning during operation [2]. Considering these data, it seems that the discipline of architecture should play a leading role in designing energy-efficient buildings. In addition, based on these data, the potential of built environments to reduce the negative effects resulting from energy consumption has made energy-efficient buildings, as well as energy-efficient houses, a topic of discussion in the discipline of architecture.

Energy efficiency generally means producing more goods and services using less energy without compromising comfort conditions. It enables the reduction of energy used per unit amount of service or product without reducing the production quantity, quality level, performance, and social welfare level. Reducing energy consumption or ensuring energy efficiency is achieved without reducing the standard of living, quality, and quantity of service or production in buildings, transportation, industry, and lighting. The energy consumption by residential buildings is the amount of energy that users generally spend to meet their basic needs, such as heating, cooling, and lighting. Ensuring energy efficiency in the housing sector means maintaining comfort and minimizing energy-related damages by reducing energy consumption [3]. Energy-efficient housing can provide many environmental, economic, and social benefits.

These residences play an essential role in meeting the needs of future generations by reducing negative environmental impacts, reducing energy consumption, and contributing to the protection of natural resources with less carbon emissions. Energy-efficient residences minimize energy consumption through appropriate architectural design, efficient heating, ventilation, air conditioning systems, lighting, water-saving systems, and renewable energy sources. In addition, the building design is made following the regional climatic conditions. However, the materials used in energy-efficient housing must be environmentally friendly, recyclable, and renewable [4]. Energy efficiency in residences can be achieved through active methods and passive strategies that aim to increase building performance in heating, cooling, ventilation, and lighting and reduce energy consumption without compromising comfort [5].

However, the fact that low-income groups have limited access to energy-efficient housing and that energy-efficient housing is made available to a certain segment of society causes energy efficiency to cause problems arising from social inequality rather than sustainability. This is generally due to the limited financial capacity of low- and sometimes middle-income groups and the fact that energy efficiency is adopted to a certain level only by high-income groups. However, the share of residences in energy consumption and the impact of energy consumption on other areas is an essential issue in terms of energy efficiency applying to all income levels.

Although sheltering is one of the basic needs of humans, in today's world, many people face housing problems. In addition, it is known that many existing shelters are of low quality and have a low standard of welfare. In response to this need, affordable housing for low- and middle-income groups has emerged, and the feasibility of low-cost housing units for this social class has been brought to the agenda.

Low-cost housing is generally defined in economic terms since it is affordable and ideal housing for middle- and low-income groups. More importance is given to constructing this type of housing, and state or non-profit organizations support it, and those houses are now adopted as an important component of housing stocks. Additionally, research shows that implementing energy-efficient strategies in low-cost housing can save 80% in energy consumption in these residences [42]. This means that low-income families can improve their living conditions and contribute to their well-being by significantly reducing their total cost of living [45]. Ensuring energy efficiency in low-cost housing can provide many environmental, economic, and social benefits. One of these is that the social class of this housing type reduces the problems caused by energy poverty.

Failure to ensure energy efficiency in residences can lead to energy poverty. Energy poverty can cause many negative effects on human health. For example, inadequate heating or cooling, low temperatures, and humid environments can cause respiratory diseases and lung infections. Additionally, low temperatures can increase the risk of cardiovascular disease, hypertension, and stroke. Energy poverty can also cause mental health problems. Living in a cold and dark house can lead to issues such as depression, anxiety, insomnia, and stress. Therefore, it is

82 *Clean energy for low-income communities*

crucial to take the necessary precautions to minimize the effects of energy poverty and increase energy efficiency in residences [6]. In addition, energy poverty in residences increases poverty, marginalization, and exclusion due to high energy loads, health problems, and environmental impacts [7]. This has made energy efficiency in residences and meeting energy needs at a low cost more critical.

Making low-cost housing energy efficient can help achieve environmental goals by contributing to reducing environmental damage. The decrease in energy costs also helps families meet other needs. Investments in energy efficiency can contribute to revitalizing local economies and promoting the marketing of energy efficiency services. In short, energy efficiency in low-cost housing can increase household incomes, reduce operating costs, and support economic growth [8]. Moreover, considering the harms of not being energy efficient in houses on a family and society scale, the importance of ensuring the energy efficiency of low-cost houses increases even more. Especially considering the financial capacity and fragility of the target audience of this housing type, the side effects of social problems arising from energy poverty show that it is inevitable to focus on ensuring energy efficiency in these housing types.

For this reason, the study focused on how comfortable, energy-efficient housing can be provided for low-income groups. In this context, a comprehensive literature review was conducted within the scope of the study, and an attempt was made to identify research trends on the subject.

3.2   Method

Bibliometrics means measuring various documents (texts, books, documents, etc.) based on numerical data. Pritchard [9] defines bibliometric studies as a concept that includes applying mathematical and statistical techniques and ranking in different communication media. Although bibliometrics is particularly prevalent in the library and information sciences fields, many research fields use bibliographic studies to explore their research trends and the impacts of those trends. These studies may use data such as reading numbers, publication content changes over the years, publishing groups, keywords, and citations. There has been a significant increase in bibliometric studies in recent years, and this increase is associated with the development of computer use, internet connections, databases, and existing statistical algorithms [10]

This study conducted a bibliometric analysis to reveal the current trends and progress of studies on low-cost, energy-efficient housing. The flow chart of the method used in this research is shown in Figure 3.2.

Clarivate Analytics' WoS database was used in this study because it provides high-quality data and allows the option to filter search queries and perform searches on the search string. Web of Science is considered the world's leading research, analytical information, and scientific citation platform [11,12].

Bibliometric studies use mapping algorithms and techniques to visualize similarities and differences in compiled data from one or more sources. Network analysis

Low-cost and energy-efficient housing design

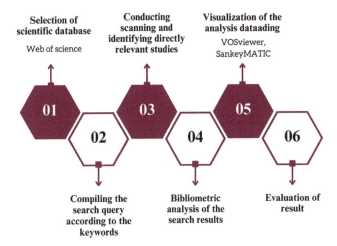

Figure 3.2 Flow of the method

is a set of techniques and algorithms derived from network theory to scientifically demonstrate the interactions of research and research trends, supported by the advent and advanced use of the Internet and computers [13]. With recent advances in computer technology, scientific indexes, and information visualization techniques, researchers can discover hidden connections and trends in the literature [14].

Various tools, such as Citespace, Pajek, VOSviewer, nodeXL, Gephi, Citespace, and SankeyMATIC have been developed to provide network analysis, visualization, and a better understanding of large amounts of data and information. This study used VOSviewer software and SankeyMATIC, which also allow working with different databases.

VOSviewer is a software tool for creating and visualizing similarities and relationships between bibliographic data through the production of visually understandable and interpretable bibliometric graphs and maps. In VOSviewer, circles and labels present the elements in the visualization network. The strength of the interrelationships of elements is visualized in connections. The clusters created are expressed in different colors. It is also widely used, especially in graphic and metadata metric studies [15–17].

SankeyMATIC makes it easy to see data flow by creating Sankey diagrams. With the Sankey diagram, many different pieces of information can be expressed in a single visual with the help of colors and arrows expressing flow [18].

In this study, "energy efficient hous*" and "low-cost*" OR "energy efficient hous*" and "low-income*" OR "low" were first searched in the title, abstract and/or keywords in the Web of Science database. Posts containing "-income hous*" and "energy efficien*" OR "low-cost hous*" and "energy efficien*" have been filtered. As a result of filtration, 230 publications were reached. It was determined that 195 of the 230 publications obtained were directly related to the subject (Figure 3.3). These publications are, respectively: their numbers (publications and citations) by

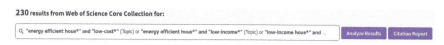

Figure 3.3 Number of results

years, types, and research areas, in addition to architecture, sources, numbers by country, author and co-authorship analysis, analysis by institutions, and keyword analysis. When the same filter is applied, the direction in which the research trend has evolved today is interpreted by presenting the details of the most cited publications between the years 2020 and 2023.

3.3 Findings

In this section, the analysis results of the data obtained from the Web of Science database through VOSviewer and SankeyMATIC are shared.

3.3.1 Publication and citation numbers by year

Figure 3.4 shows the publication and citation numbers of the 195 publications obtained as a result of the filtering process in the study by year. The first publication was made in 1993. It is stated that between 1993 and 2023, the most publications were made in 2019 and 2020 with 25 publications. These two years are followed by 2022 with 23 publications. The years 1998, 2001, 2002, and 2003 are the years with the least publications, with zero publications. The number of publications started to increase after 2011. Although there was a decrease in 2014, 2018, and 2021 compared to the previous year, the increase continued in the next year.

Citations refer to the number of times a scientific article or document is cited in another article to reflect its scientific impact [19]. When Figure 3.4 is examined, it appears that the number of citations in publications has increased since 2015. The highest number of citations was made in 2021–2022.

3.3.2 Research areas

The determined keywords were scanned "in all areas". It seems that publications are available in different fields. These fields are *energy fuels, environmental sciences ecology, engineering, construction building technology, science technology other topics, business economics, public environmental occupational health, urban studies, architecture, meteorology atmospheric sciences, public administration, general internal medicine, materials science, water resource, biomedical social sciences, computer science, development studies, geography, physics, social sciences other topics, thermodynamics, agriculture, automation, control systems, chemistry, family studies.* According to Figure 3.5, energy fuels rank first with a rate of 45%, environmental sciences, and ecology rank second with a rate of 33%, engineering ranks third with a rate of 29%, construction technology ranks fourth with a rate of 26%, and science and technology other subjects rank fifth with a rate of 20%.

Low-cost and energy-efficient housing design 85

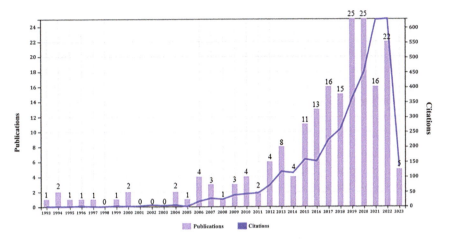

Figure 3.4   Publication and citation numbers by year

Field: Research Areas	Record Count	% of 195
Energy Fuels	88	45.128%
Environmental Sciences Ecology	65	33.333%
Engineering	57	29.231%
Construction Building Technology	51	26.154%
Science Technology Other Topics	40	20.513%
Business Economics	36	18.462%
Public Environmental Occupational Health	12	6.154%
Urban Studies	11	5.641%
Architecture	6	3.077%
Meteorology Atmospheric Sciences	4	2.051%
Public Administration	4	2.051%
General Internal Medicine	3	1.538%
Materials Science	3	1.538%
Water Resources	3	1.538%
Biomedical Social Sciences	2	1.026%
Computer Science	2	1.026%
Development Studies	2	1.026%
Geography	2	1.026%
Physics	2	1.026%
Social Sciences Other Topics	2	1.026%
Thermodynamics	2	1.026%
Agriculture	1	0.513%
Automation Control Systems	1	0.513%
Chemistry	1	0.513%
Family Studies	1	0.513%

Figure 3.5   Research areas

86   *Clean energy for low-income communities*

3.3.3   Publication type

Publication types in the Web of Science database are articles, conference papers/ proceedings papers, review articles, book chapters, early access publications, and meeting abstracts. It is divided into six subheadings. When Figure 3.6 is examined, it is seen that the majority of the 195 publications analyzed were published in the form of articles (82% rate) or conference papers (14% rate).

3.3.4   Sources of publications

As stated in the previous analysis, publications were mostly written in the form of articles or conference papers/papers. When Figure 3.7 is examined in support of this analysis, all the top 10 sources according to the number of publications are journal articles. Energy Policy magazine ranks first with 11.282%. Energy and

Select All	Field: Document Types	Record Count	% of 195
☐	Article	160	82.051%
☐	Proceeding Paper	28	14.359%
☐	Review Article	8	4.103%
☐	Book Chapters	1	0.513%
☐	Early Access	1	0.513%
☐	Meeting Abstract	1	0.513%

Figure 3.6   Publication type

Field: Publication Titles	Record Count	% of 195
ENERGY POLICY	22	11.282%
ENERGY AND BUILDINGS	16	8.205%
SUSTAINABILITY	10	5.128%
ENERGY RESEARCH SOCIAL SCIENCE	7	3.590%
APPLIED ENERGY	6	3.077%
BUILDING AND ENVIRONMENT	6	3.077%
ENERGY PROCEDIA	6	3.077%
RENEWABLE ENERGY	6	3.077%
INDOOR AND BUILT ENVIRONMENT	5	2.564%
ENERGY FOR SUSTAINABLE DEVELOPMENT	4	2.051%
IMPROVING RESIDENTIAL ENERGY EFFICIENCY INTERNATIONAL CONFERENCE IREE 2017	4	2.051%
RENEWABLE SUSTAINABLE ENERGY REVIEWS	4	2.051%
BUILDING RESEARCH AND INFORMATION	3	1.538%
ENERGIES	3	1.538%
ENVIRONMENTAL RESEARCH LETTERS	3	1.538%

Figure 3.7   Sources of publications and publication rates

Buildings 8.205%, Sustainability 5.128%, Energy Research Social Science 3.590%, Applied Energy 3.077%, Building and Environment 3.077%, Energy Procedia 3.077%, Renewable Energy 3.077%, Indoor and Built Environment 2.564%, and Energy for Sustainable Development 2.051% are among the sources with the most publications.

3.3.5 Publication numbers by country

Figure 3.8 shows the distribution of the publications obtained as a result of filtering according to the countries where they were made. According to the data obtained, America (53 publications), England (31 publications), Austria (20 publications), South Africa (17 publications), and Brazil (10 publications) are among the top five countries that publish the most on the subject under investigation.

3.3.6 Authorship and co-authorship analysis

Co-authorship analysis is an analytical tool used to identify and evaluate leading organizations, countries, scientific collaboration trends, or individual scientists [28]. For author and co-authorship analysis, the number of publications per author was determined as a minimum of three as a parameter in the VOSviewer program. Thirty-one publications and 10 authors met this criterion. Figure 3.9 lists the top 10 most prolific authors publishing in the research field mentioned at the outset. On this subject, Mathews, E. has four publications and other authors have three publications. If the citations to the authors' publications are considered in order, Boardman, B. ranks first with 223 citations; Hernandez, D. 108 citations; Maller, C. 64 citations; Chen, Ca. 62 citations; Xu, X. 62 citations; Ridley, I. 59 citations; Willand, N. 59 citations; McCoy, A., P. 26 citations; Mathews, E. 24 citations; and Hankey, S. 11 citations, following Boardman, B.

In Figure 3.10, a collaborative authorship network for authors has been developed. Interauthor lines refer to collaborations and define an author's

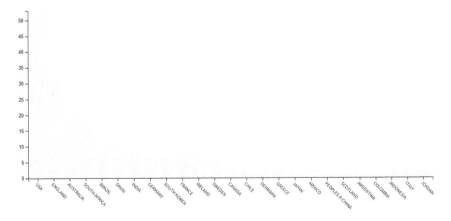

Figure 3.8 Publication numbers by country

88 Clean energy for low-income communities

Author	Documents	Citations
mathews, eh	4	24
boardman, b	3	223
hernandez, diana	3	108
maller, cecily	3	64
chen, chien-fei	3	62
xu, xiaojing	3	62
ridley, ian	3	59
willand, nicola	3	59
maccoy, Andrew p.	3	26
hankey, steve	3	11

Figure 3.9   Prolific authors publishing in the field of research and their number of citations

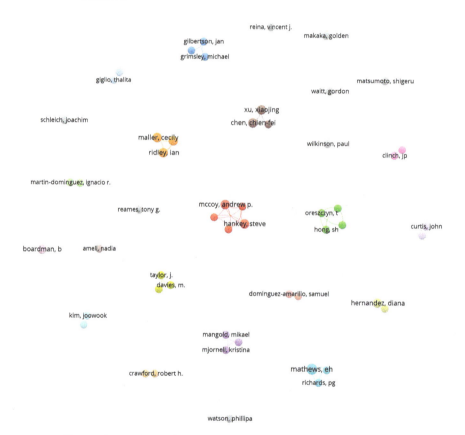

Figure 3.10   Co-authorship network/inter-author collaboration

connection with other authors [20]. For co-authorship analysis, the number of publications and citations per author was determined as a minimum of two as parameters in the VOSviewer program. In the table, 53 authors and 26 co-authorship clusters are identified. According to the table, there is an active

Low-cost and energy-efficient housing design 89

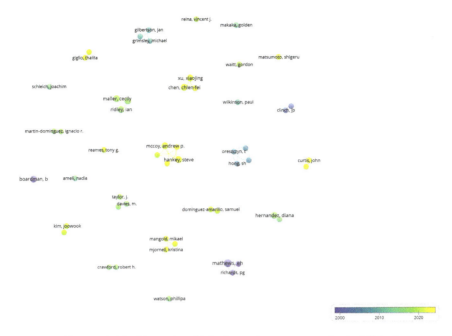

Figure 3.11 Years in which the authors were active/published

collaboration network among the following authors: Hankey, S., Maccoy, A., Reichard, G., Schenk, T., Yeganeh, A.; Hong, Sh., Oreszczyn, T., Ridley, I., Wikinsons, P.; Gilbertson, J., Green, G., Grimsley, M.; Davies, M., Hamilton, I. G., Taylor, J.; Mangold, M., Mjornell, K., Von Platten, J.; Mathews, E., Richards, P., Vanwyk, S.; Maller, C., Ridley, İ., Willand, N.; Chen, C., Nelson, H., Xu, X.; Clinch, J., Jealy, J.; Dominguez-amarilli, S., Fernandez-Aguera, J.; Martin-dominguez, İ., Rodriguez-Muanoz, N.; Giglio, T., Nunes, G.; Hernandez, D., Philipse, D.; Curtis, J., Pillai, A; Kim, J., Song, D.; Crawford, R., Wrigley, K.

Figure 3.11 visually expresses the years in which authors with a collaboration network published. Accordingly, the authors with the most recent publications include Curtis, J. and Pillai, A.

3.3.7 Publication and citation numbers by institutions

The publication and citation numbers of the publishing institutions are examined in this section. Among the institutions where the filtered publications were made, those with at least four publications are included in Figure 3.12. Accordingly, in the first six places according to the number of publications, the University of College London (six publications, 361 citations), Oxford University (five publications, 261 citations), Sheffield Hallam University (four publications, 365 citations), London School of Hygiene and Tropical Medicine University (four publications, 403 citations), Columbia University (four publications, 163 citations), and the University of Johannesburg (four publications, 54 citations).

90 *Clean energy for low-income communities*

Organization	Documents	Citations
UCL	6	361
Univ oxford	5	261
Sheffield hallam univ	4	375
Univ London London sch hyg & trop...	4	403
Columbia univ	4	163
Univ Johannesburg	4	54

Figure 3.12 Publication and citation numbers by institutions

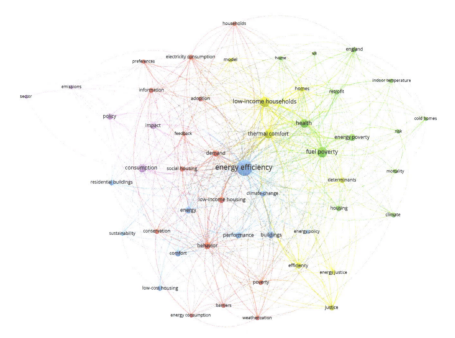

Figure 3.13 Keyword network analysis

3.3.8 *Keyword analysis*

Keywords are expressions, terms, and names that reflect the essence of a publication [21]. The contents of 195 publications were analyzed with the co-occurrence option in the VOSviewer program and keywords were found. Co-occurrence means the common presence or proximity of similar keywords. Co-occurrence can also include similar keywords that occur in the same topic but are not the same in meaning, the closeness of keywords is directly related to the degree of co-occurrence [22].

In the analysis, the minimum number of repetitions of keywords was determined as five. The keywords listed with the co-occurrence option were filtered considering their repetition and 51 keywords were found. These 51 keywords were divided into five separate clusters by the VOSviewer program according to their relationships with each other in the publications. Each cluster is expressed with a different color in Figure 3.13.

Cluster 1 is shown in red in Figure 3.14. In the cluster, there are 15 keywords like adoption, barriers, behavior, conservation, demand, electricity consumption, energy consumption, feedback, household, information, low-income housing, poverty, preferences, social housing, and weatherization. After examining the articles of the cluster containing these keywords, they were grouped under the title Energy Efficiency Policy and Energy Behavior.

Cluster 2 is expressed in green and with 13 keywords. Keywords in this cluster: climate, cold homes, energy poverty, England, fuel poverty, health, home, housing, indoor temperature, mortality, retrofit, risk, United Kingdom (UK). After examining the articles of the cluster containing these keywords, they were grouped under the title Energy Poverty and Health.

There are 10 keywords in Cluster 3, shown in blue. These keywords: buildings, climate-change, comfort, energy, energy efficiency, energy policy, low-cost housing, performance, residential buildings, and sustainability. After examining the articles of the cluster containing these keywords, they were grouped under the heading Energy Policy and Energy Efficiency.

Cluster 4 is defined in yellow and with eight keywords. Determinants in this cluster: efficiency, energy justice, homes, justice, low-income households, model, and thermal comfort. After examining the articles of the cluster containing these keywords, they were collected under the title Energy Justice and Thermal Comfort.

Cluster 5 is shown in purple and there are five keywords in the cluster: consumption, emissions, impact, policy, and sector. After examining the articles of the cluster containing these keywords, they were summarized under the title Energy Consumption and Emissions.

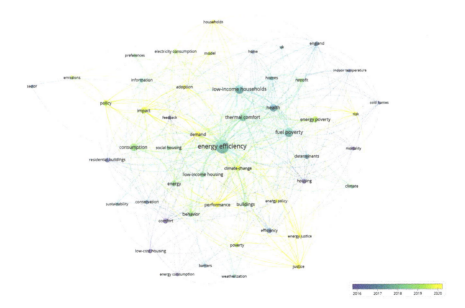

Figure 3.14 Change of the frequently used keywords through the years

92 *Clean energy for low-income communities*

When the changes in the most used keywords in publications are examined over the years, it can be said that this change can be visualized in four separate categories. Categorization is made regarding dark blue, turquoise, green, and yellow color groups. Accordingly, in the first category when listed from past to present: mortality rate, low-cost housing, energy consumption, housing, comfort, conservation, feedback, cold homes, England, home, indoor temperature, residential buildings, efficiency, and sector.

The second category consists of the words: barriers, information, weatherization, fuel poverty, health, United Kingdom (UK), energy efficiency, sustainability, determinants, homes, low-income households, behavior, low-income housing, preferences, social housing, climate, retrofit, buildings, energy, thermal comfort, and consumption.

The third category: electricity consumption, poverty, energy poverty, risk, climate-change, energy policy, performance, model, emissions, impact, and policy.

In the fourth category: adoption, demand, household, energy justice, and justice.

When the keywords of the most current publications are examined, energy policy, energy efficiency, and energy poverty can be seen.

In Figure 3.15, keywords are visualized under the cluster headings to which they are linked according to their number of repetitions. Among the keywords, energy (133 repetitions), efficiency (80 repetitions), and energy efficiency (79 repetitions) were repeated the most. When the clusters were examined, most keywords were collected under the subject heading Energy Policy and Energy Efficiency.

If the results of the analyzes are considered, the researches:

- It is mostly concentrated in the field of Energy Policy and Energy Efficiency, and these studies are related to concepts related to energy, energy efficiency, and the building sector.
- In second place are the studies handled in the context of Energy Justice and Thermal Comfort. Efficiency, Fairness, and Thermal Comfort are concepts that studies in this field are related to.
- Energy Poverty and Health problems were taken into account in the third place. These studies are most associated with the concepts of housing, fuel poverty, health, and retrofit
- In the fourth place, Energy Efficiency Policy and Energy Behavior were studied. Studies in this field are related to the concepts of poverty, household, and low-income housing.
- Finally, Energy Consumption and Emissions are included. These studies are also associated with consumption and emissions.

Another result of the analysis is that all studies are directly or indirectly related to each other.

If we look at the associated concepts of the main headings created based on content analysis in the studies: studies that holistically evaluate the issue of low-cost energy energy-efficient housing with all effective parameters (energy policy,

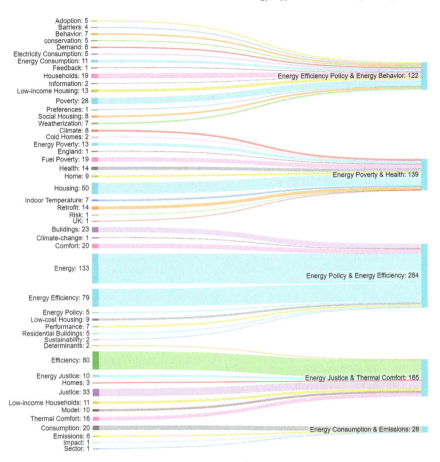

Figure 3.15 Keywords and the clusters they are affiliated with according to their number of repetitions

social class, energy poverty, energy justice, and health problems) are not sufficient. However, it is thought that it is necessary to conduct a content analysis of the most cited studies in order to identify gaps in this field and a potential area for further research.

3.3.9 Most cited studies published between 2000 and 2023

In this section, the most cited publications are discussed. An attempt was made to find out what the focus of these publications was. Publications, content, and number of citations are shown in Table 3.1.

When the contents of these publications are examined, it is seen that the relationship between fuel poverty and health problems in the built environment is discussed and suggestions are made to reduce these problems in energy policies. The role of energy efficiency programs in reducing fuel poverty and reducing

94　*Clean energy for low-income communities*

Table 3.1　Most cited publications

Ref.	Publications	Content	Citation
[6]	"Housing Improvements for Health and Associated Socio-Economic Outcomes: A Systematic Review," Thomson, H., Thomas, S., Sellström, E. Petticrew, M. 2013 | *Campbell Systematic Reviews* 9(1), pp. 1–348.	The health and social effects of improvements in the physical structure of the residences and the provision of comfort to the residents were evaluated.	170
[31]	"Quantifying the severity of fuel poverty, its relationship with poor housing and reasons for non-investment in energy-saving measures in Ireland," Healy, J. D. and Clinch, J. P. 2004 | *Energy Policy* 32(2), pp. 207–220	Fuel poverty, health problems caused by unsuitable housing, and the reasons why homeowners invest in energy efficiency were examined, and suggestions were made to reduce fuel poverty in energy policies.	170
[3]	"Making cold homes warmer: the effect of energy efficiency improvements in low-income homes," Milne, G. and Boardman, B. 2000 | *Energy Policy* 28 (6–7), pp. 411–424	It has been calculated to provide thermal comfort and energy savings in low-income housing with energy efficiency improvement strategies.	155
[36]	"Determinants of winter indoor temperatures in low-income households in England," Oreszczyn, T., Ridley, I., Hong, S. H., and Wilkinson, P. 2006 | *Indoor and Built Environment* 15(2), pp. 125–135	The relationship between indoor temperature of houses in cold seasons and factors such as housing characteristics, social class of low-income families, and the number of people living in the house was examined.	145
[44]	"Energy and Health 4 - Energy, energy efficiency, and the built environment," Wilkinson, P., Smith, K. R., Beevers, S., Tonne, C., and Oreszczyn, T. 2007 | *Lancet* 370(9593), pp. 1175–1187	Fuel poverty in built environments and the relationship between fuel poverty and health problems are discussed.	112
[41]	"Fuel poverty, affordability, and energy justice in England: Policy insights from the Warm Front Program," Sovacool, B. K. 2015 | *Energy* 93, pp. 361–371	The role of energy efficiency programs in reducing fuel poverty and greenhouse gas emissions and the savings in families' income costs are discussed.	101
[26]	"Living in cold homes after heating improvements: Evidence from Warm-Front, England's Home Energy Efficiency Scheme," Critchley, R., Gilbertson, J., Grimsley, M., and Green, G. 2007 | *Applied Energy* 84(2), pp. 147–158	Health problems in houses with heating problems due to fuel poverty and the potential of support programs to solve these problems were examined.	96

(Continues)

Low-cost and energy-efficient housing design 95

Table 3.1 Most cited publications (Continued)

Ref.	Publications	Content	Citation
[23]	"Determinants of households' investment in energy efficiency and renewable: evidence from the OECD survey on household environmental behavior and attitudes," Ameli, N. and Brandt, N. 2015 | *Environmental Research Letters* 10(4), pp. 44–59	In investments in energy efficiency and renewable energy technologies, the impact of families' income level and social class on consumer preferences is mentioned.	92
[34]	"Reducing energy consumption in low-income public housing: Interviewing residents about energy behaviors," Langevin, J., Gurian, P. L., and Wen, J. 2013 | *Applied Energy* 102, pp. 1358–1370	An attempt was made to determine energy behavior and energy knowledge gaps and attitudes among low-income public housing residents.	86
[39]	"Excess winter morbidity among older people at risk of cold homes: A population-based study in a London borough," Rudge, J. and Gilchrist, R 2005 | *Journal of Public Health* 27(4), pp. 353–358	The relationship between older people's health and fuel poverty in low-income households has been examined.	85
[7]	"Psychosocial routes from housing investment to health: Evidence from England's home energy efficiency scheme," Gilbertson, J; Grimsley, M and Green, G. 2012 | *Energy Policy* 49, pp. 122–133	The improvement of health status by eliminating fuel poverty was examined, and the financial and psychosocial benefits of energy efficiency were mentioned.	85
[30]	"Health effects of home energy efficiency interventions in England: a modeling study," Hamilton, I., Milner, J., Chalabi, Z., Das, P., Jones, B., Shrubsole, C., and Wilkinson, P. 2015 | *BMJ Open* 5 (4), pp. 7--84	The effects of improving energy efficiency and indoor air quality on household health are discussed.	61
[24]	"New directions for household energy efficiency: evidence from the UK," Boardman, B. 2004 | *Energy Policy* 32 (17), pp. 1921–1933	Developing a market-oriented policy regarding new energy-efficient products and providing support to those who produce these products are mentioned.	58
[27]	"Indoor air quality in green-renovated vs. non-green low-income homes of children living in a temperate region of US (Ohio)," Coombs, K. C., Chew, G. L., Schaffer, C., Ryan, P. H., Brokamp, C., Grinshpun, S. A., and Reponen, T. 2016 | *Science of the Total*	Differences in indoor air quality between green and non-green houses in low-income housing sites were determined.	58

(Continues)

96 *Clean energy for low-income communities*

Table 3.1   Most cited publications (Continued)

Ref.	Publications	Content	Citation
	Environment 554, pp. 178–185		
[38]	"Effects of customized consumption feedback on energy efficient behavior in low-income households," Podgornik, A., Sucic, B., and Blazic, B. 2016 | *Journal of Cleaner Production* 130, pp. 25–34	The impact of customized consumption feedback and other information interactions on energy consumption behavior and energy saving in low-income households was evaluated.	57
[40]	"Energy efficient technology adoption in low-income households in the European Union - What is the evidence?," Schleich, J. 2019 | *Energy Policy,* 125, pp. 196–206	The relationship between whether energy efficient technologies are adopted or not and household income is discussed, and the role of support programs in this regard is mentioned.	54
[37]	"Mould and winter indoor relative humidity in low-income households in England," Oreszczyn, T., Ridley, I., Hong, S. H., and Wilkinson, P. 2006 | *Indoor and Built Environment* 15(2), pp. 125–135	The relationship between moisture and mold formation in houses during cold seasons, housing characteristics, social class of the family, provision of thermal comfort and health was examined.	52
[25]	"A study of residential energy use in Hong Kong by decomposition analysis, 1990–2007," Chung, W., Kam, M. S., and Ip, C. Y. 2011 | *Applied Energy* 88 (12), pp. 5180–5187	Factors affecting energy consumption in residences were discussed, and it was shown that the increase in low-income housing in Hong Kong caused an increase in energy consumption in the residential sector.	51
[43]	"Increasing energy efficiency in low-income households through targeting awareness and behavioral change," Vassileva, I. and Campillo, J. 2014 | *Renewable Energy* 67, pp. 59–63	Informing households about the benefits of energy efficiency and the impact of this awareness on consumers' energy consumption behavior and ensuring energy efficiency are mentioned.	50

greenhouse gas emissions has also been highlighted. The relationship between the physical features of the house and the social class of the families was examined, and the importance of support programs in improving living standards in houses was mentioned. However, it has been shown that investments in energy efficiency, renewable energy technologies, and raising living standards are directly related to the income levels and social classes of families. Additionally, energy behavior and energy knowledge gaps have been shown to be an influencing factor among low-income residents. To summarize, the focus of the most cited publications is on energy poverty and health problems, the development of energy-efficient policies for low-income families, the reduction of greenhouse gas emissions, and the provision of benefits at different levels for low-income groups thanks to these policies.

3.4   Results and potential areas for future studies

In this chapter, a bibliometric analysis of published research was conducted in order to determine how researchers collaborate on low-cost energy-efficient housing and where the gaps and potential in the field lie. Therefore, "energy efficient hous*" and "low-cost*" OR "energy efficient hous*" and "low-income*" OR "low-income hous*" and "energy efficien*" in the Web of Science database Information on 195 publications published at all times, including the subject headings " OR "low-cost hous*" and "energy efficient*", was examined.

The number of publications by year and the number of citations, common research areas with architecture, type, sources, number by country, author and co-authorship analysis, number, and number of citations by institutions, and keyword analysis are interpreted. The 19 most cited publications between 2000 and 2023 were identified and their contents were examined to observe research trends. The results are as follows:

- Publications were made between 1993 and 2023. While there was an increase compared to 2015, most of the publications were made in 2020 and 2021.
- Apart from architecture, other areas of publications include engineering, construction technology, material science, urban studies, energy fuels, environmental engineering, and social science research. It is possible to say that the publications are mostly related to architecture and the field of energy engineering.
- The majority of the 195 publications analyzed are in the type of articles or conference papers.
- All of the top 10 sources by number of publications are journal articles.
- When the number of publications by country is examined, America and England are in the first two countries among the countries that publish the most on the subject under investigation. They are followed by Austria, South Africa, and Brazil, respectively.
- The most productive author in the field studied between 1993 and 2023 is Mathews, E.. 26 co-authorship clusters with 53 authors were identified. Most recent authors include Curtis, J. and Pillai, A.
- According to the number of publications, the top three institutions are College London University, Oxford University, and Sheffield Hallam University, respectively.
- 51 keywords divided into five clusters were used for keyword analysis. Apart from the keywords energy efficiency and thermal comfort, the most recurring

98 *Clean energy for low-income communities*

words in the examined publications were fuel poverty, energy poverty, social housing, and health.

- The keywords of the most current publications were determined as energy policy, energy efficiency, and energy poverty.
- Among the 19 most cited publications, one is a review article and the others are research articles.
- Studies on energy poverty, energy justice, and energy policies aim to identify analysis and potential study areas.

As a result of the evaluations, the issues that can be determined as academic study areas can be summarized as the social dimensions of energy injustice and the place of energy efficiency of low-income housing in energy politics. In this context, the importance of considering the concept of energy efficiency from the perspective of energy justice and therefore reducing inequalities should be emphasized. For this purpose, it is necessary to provide more information on the health consequences of energy efficiency interventions, to focus energy policies on poverty, marginalization, and exclusion, to increase research on these issues, and to contribute to the development of policies by raising public awareness. In summary, energy efficiency and energy-efficient housing should not be considered as a goal but perhaps as a tool to ensure social justice, and this issue should be emphasized more.

References

[1] IEA. (2022). Energy Efficiency, 21. Retrieved from www.iea.org.

[2] Zaid, N., and Graham, P. (2011). Low-cost housing in Malaysia: A contribution to sustainable development. *Energy, Environment and Sustainability*, 20(11), 82–87.

[3] Aydın, M. (2016). The role of energy efficiency in sustainable development: An assessment of Turkey. *Journal of Management Sciences*, 14(28), 409–441.

[4] Pacheco, R., Ordóñez, J., and Martínez, G. (2012). Energy efficient design of building: A review. *Renewable and Sustainable Energy Reviews*, 16(6), 3559–3573.

[5] Kibert, C. J. (2007). The next generation of sustainable construction. *Building Research and İnformation*, 35(6), 595–601.

[6] Thomson, H., Thomas, S., Sellstrom, E., and Petticrew, M. (2013). Housing improvements for health and associated socio-economic outcomes. *Cochrane Database of Systematic Reviews*, 9(1), 1–348.

[7] Brown, M. A., Soni, A., Doshi, A. D., and King, C. (2020). The persistence of high energy burdens: A bibliometric analysis of vulnerability, poverty, and exclusion in the United States. *Energy research and Social Science*, 70, 101756.

[8] Figus, G., Turner, K., McGregor, P., and Katris, A. (2017). Making the case for supporting broad energy efficiency programmes: Impacts on household incomes and other economic benefits. *Energy Policy*, 111, 157–165.

[9] Pritchard, A. (1969). Statistical bibliography or bibliometrics. *Journal of Documentation*, 25(4), 348–349.

[10] Sajovic, I., Tomc, H.G., and Podgornik, B.B. (2018). Bibliometric study and mapping of a journal in the field of visualization and computer graphics. *COLLNET Journal of Scientometrics and Information Management*, 12(2), 263–287.

[11] Li, K., Rollins, J., and Yan, E. (2018), Web of science use in published research and review papers 1997–2017: a selective, dynamic, cross-domain, content-based analysis. *Scientometrics*, 115(1), 1–20.

[12] Chavarro, D., Rafols, I. and Tang, P. (2018). To what extent is inclusion in the web of science an indicator of journal quality? *Research Evaluation*, 27(2), 106–118.

[13] Smiraglia, R. P. (2015). Domain analysis of domain analysis for knowledge organization: Observations on an emergent methodological cluster. *Knowledge Organization*, 42(8), 602–614.

[14] Li, X., Wu, P., Shen, G.Q., Wang, X., and Teng, Y. (2017). Mapping the knowledge domains of Building Information Modeling (BIM): A bibliometric approach. *Automation In Construction*, 84, 195–206.

[15] Aghimien, D.O., Aigbavboa, C.O., Oke, A.E. and Thwala, W.D. (2019). Mapping out research focus for robotics and automation research in construction-related studies. *Journal of Engineering, Design and Technology*, 18(5), 1063–1079.

[16] Akinlolu, M., Haupt, T.C., Edwards, D.J. and Simpeh, F. (2020). A bibliometric review of the status and emerging research trends in construction safety management Technologies. *International Journal of Construction Management*, 22(14), 1–13.

[17] Wu, Z., Yang, K., Lai, X., and Antwi-Afari, M.F. (2020). A scientometric review of system dynamics applications in construction management research. *Sustainability*, 12(18), 7474.

[18] Riehmann, P., Hanfler, M., and Froehlich, B. (2005). Interactive Sankey diagrams. *IEEE Symposium on Information Visualization*. October 23–25, 2005, Minneapolis, MN, USA, pp. 233–240.

[19] Guo, Y.M., Huang, Z.L., Guo, J., Li, H., Guo, X.R., and Nkeli, M.J. (2019). Bibliometric analysis on smart cities research. *Sustainability*, 11(13), 36–54.

[20] Van Eck, N.J., and Waltman, L. (2010), Software survey: VOSviewer, a computer program for bibliometric mapping. *Scientometrics*, 84(2), 523–538.

[21] Xiang, C., Wang, Y., and Liu, H. (2017). A scientometrics review on non-point source pollution research. *Ecological Engineering*. 99, 400–408.

[22] Lozano, S., Calzada-Infante, L., Adenso-Díaz, B., and García, S. (2019). Complex network analysis of keywords co-occurrence in the recent efficiency analysis literatüre. *Scientometrics*, 120(2), 609–629.

[23] Ameli, N., and Brandt, N. (2015). Determinants of households' investment in energy efficiency and renewables: Evidence from the OECD survey on

household environmental behaviour and attitudes. *Environmental Research Letters*, 10(4), 44–59.

[24] Boardman, B. (2004). New directions for household energy efficiency: evidence from the UK. *Energy Policy*, 32(17), 1921–1933.

[25] Chung, W., Kam, M. S., and Ip, C. Y. (2011). A study of residential energy use in Hong Kong by decomposition analysis, 1990–2007. *Applied Energy*, 88(12), 5180–5187.

[26] Critchley, R., Gilbertson, J., Grimsley, M., Green, G., and Warm Front Study Group. (2007). Living in cold homes after heating improvements: evidence from warm-front, England's home energy efficiency scheme. *Applied Energy*, 84(2), 147–158.

[27] Coombs, K. C., Chew, G. L., Schaffer, C., *et al.* (2016). Indoor air quality in green-renovated vs. non-green low-income homes of children living in a temperate region of US (Ohio). *Science of the Total Environment*, 554, 178–185.

[28] Fonseca, B.D.P.F., Sampaio, R.B., de Araújo Fonseca, M.V. and Zicker, F. (2016). Co-authorship network analysis in health research: method and potential use. *Health Research Policy and Systems*, 14(1), 1–10.

[29] Gilbertson, J., Grimsley, M., Green, G., and Warm Front Study Group. (2012). Psychosocial routes from housing investment to health: Evidence from England's home energy efficiency scheme. *Energy Policy*, 49, 122–133.

[30] Hamilton, I., Milner, J., Chalabi, Z., *et al.* (2015). Health effects of home energy efficiency interventions in England: A modelling study. *BMJ Open*, 5 (4), 72–84.

[31] Healy, J. D., and Clinch, J. P. (2004). Quantifying the severity of fuel poverty, its relationship with poor housing and reasons for non-investment in energy-saving measures in Ireland. *Energy Policy*, 32(2), 207–220.

[32] Internet: World Energy Consumption by Energy Source. Retrieved from https://www.eia.gov/todayinenergy/detail.php?id=32912

[33] Intergovernmental Panel on Climate Change. (2021). *Climate Change 2021: Synthesis Report*. Retrieved from https://www.ipcc.ch/report/sixth-assessment-report-cycle/

[34] Langevin, J., Gurian, P. L., and Wen, J. (2013). Reducing energy consumption in low-income public housing: Interviewing residents about energy behaviors. *Applied Energy*, 102, 1358–1370.

[35] Milne, G., and Boardman, B. (2000). Making cold homes warmer: the effect of energy efficiency improvements in low-income homes: A report to the Energy Action Grants Agency Charitable Trust. *Energy Policy*, 28(6–7), 411–424.

[36] Oreszczyn, T., Hong, S. H., Ridley, I., Wilkinson, P., and Warm Front Study Group. (2006). Determinants of winter indoor temperatures in low-income households in England. *Energy and Buildings*, 38(3), 245–252.

[37] Oreszczyn, T., Ridley, I., Hong, S. H., Wilkinson, P., and Warm Front Study Group. (2006). Mould and winter indoor relative humidity in low-income households in England. *Indoor and Built Environment*, 15(2), 125–135.

[38] Podgornik, A., Sucic, B., and Blazic, B. (2016). Effects of customized consumption feedback on energy efficient behaviour in low-income households. *Journal of Cleaner Production*, 130, 25–34.

[39] Rudge, J., and Gilchrist, R. (2005). Excess winter morbidity among older people at risk of cold homes: a population-based study in a London borough. *Journal of Public Health*, 27(4), 353–358.

[40] Schleich, J. (2019). Energy efficient technology adoption in low-income households in the European Union–What is the evidence?. *Energy Policy*, 125, 196–206.

[41] Sovacool, B. K. (2015). Fuel poverty, affordability, and energy justice in England: Policy insights from the Warm Front Program. *Energy*, 93, 361–371.

[42] Sunday, D., Shukor Lim, N. H. A., and Mazlan, A. N. (2021). Sustainable affordable housing strategies for solving low-income earners housing challenges in Nigeria. *Studies of Applied Economics*, 39(4), 1–15.

[43] Vassileva, I., and Campillo, J. (2014). Increasing energy efficiency in low-income households through targeting awareness and behavioral change. *Renewable Energy*, 67, 59–63.

[44] Wilkinson, P., Smith, K. R., Beevers, S., Tonne, C., and Oreszczyn, T. (2007). Energy and Health 4 - Energy, energy efficiency, and the built environment. *The Lancet*, 370(9593), 1175–1187.

[45] Yeganeh, A., Agee, P. R., Gao, X., and McCoy, A. P. (2021). Feasibility of zero-energy affordable housing. *Energy and Buildings*, 241, 110–119.

Chapter 4

Enabling solar energy production for low-income communities

Lutfu S. Sua[1] and Figen Balo[2]

Solar power has a lot of significance among renewable power resources. Electric production from solar power has become a priority for many nations around the world in the last few years. It is very important to choose the right solar panel when it comes to producing electricity through sunlight. If the selected solar panel is not suitable for the system, there may be many different disadvantages, such as waste of energy and capital.

In this study, it is attempted to determine the best solar panel in terms of technical features and cost in a solar facility planned to be used to enable widespread use of solar energy for low-income communities.

While determining the solar panels studied worldwide among the alternative brands, the evaluation of some characteristics of the brands for low-income communities was taken into consideration simultaneously. Such characteristics include, long "product warranty length," minimum "warranted yearly efficiency reduction," and maximum "warranted energy output," etc. For the case study, 415 W monocrystalline solar panel brands were analyzed. The results were evaluated using Analytic Hierarchy Process (AHP) among Multi-Criteria Decision Making (MCDM) techniques.

Keywords: Green housing; MCDM; Clean energy; Solar power; Low-income communities

Nomenclature

AHP	Analytic Hierarchy Process
ANP	Analytic Network Process
ARAS	Additive Ratio Assessment
CRITIC	Criteria Importance Through Intercriteria Correlation
EDAS	Evaluation Based on Distance from Average Solution

(Continues)

[1]Management Department, Southern University and A&M College, USA
[2]Department of METE, Firat University, Türkiye

104 *Clean energy for low-income communities*

(*Continued*)

MCDA	Multi-Criteria Decision Analysis
MCDM	Multi-Criteria Decision Making
MULTIMOORA	Multi-Objective Optimization on the basis of Ratio Analysis plus full multiplicative form
PROMETHEE	Preference Ranking Organization Method for Enrichment Evaluation
SWARA	StepwiseWeight Assessment Ratio Analysis
TOPSIS	Technique for Order Performance by Similarity to Ideal Solution
VIKOR	VlseKriterijumska Optimizcija I Kaompromisno Resenje
WASPAS	Weighted Aggregated Sum Product Assessment

4.1 Introduction

Energy is known as the driving force behind financial growth. Energy sourced from renewable and non-fossil resources, like sun, wind, geothermal, hydropower, ocean, biogas, and biomass, is what the "Directive 2009-28-EC Additional 1" defines as renewable power sources [1]. Due to the increasing demand, a considerable portion of the power resources being used will finally run out [2]. The improvement of sustainable power is a policy used by governments to lessen citizens' reliance on fossil fuels, particularly when supplying electricity [3]. Because of the air pollution caused by fossil fuels, many people worldwide suffer from fatal diseases, including cancer and asthma, as well as other catastrophic illnesses. When fossil fuels are used, they often produce a variety of SO_2, CO_2, NO_x, and Particular Materials (PM) emissions [4].

For several reasons, "Renewable Energy Sources" is the first and most sensible option that people consider at this time. People have started producing energy using renewable energy sources, mainly because they are clean, which means they lower air pollution and do not harm the environment. As a result, it is simple to predict that there will be fewer threats to nature and public safety. Another justification is that individuals can always use clean energy to complete their everyday needs because, unlike fossil fuels, renewable energy sources do not deplete over time.

The sun can sometimes provide renewable energy sources directly (like solar energy) and sometimes indirectly (like wind and hydropower) [5]. Hydro, biomass, geothermal, wind, and solar energy make up the majority of renewable energy sources. Some forms of energy are simple to replenish because they come from non-depleting natural sources, so they are renewable [6]. It is anticipated that this renewable energy will also be obtained at the individual or household level, in addition to the technological level (for instance, by a private or sizable state enterprise) [7]. However, developments in renewable and alternative energy technologies are slowly turning them into a common option in most developing nations, and they have significant potential to reduce dependency on non-renewable energy sources [8]. Based on geography and available resources, many nations have begun to move toward renewable resources [9]. Solar photovoltaic mechanisms, wind energy generating mechanisms, microturbines, fuel cells, and other renewable energy-based systems are the main sources.

Sun power is considered a promising, profitable, and reliable power resource. Some advantages involve a long lifespan, low maintenance, and being pollution-free [10]. The most abundant energy resource that can fulfill community requirements arising from sustainable financial development is solar power [11]. Sun power, solar thermal energy, and photovoltaic energy all have numerous favorable impacts on the ecology, enhancing societal sustainability and raising the standard of living [12]. Due to the decreasing cost of solar panels, establishing solar power plants is getting simpler today [13,14]. The benefits of solar energy include enhanced CO_2 mitigation, a lack of noise, reduced toxic waste, and a lack of environmental remediation procedures [15]. Heat and light from solar radiation can be transformed into usable thermal energy or used to generate electricity utilizing solar thermal or photovoltaic processes [16]. The field of solar photovoltaic technology has been growing technologically, and today several commercially viable solar technologies have arisen. Developed nations are mindful of protecting the environment even as they strive for quicker economic progress and industrial expansion. There are legitimate concerns about environmental preservation in the middle of all the succeeding industrial growth since a nation's industrial development cannot be compromised at any cost. Numerous industrialized nations have made great achievements toward the forefront of green industrial operations and have been extremely eco-friendly in their industrial endeavors without negatively compromising their earnings or other organizational goals. When compared to other production steps, photovoltaic modules require more labor than money and technology. This is a fantastic chance for developing nations.

In recent years, solar panels have been utilized to generate electricity on a smaller scale, particularly for residential or commercial use in apartment buildings or single structures, with efficiencies ranging from 18% to 21%. The effectiveness of solar energy projects across the nation depends on the availability of accurate data on solar radiation [17]. Polycrystalline and monocrystalline crystals are two of the crystal kinds used in solar panels [18]. Due to the presence of polycrystalline panels, which are generally less efficient but also more affordable, monocrystalline panels, while slightly more expensive, tend to be more efficient [19].

It can be challenging to select the precise required solar panel because of the large number of possibilities and the diversity of selection criteria. Multi-attribute resolution analysis, which ranks options depending on available information, is the most appropriate method for resolution in this situation [20]. It will be challenging to choose the best solar panel manually or based just on a small number of parameters because there are many different factors to consider, such as pricing and technical properties [21]. Problems relating to sustainability are resolved using Multi-Criteria Decision Analysis (MCDA) approaches, which have demonstrated their efficacy in evaluating sustainability. The MCDA approach is therefore required because it accepts all criteria for all alternatives. For instance, the Preference Ranking Organization Method for Enrichment Evaluation (PROMETHEE) approach with steadiness evaluation was used to evaluate offshore wind facility areas [22,23] to

106 *Clean energy for low-income communities*

establish the elements of the decision-making process, perform the sensitivity analysis, rank the actions, and choose the best action. AHP and Analytic Network Process (ANP) were used to build wind farms [24], and novel enlargements of the AHP methodology are being suggested [25]. The main difference between AHP and ANP is that while ANP is used to structure a decision problem as a network, AHP structures it into a hierarchy with an objective, decision criteria, and decision alternatives.

One of the key issues during the multiple-attribute decision-support process is the significance levels/weights of the criterion. Objective and subjective weights for criteria are the two forms that are covered in the literature [26]. The information in decision matrices provides the objective weights, while the knowledge provided by the decisions is used to estimate the subjective weights [27]. Many authors have devised various methods for evaluating the weights of objective attributes [28–30]. For the assessment of the issue of selecting a sustainable power industry, Maghsoodi *et al.* proposed a combined technique by merging Multi-Objective Optimization on the basis of Ratio Analysis plus full multiplicative form (MULTIMOORA) and StepwiseWeight Assessment Ratio Analysis (SWARA) approach [31]. The Weighted Aggregated Sum Product Assessment (WASPAS) and SWARA techniques were thoroughly reviewed by Mardani *et al.* along with their uses in various fuzzy contexts [32]. SWARA and WASPAS methods are usually used in conjunction because SWARA is a method that is used in ranking the decision criteria based on their relative importance, while WASPAS is utilized in evaluating and ranking the decision alternatives. Karabasevic *et al.* addressed a framework for evaluation based on the SWARA and Additive Ratio Assessment (ARAS) methodologies [33]. Kersuliene *et al.* introduced the SWARA methodology as a novel and efficient way of calculating subjective criteria weights [34]. In order to solve MCDM challenges, this method was then employed to evaluate building equipment in light of sustainability dimensions. The eco-technological thermal energy facility choice issue was studied by Mishra and Rani using a hybrid approach that combined the VlseKriterijumska Optimizcija I Kaompromisno Resenje (VIKOR) and SWARA methodologies in a neutrosophic-fuzzy environment [35]. Ghorabaee *et al.* improved fuzzy combined techniques based on the Evaluation Based on Distance from Average Solution (EDAS), SWARA, and Criteria Importance Through Intercriteria Correlation (CRITIC) methodologies in calculating the weights of the criteria and ranking the alternatives [36]. EDAS is a method used for classification and decision-making for conflicting decision criteria.

The following is a view of the literature on solar power research using MCDM techniques: The carbon loans obtained and solar power produced through the solar panels of the multi-crystalline photovoltaic panel were estimated by Maheshwari *et al.*, in their research on the photovoltaic systems' economic viability. The payback period and carbon footprint have also been analyzed [37]. In order to categorize and choose the top 100 W brand PV models, Sasikumar and Ayyappan employed Fuzzy AHP and Technique for Order Performance by Similarity to Ideal Solution (TOPSIS) techniques [38]. To assist the expert in selecting the best photovoltaic in the market given a series of attributes, El-Bayeh *et al.* have suggested an algorithm based on weightiness-ratio.

Enabling solar energy production for low-income communities 107

They employ TOPSIS for the outcome validation process. Moreover, when using the TOPSIS approach, the authors did not consider the opinions of experts [39]. Kharseh and Wallbaum gathered numerous aspects and compared various PV technologies. They conducted a good cross-check, but they made no optimization recommendations or used any multi-attribute decision-support techniques to choose the best photovoltaic panels [40]. Balo and Sagbansua used the AHP multi-criteria decision-support approach to try to get the optimum performance out of six distinct PV panels. Many factors, including mechanical, electrical, consumer, and ecological norms, were taken into consideration in the suggested method for choosing solar panels. The application was based on 200 W brand names, and it was a unique situation that might not accurately represent the requirements of other consumers. Nonetheless, the suggested modeling can be modified and performed as a guide when choosing the best solar panels for numerous situations in various locations or nations.

A multi-attribute decision-making method has the capability of incorporating both quantitative and qualitative aspects in selecting solar panels for a photovoltaic system. Owing to the fact that it uses both quantitative and qualitative factors, MCDM is a sophisticated decision-making technique. Several MCDM methodologies and methods have been proposed in recent years to address issues with energy planning. This paper's goal is to find the most effective solar panel by AHP as one of the multiple attribute decision support approaches in accordance with the best technical characteristics and the minimum costs for low-income communities.

4.2 Solar power at low-income communities

As roughly three-quarters of residential rooftops worldwide cannot be used for residential solar systems due to shade and poor roof conditions, community solar is a desirable alternative for everyone, regardless of economic level. Nonetheless, numerous governments have started looking into community solar within the larger shared solar market as a way to increase the acquisition of solar power for low-income communities.

Governments at all levels are utilizing different strategies and finance schemes to increase the acquisition of solar energy, which is causing the low-income communities' solar policy environment to change quickly. However, despite this increased focus, it is still difficult to expand solar access to low-income populations.

In terms of solar, many countries have run into problems fulfilling market needs. This has resulted in program restructuring and, in some cases, a de-emphasis on the low-income community market. A way forward might be provided by multifaceted strategies that simultaneously address many difficulties that low-income communities struggle with when attempting to use solar energy.

4.2.1 Solar energy implementation challenges in low-income communities

Emerging policies and programs aim to remove a number of barriers preventing low- and moderate-income populations from accessing solar energy. The most noticeable

108   *Clean energy for low-income communities*

examples are the National Renewable Energy Laboratory (NREL) [41–43], solar programs in California, Colorado, and Louisiana [44], the Interstate Renewable Energy Council (IREC) [45], the U.S. Department of Energy [46], the Center for Social Inclusion [47], and the Clean Energy Savings for All Americans Initiative [48].

To date, public sector support and involvement have been crucial in increasing low-income populations' access to solar energy, even if the processes and financing sources used have varied greatly. The financing strategies used in the low-income solar market are probably going to change as well. Although there is a limited amount of literature and research on low-income solar applications, a number of prospective funding channels and sources are beginning to emerge.

It can be difficult to communicate the long-term advantages of solar power because low-income communities may not be instantly aware of these advantages and have typically not been aimed at solar-based investments.

In some places, the electric ratio provided through low-income community subsidized photovoltaic installations or solar participation may be more than what customers are already paying if they are eligible for reduced electricity rates through energy assistance programs.

Low-income neighborhoods typically have lower homeownership rates and a higher possibility of residing in multiple families and economic building units, which translates into having less influence on utility and rooftop solar policy. Even where lower-income people own their own homes, rooftop solar may not be a realistic choice.

Direct incentives can be paid to community solar facilities that serve low-income customers directly in cash or can lower the cost of PV systems or the subscription price for community solar. They can also increase the bill credit that users will receive.

Public money is maintained in reserve under loan loss reserve programs to compensate for potential losses that credit suppliers may experience if a borrower defaults on a credit. This can reduce the detected risk and ease the loan application process for those with bad credit scores. With a few lending programs, the city allows lenders to take advantage of government-funded credit loss resource calculations to reduce loan risk while also providing credits to low-income consumers to buy society-sun-sourced revenues.

Lowering minimal credit score requirements or focusing on prior invoice expense history more than credit score alone are two more strategies for enhancing low-income consumers' access to loans and credit.

Placing energy-sourced loan expenses straight onto the electricity bills of customers is a popular practice in the energy efficiency sector and may provide benefits for both institutions and consumers, such as transferability and lower bills. On-bill financing can also be more understandable and less reliant on credit ratings when addressing low-income solar access because clients can see beneficial bill conserving from solar power offsetting their loan expenses.

If low-income persons have a sufficiently high tax burden or do not fall into an appropriate tax bracket, they may not be beneficiaries of tax benefits for sun-sourced mechanisms. Tax credits are the main government incentive for solar PV installations; however, they have not been effective in expanding affordability to lower-income areas.

Enabling solar energy production for low-income communities 109

Higher upfront installation expenses or community solar subscription charges are unaffordable in low-income communities since there is less disposable cash in those areas. Also, low-income people sometimes have worse credit scores, which can make getting credit for sun-based investments challenging. Even in circumstances where credits are offered for society through sun-sourced buys, they might not provide access for people with bad credit.

4.2.2 *Potential funding resources for cities*

Placement of energy-sourced credit payments straight onto clients' electricity invoices is also a common practice and may provide similar benefits for both financial institutions and consumers. Although the processes and funding sources have varied substantially, public sector involvement and assistance have been essential in spreading solar access to low-income groups. The financing strategies used in the low-income solar market are probably going to change as well. Although there is a limited amount of literature and research on low-income solar-based best applications, a number of prospective financing options and funding sources are beginning to emerge [49].

Direct incentives can be paid to community solar facilities that serve low-income customers directly in cash or can lower the cost of PV systems or the subscription price for community solar. They can also increase the bill credit that users will receive.

Solar might be a crucial component of these kinds of place-sourced programs in relation to encouraging both financial and environmental advantages. Some communities are establishing more strategic partnerships to engage in solar energy projects as a way to create regional job opportunities and enhance local prosperity.

The growth of solar access in low-income communities can be significantly facilitated by community improvement entities and community improvement institutions, which are major actors in the development of economic homes and enterprises in these areas. This can help communities gain access to other financial tools to extend local solar power use. In other instances, these can proactively assist their projects' solar integration.

In this study, global brands of solar panels, one of the most costly pieces of equipment for a solar farm, were investigated. It is aimed at determining the solar panel brands that can be preferred among the high-efficiency but low-cost solar panel brands for low-income communities.

4.3 Methodology

According to the literature, the following factors make the decision-making process for solar panel selection both challenging and complex:

- Numerous criteria that are both qualitative and quantitative; conflicts between criteria; and criteria with competing aims.
- Due to the intense global business competition, there are several external and internal restraints placed on the purchasing operation.

110   *Clean energy for low-income communities*

- The AHP approach is typically employed in applications requiring nearly precise decisions. The AHP implies that the notional relevance of the attributes impacting the option's efficiency is definite; hence, it does not explain risk and uncertainty when evaluating the option's potential efficiency. Decision-makers' subjective assessment, choice, and preference have a significant impact.

The AHP process consists of six stages: identifying the decision to be made, determining the criteria to be considered in the decision-making process, creating a binary crosscheck matrix for pairwise comparison of the criteria, determining the relative importance of each criterion based on expert opinions, an accuracy check indicator, getting the final score, and choosing what to do. The process follows the steps below:

Step 1: Creating the hierarchical structure based on the alternatives and the listed characteristics.

Step 2: Construct the pairwise comparison matrix in relation to relative significance values.

Step 3: Calculate attribute weights (W_j).

Step 4: Create the weighted matrix as the product of normalized technical properties and weight vectors and check for consistency through indication consistency values, rate, and index. Consistency values are the product of index-weighted and attribute-weighted values.

Consistency index (*CI*) is computed through,

$$CI = (\lambda_{\max} - n) / (n - 1) \tag{4.1}$$

Here, n is the attribute number and $\lambda_{\max}$ is the mean of consistency values.

CR (consistency rate) is calculated by,

$$CR = CI / RI \tag{4.2}$$

Random index (*RI*) is found with respect to earlier research. Eventually, to evaluate the weight value consistency, the obtained consistency rate is crosschecked with an upper limit accepted of 0.1.

4.4   Results and discussion

Tables 4.1 and 4.2 present the selected characteristics of 15 solar panel brands compared in this study. Panel efficiency (%), approximate cost per Watt ($), approximate cost per panel ($), panel weight (kg), number of busbars, temperature coefficient (%/°C), front load resistance (Pa), rear load/wind resistance (Pa), product warranty length (years), warranted annual performance degradation year 2–25 (% per year), power output warranted at year 25 (%) are the technical properties coded as TP# in the subsequent tables. The characteristics provided in Table 4.1 are used within the multi-criteria decision-making methodology, while the remaining characteristics in Table 4.2 are provided as supplementary information about the panel brands.

Table 4.1   Solar panel (415 W) technical properties

Brand	panel efficiency (%)	App. cost per Watt ($)	App. cost per panel ($)	Panel weight (kg)	Number of busbars	Temperature coefficient (%/°C)	Front load resistance (Pa)	Rear load/ wind resistance (Pa)	Product warranty length (years)	Warranted annual performance degradation year 2–25 (% per year)	Power output warranted at year 25 (%)
1	20.4	0.70	280	22	9	−0.34	5,400	2,400	25	0.55	84.8
2	21.3	0.65	269	22	9	−0.35	5,400	2,400	12	0.6	83.1
3	20.8	0.60	264	24	9	−0.35	5,400	2,400	12	0.55	84.8
4	21.3	0.62	257	21	9	−0.34	5,400	2,400	15	0.55	84.8
5	21	0.73	299	22	5	−0.38	5,400	2,400	20	0.45	87.2
6	21.3	0.77	320	21	12	−0.34	5,400	3,600	25	0.5	86
7	22.3	0.70	393	22	16	−0.26	7,000	4,000	25	0.25	92
8	21.6	0.67	261	21	9	−0.34	5,400	2,400	25	0.55	84.8
9	21.3	0.61	253	19	9	−0.34	5,400	2,400	30	0.55	84.8
10	21.3	0.70	290	21	9	−0.34	5,400	2,400	25	0.55	84.8
11	21.1	0.79	328	21	9	−0.34	5,400	2,400	25	0.45	87.2
12	21.3	0.75	311	21	9	−0.36	5,400	3,800	25	0.4	89.4
13	20.1	0.75	250	22	9	−0.34	3,600	2,400	25	0.54	84
14	20.8	0.60	250	22	9	−0.34	5,400	2,400	25	0.55	84.8
15	21.2	0.82	311	22	9	−0.35	5,400	2,400	25	0.53	85.28

112   *Clean energy for low-income communities*

Table 4.2   The solar panel (415 W) general information (operation temperature: −40 °C/−85 °C)

Panel brand	Panel technology	Country of manufacture	Annual revenue of the company (USD)	Company origin
1	Shingled	China	$39 billion	South Korea
2	Monocrystalline	China	$2.41 billion	China
3	Monocrystalline	China	$3.2 billion	China
4	Monocrystalline	China	$876 million	China
5	Monocrystalline	China	$800 million	China
6	Monocrystalline	China	$2.42 billion	Germany
7	Monocrystalline Heterojunction	Singapore	$820 million	Norway
8	Monocrystalline	China	$808 million	China
9	Monocrystalline	China	$780 million	China
10	Monocrystalline	China	$489 million	Israel
11	Monocrystalline	Malaysia// Philippines// Mexico	$2.7 billion	USA
12	Monocrystalline	China	$1.24 billion	China
13	Monocrystalline PERC	Australia	$820 million	Australia
14	Monocrystalline	China	$3.15 billion	China
15	Monocrystalline	China	$80 million	Taiwan

Table 4.3   Comparison scale

Intensity	Explanation	Definition
9	The strongest potential order of affirmation can be found in the information supporting one activity more than another.	Extreme significance
7	One activity is greatly preferred than other, and this dominance is put into action.	Demonstrated or strong significance
5	Judgment and experience favor one activity more than the other quite strongly.	Strong significance
3	Judgment and experience favor one activity mildly more than the other.	Moderate significance
1	The goal is equally benefited by the two activities.	Equal significance
2, 4, 6, 8	There are occasions when compromise is needed.	Intermediate values

The scale used for comparison of the criteria is provided in Table 4.3.

Based on the average score of experts' views using the scale in Table 4.3, the decision matrix in Table 4.3 is developed to provide pairwise comparisons of the decision criteria used in the study.

The correlation matrix in Table 4.4 indicates that *"Panel Efficiency"* is the most significant factor contributing to the overall attractiveness of solar panels.

Table 4.4   Correlation matrix

	TP1	TP2	TP3	TP4	TP5	TP6	TP7	TP8	TP9	TP10	TP11
TP1	1	1	3	5	7	4	6	3	2	2	3
TP2	1	1	3	6	2	1/3	1/2	1/4	1/5	1/5	1/3
TP3	1/3	1/3	1	3	4	¼	1/3	1/5	1/6	1/6	¼
TP4	1/5	1/6	1/3	1	1/3	¼	1/3	1/5	1/6	1/6	¼
TP5	1/7	1/2	1/4	3	1	1/5	¼	1/6	1/7	1/7	1/5
TP6	1/4	3	4	4	5	1	2	1/2	2	2	1
TP7	1/6	2	3	3	4	1/2	1	1/3	1/3	1	1
TP8	1/3	4	5	5	1/4	2	3	1	2	2	1
TP9	1/2	5	6	1/4	1/7	1/2	3	1/2	1	1	1/2
TP10	1/2	5	6	1/4	1/7	1/2	1	1/2	1	1	2
TP11	1/3	3	4	4	5	1	1	1	2	1/2	1
Weight	**21,4%**	**6,1%**	**3,5%**	**1,9%**	**2,2%**	**11,1%**	**6,5%**	**14,2%**	**11,5%**	**11,8%**	**10,0%**

114 *Clean energy for low-income communities*

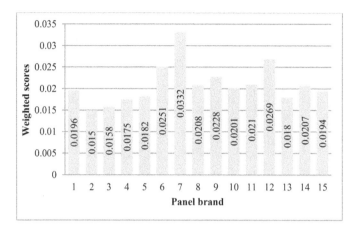

Figure 4.1   Total scores of the alternatives

Figure 4.1 provides a comparison of the alternative solar panels based on the 11 criteria. The results show that Panel 7 has the highest score among the 15 brands compared in this study.

4.5 Conclusions

The pace of global energy consumption is rising quickly, along with the effects that this consumption has on the environment. This trend is acknowledged by a variety of communities, including scientists, engineers, and even politicians [50].

Compared to wind and tidal energy, solar energy has been one of the most broadly utilized non-conventional power resources. Solar energy is produced through photovoltaics using a photovoltaic system. The development of solar power generation has led to a more affordable alternative to traditional gas- or coal-sourced thermal energy facilities, which are not only expensive but also environmentally unfriendly. With a photovoltaic system, the solar panel is essential, and extensive research is being conducted worldwide to lower costs while improving efficiency. The choice of a solar panel is influenced by a variety of complex elements, including both subjective and objective criteria. The right solar panel must be chosen by balancing both subjective and quantitative factors. For choosing solar panels, research was carried out to support the suggested approach.

This study contributes to both the current literature and the industry by proposing a structured assessment of monocrystalline solar panels using an MCDM method. Special consideration was given to the lower-income communities by considering a number of properties that are more relevant to them. As a result of the analysis made in this study, "Panel 7" was found to be the most effective panel. Compared to Panel 2, which has the lowest power output, the power output that can be achieved with Panel 7 is 9.67% higher. Likewise, the profit rates that can be

Enabling solar energy production for low-income communities 115

achieved with Panel 7 compared to Panel 2 are more than 7.14% and 31.55% for "application cost per Watt" and "application cost per panel," respectively.

References

[1] Sholikha, M. *Hambatan malta mencapaı target energı terbarukan dalam kerangka renewable energy directive uni eropa periode 1st interim 2010– 2014.* Universitas Airlangga (2019).

[2] Zoghi, M., Houshang E., Sadat, M., Amiri, M.J., and Karimi, S. Optimization solar site selection by fuzzy logic model and weighted linear combination method in arid and semi-arid region: A case study Isfahan-IRAN. *Renewable and Sustainable Energy Reviews* (2017) 68:986–996. https://doi.org/10.1016/j.rser.2015.07.014.

[3] Widjaja, M.R. *Investment-State Dispute Settlement Dalam Pemberian Insentif Bagi Investasi Asing Di Sektor Energi Terbarukan Indonesia.* Universitas Aırlangga (2020).

[4] Paraschiv, S., and Paraschiv, L.S. Trends of carbon dioxide (CO_2) emissions from fossil fuels combustion (coal, gas, and oil) in the EU member states from 1960 to 2018. *Energy Reports* (2020) 6:237–242.

[5] Ellabban, O., Abu-Rub, H., and Blaabjerg, F. Renewable energy resources: Current status, future prospects, and their enabling technology. *Renewable and Sustainable Energy Reviews* (2014) 39:748–764.

[6] Mohtasham, J. Renewable energies. *Energy Procedia* (2015) 74:1289–1297.

[7] Strielkowski, W., Streimikiene, D., Fomina, A., and Semenova, E. Internet of energy (IoE) and high-renewables electricity system market design. *Energies* (2019) 12(24):1–17. https://doi.org/10.3390/en12244790.

[8] Vaka, M., Walvekar, R., Rasheed, A.K., and Khalid, M. A review on Malaysia's solar energy pathway towards carbon-neutral Malaysia beyond Covid'19 pandemic. *Journal of Cleaner Production* (2020) 273:122834. https://doi.org/10.1016/j.jclepro.2020.122834.

[9] Akrami, M., Gilbert, S.J., Dibaj, M., *et al.* Decarbonisation using hybrid energy solution: Case study of Zagazig, Egypt. *Energies* (2020) 13(18). https://doi.org/10.3390/en13184680.

[10] Gupta, A., Chauhan, Y.K., and Pachauri, R.K. A comparative investigation of maximum power point tracking methods for solar PV system. *Solar Energy* (2016) 136:236–253. https://doi.org/10.1016/j.solener.2016.07.001.

[11] Husain, A.A.F., Hasan, W.Z.W., Shafie, S., Hamidon, M.N., and Pandey, S.S. A review of transparent solar photovoltaic technologies. *Renewable and Sustainable Energy Reviews* (2017) 94:779–791. https://doi.org/10.1016/j.rser.2018.06.031.

[12] Tsoutsos, T., Frantzeskaki, N., and Gekas, V. Environmental impacts from the solar energy technologies. *Energy Policy* (2005) 33:289–296.

[13] Candelise, C., Winskel, M., and Gross, R.J. The dynamics of solar PV costs and prices as a challenge. *Renewable and Sustainable Energy Reviews* (2013) 26:96–107.

116 *Clean energy for low-income communities*

[14] Honrubia-Escribano, A., Javier Ramirez, F., Gómez-Lázaro, E., Garcia-Villaverde, P.M., Ruiz-Ortega, M.J., and Parra-Requena, G. Influence of solar technology in the economic performance of PV power. *Renewable and Sustainable Energy Reviews* (2018) 82:488–501.

[15] Alsema, E.A. Energy requirements and CO2 mitigation potential of PV systems. In *Proceedings of the BNL/NRELWorkshop*, Colorado, CO, USA, 23–24 July 1998.

[16] Soda, S., Sachdeva, A., and Garg, R.J. GSCM: Practices, trends, and prospects in Indian context, Department of Industrial and Production Engineering, National Institute of Technology, Jalandhar, India (2015).

[17] Byrne, J.A., Fernandez-Ibanez, P.A., Dunlop, P.S.M., Alrousan, D.M.A., and Hamilton, J.W.J. Photocatalytic enhancement for solar disinfection of water: A review. *International Journal of Photoenergy* (2011) 2011:798051.

[18] Wander, J. Stimulating the diffusion of photovoltaic systems: A behavioural perspective. *Energy Policy* (2006) 34(14):1935–1943.

[19] Becchio, C., Bottero, M., Corgnati, S.P., and Dell'Anna, F. A MCDA-based approach for evaluating alternative requalification strategies for a net-zero energy district (NZED). In *Multiple Criteria Decision Making*. Berlin: Springer (2017), pp. 189–211.

[20] Alsadi, S., and Khatib, T. Photovoltaic power systems optimization research status: A review of criteria, constraints, models, techniques, and software tools. *Applied Sciences* (2018) 8:1761.

[21] Ziemba, P. Towards strong sustainability management—A generalized prosa method. *Sustainability* (2019) 11:1555.

[22] Ziemba, P., Watróbski, J., Zioło, M., and Karczmarczyk, A. Using the prosa method in offshore wind farm location problems. *Energies* (2017) 10:1755.

[23] Rehman, A., Shekhovtsov, A., Rehman, N., Faizi, S., and Sałabun, W. On the analytic hierarchy process structure in group decision-making using incomplete fuzzy information with applications. *Symmetry* (2021) 13:609.

[24] Shekhovtsov, A., Kizielewicz, B., and Sałabun, W. New rank-reversal free approach to handle interval data in mcda problems. *International Conference on Computational Science*. Berlin: Springer (2021).

[25] Liu, J., Liu, P., Liu, S.F., Zhou, X.Z., and Zhang, T. A study of decision process in MCDM problems with large number of criteria. *International Transactions in Opererational Research* (2015) 22:237–264.

[26] Diakoulaki, D., Mavrotas, G., and Papayannakis, L. Determining objective weights in multiple criteria problems: The CRITIC method. *Computers and Operations Research* (1995) 22:763–770.

[27] Zavadskas, E.K., and Podvezko, V. Integrated determination of objective criteria weights in MCDM. *International Journal of Information Technology and Decision Making* (2016) 15:267–283.

[28] Mishra, A.R., Rani, P., Pardasani, K.R., and Mardani, A. A novel hesitant FuzzyWASPAS method for assessment of green supplier problem based on exponential information measures. *Journal of Cleaner Production* (2019).

[29] Rani, P., Mishra, A.R., Pardasani, K.R., Mardani, A., Liao, H., and Streimikiene, D. A novel VIKOR approach based on entropy and divergence measures of Pythagorean fuzzy sets to evaluate renewable energy technologies in India. *Journal of Cleaner Production* (2019) 238:1–17.

[30] Maghsoodi, A.I., Maghsoodi, A.I., Mosavi, A., Rabczuk, T., and Zavadskas, E.K. Renewable energy technology selection problem using integrated H-SWARA-MULTIMOORA approach. *Sustainability* (2018) 10:448.

[31] Mardani, A., Nilashi, M., Zakuan, N., *et al.* Asystematic review and meta-analysis of SWARA and WASPAS methods: Theory and applications with recent fuzzy developments. *Applied Soft Computing* (2017) 57:265–292.

[32] Karabasevic, D., Paunkovic, J., and Stanujkic, D. Ranking of companies according to the indicators of corporate social responsibility based on SWARA and ARAS methods. *Serbian Journal of Management* (2016) 11:43–53.

[33] Kersuliene, V., Zavadskas, E.K., and Turskis, Z. Selection of rational dispute resolution method by applying new step-wise weight assessment ratio analysis (SWARA). *Journal of Business Economics and Management* (2010) 11:243–258.

[34] Rani, P., and Mishra, A.R. Single-valued neutrosophic SWARA-VIKOR framework for performance assessment of eco-industrial thermal power plants. *ICSES Transactions on Neural and Fuzzy Computing* (2020) 3:1–9.

[35] Ghorabaee, M.K., Amiri, M., Zavadskas, E.K., Turskis, Z., and Antuchevicience, J. An extended step-wise weight assessment ratio analysis with symmetric interval type-2 fuzzy sets for determining the subjective weights of criteria in multi-criteria decision-making problems. *Symmetry* (2018) 10:91.

[36] Maheshwari, H., and Jain, K. Financial viability of solar photovoltaic system: A case study.*International Journal of Civil Engineering and Technology (IJCIET)* (2017) 8(11):180–190.

[37] Gnanasekaran S., and Sivasangari A. Multi-criteria decision making for solar panel selection using fuzzy analytical hierarchy process and technique for order preference by similarity to ideal solution (TOPSIS): An empirical study. *Journal of The Institution of Engineers (India): Series C* (2019) 100 (4):707–715.

[38] El-Bayeh,C.Z., Alzaareer,K., Brahmi, B., Zellagui, M., and Eicker, U. An original multi-criteria decision-making algorithm for solar panels selection in buildings. *Energy* (2021) 217:1–28.

[39] Kharseh, M., and Wallbaum, H. Comparing different PV module types and brands under working conditions in the United Kingdom. In *Reliability and Ecological Aspects of Photovoltaic Modules*. IntechOpen (2020).

[40] Balo, F., and Şağbanşua, L. The selection of the best solar panel for the photovoltaic system design by using AHP. *Energy Procedia.* (2016) 100:50–53.

[41] Gagne, D., and Aznar, A. *Low-Income Community Solar: Utility Return Considerations for Electric Cooperatives*. Golden, CO: National Renewable Energy Laboratory (2018). NREL/TP-7A40-70536.

[42] Kiatreungwattana, K., Bird, L., and Heeter, J. *Modeling the Cost of LMI Community Solar Participation: Preliminary Results*. Golden, CO: National

118   *Clean energy for low-income communities*

Renewable Energy Laboratory (2018). Available from: https://www.nrel.gov/docs/fy18osti/72135.pdf.

[43] Heeter, J., Bird, L., O'Shaughnessy, E., and Koebrich, S. *Design and Implementation of Community Solar Programs for Low- and Moderate-Income Customers.* Golden, CO: National Renewable Energy Laboratory (2018). NREL/TP-6A20- 71652.

[44] Bovarnick, B., and Banks, D. *State Policies to Increase Low-Income Communities' Access to Solar Power.* Center for American Progress. Policy brief discussing broad policy recommendations and specific LMI solar programs in California, Colorado, and Louisiana (2014).

[45] Interstate Renewable Energy Council (IREC). *Shared Renewable Energy for Low-to Moderate-Income Consumers: Policy Guidelines and Model Provisions* (2016). A report on implementing shared renewables programs with a specific focus on providing benefits to LMI customers.

[46] Department of Energy Office of Energy Efficiency & Renewable Energy. *National Community Solar Partnership* (2015). The National Community Solar Partnership is being led by the Department of Energy, in partnership with the Department of Housing and Urban Development, the Environmental Protection Agency, and the Department of Agriculture, to promote solar access, with specific attention to increasing access for LMI communities.

[47] Center for Social Inclusion, GRID Alternatives, & Vote Solar. *Low-Income Solar Policy Guide* (2016). Guide outlines both the barriers to solar access in low-income communities and potential strategies and solutions for overcoming those hurdles.

[48] The White House, Office of the Press Secretary. *Fact Sheet: Obama Administration Announces Clean Energy Savings for All Americans Initiative* (July 19, 2016). The Clean Energy Savings for All Initiative is a multi-agency effort to expand solar power and energy efficiency measures to LMI families.

[49] Heeter, J., Sekar, A., Fekete, E., Shah, M., and Cook, J. *Affordable and Accessible Solar for All: Barriers, Solutions, and On-Site Adoption Potential.* Golden, CO: National Renewable Energy Laboratory (2021). NREL/TP-6A20-80532.

[50] Alirezaei, M., Noori, M., and Tatari, O. Getting to net zero energy building: Investigating the role of vehicle to home technology. *Energy and Buildings* (2016)130:465–476.

Chapter 5

Building energy efficiency improvements and solar PV systems integration

Sogo Mayokun Abolarin[1], Manasseh Babale Shitta[2], Olanrewaju G. Oluwasanya[2], Charles Asirra Eguma[2], Azizat Olusola Gbadegesin[2], Emmanuel O. Ogedengbe[3] and Louis Lagrange[4]

Energy efficiency can be applied at all levels, from energy generation to energy end-use, to achieve technical-, economic- and environmental benefits. The solar photovoltaic component size in a five-bedroom duplex residential building in Lagos State, Nigeria, was investigated. The adopted energy management strategy compared two cases—a present (base) case involving inefficient or low energy-efficient appliances, and a proposed efficient case involving more energy-efficient appliances. The energy supply infrastructure (solar PV modules, batteries, and inverters) size had to meet the energy needs of the appliances in the two cases. The appliances fall into three categories, namely lighting, air conditioners, and other appliances. An electronic energy audit and solar sizing (e-EASZ) tool was used for the energy audit, data analysis, and photovoltaic system infrastructure sizing. The study validated the results with the literature. The energy efficiency opportunities identified for retro-commissioning include installing more energy-efficient lighting, replacing existing inefficient air conditioning units with units conforming to the Minimum Energy Performance Standard (MEPS) for electrical appliances, and ensuring that appliances not in use are switched off. The retro-commissioning resulted in a significant reduction in energy demand, energy costs, and solar PV system infrastructure components, for different load scenarios. Energy efficiency measures led to a 42%, 26%, and 20% reduction in the energy demand and cost of lighting, air conditioning, and other appliances, respectively, while the solar PV peak power, battery bank and inverter

[1]Department Engineering Sciences, University of the Free State, South Africa
[2]National Centre for Energy Efficiency and Conservation, [Energy Commission of Nigeria], Faculty of Engineering, University of Lagos, Nigeria
[3]Department Mechanical Engineering, University of Lagos, Nigeria
[4]Department Engineering Sciences, University of the Free State, South Africa

120 *Clean energy for low-income communities*

capacity were reduced by 19%–42%. Correlations were also developed to predict the number of solar PV system components given the energy demand. The predicted results from the correlations excellently agreed with the results of the study. The correlations were determined using the coefficient of variation root mean square error, which ranged from 0.3% to 0.85% at a 95% confidence level. It is anticipated that this e-EASZ tool could also aid in sizing PV system components for larger energy efficiency and microgrid projects.

Keywords: Energy efficiency; Solar PV components sizing and minimization; Clean energy; Energy audit; Residential building

Nomenclature

Symbol	Description	Unit
B_u	Breadth of a unit solar module	m
B_{lf}	Battery loss factor	
B_{sc}	Battery bank capacity	Ah
$\cos\varphi$	Power factor	
CV_{RMSE}	Coefficient of variation root mean square error	%
D_{oa}	Days of autonomy for batteries	days
D_{od}	Depth of discharge for batteries	
E_y	Solar energy yield or output	kWh
EC_a	Annual energy consumption	kWh
EC_d	Daily energy consumption	kWh
EC_i	Energy consumption for day i	kWh
EC_w	Monthly energy consumption	kWh
EC_w	Weekly energy consumption	kWh
ECC	Energy consumption cost	
ECC_w	Weekly energy consumption cost	₦ or $
ECC_u	Unit energy consumption cost	cost/kWh
EER	Energy efficiency ratio	
f_{sp}	Solar PV derating factor	
G_{STC}	Solar radiation at standard testing condition	kWh/m²
G_T	Solar radiation on site	kWh/m²
L_u	Length of a unit solar module	m
N	Number of days in a week	
₦	Currency in Naira	₦
N_{dm}	Number of days in a month	days
N_{dy}	Number of days in a year	days
N_b	Number of batteries	
N_{sp}	Number of solar panels	
P_{cc}	Air conditioner cooling capacity	kW
P_{ri}	Rated input power	kW
P_{sp}	Peak solar PV array power	kW
P_{TD}	Total power demand of appliance	kW
P_{up}	Unit power of solar module	kW
Q_{ty}	Quantity	
$	Dollar	

(*Continued*)

S_{sh}	Solar sun hours	hr
$t_{adu,i}$	Daily duration of appliance use	hr
$t_{OFF,i}$	Time when appliance was turned OFF on day i	hr
$t_{ON,i}$	Time when appliance was turned ON on day i	hr
T_C	PV cell temperature is the current time step	°C
$T_{C,STC}$	Cell temperature under standard testing condition (25 °C)	°C
V_{bn}	Nominal voltage of battery	V

Subscripts

B	Battery
E	Energy efficient or alternative case
Inv	Inverter
P	Base or present case
S	Saving
Max	Maximum
Sp	Solar panel
STC	Standard testing condition
U	Unit

Greek symbol

A	Temperature coefficient of PV module in the current time step %/°C
H	Efficiency
Φ	Phase angle

Abbreviations

AC	Air conditioning	
CFL	Compact fluorescent lamp	
EC	Energy consumption	kWh
LED	Light emitting diode	
$MEPS$	Minimum Energy Performance Standards	
$e\text{-}EASZ$	Electronic Energy Audit and Solar siZing	
$RMSE$	Root mean square error	
$TAEC$	Total appliance energy consumption	kWh
$TDEC$	Total daily energy consumption	kWh
VBA	Visual Basic for Applications	

5.1 Introduction

The increasing population, industrialization, and rising demand for housing require a consistent supply of electricity for productive uses in an economy. The electricity supply industry in many developing countries has experienced multiple challenges over the years that are unabated despite interventions from the government [1,2]. The power sector continues to witness abysmal power generation, load shedding, and poor power generation, transmission, and distribution infrastructure which adversely impact the quality of the electricity supply [3]. Electricity users continue to struggle with these challenges and are uncertain as to how and when these challenges will be addressed [4,5]. Due to these challenges, consumers are now actively embracing alternative energy solutions in the form of adopting energy efficiency practices and the installation of renewable energy sources [6]. Different alternative energy options are being utilized, which include conventional, clean, or hybrid energy solutions.

The available conventional energy sources installed to power electrical appliances in residential buildings include gasoline, diesel, and natural gas. These energy sources are primarily consumed as fuel to power electrical generators and, in most cases, are available for purchase at petrol (gas) stations. The availability of these fuels may not be guaranteed during periods of scarcity due to unrest, truncation in supply, vandalization of petroleum pipelines that transport the fuel from one location to another, or an increase in price and demand resulting from industrial actions [7–10]. While individuals have resorted to providing their own energy privately, many users do not realize the adverse effect of greenhouse gas emissions associated with the use of conventional sources of fuel for electricity generation [11,12], which has been found to have an adverse impact on nature, ecology, and the environment [13–16].

For many years, the electricity sector in many developing countries has continued to depend largely on fossil fuel consumption for electricity generation [17–20]. The consumption of and reliance on fossil fuels for electricity generation have some adverse consequences, which include high carbon emissions leading to environmental degradation, inadequate maintenance of equipment leading to the subsequent collapse of installed large-scale generating stations, and a continuous increase in the price of electricity [21]. These consequences are now influencing the way energy is locally, privately, and nationally generated, as both the government and private individuals are now transitioning toward clean energy technologies for electricity generation. This transition has led to the adoption of renewable energy technology solutions such as wind, biomass, and in particular solar photovoltaics (solar PV) to provide electricity for commercial, industrial, and residential facilities [22].

Solar PV systems are becoming popular in many developing countries as private homeowners have started to install solar PV in their residences to improve their power supply and guarantee energy access. Decentralized on-grid, mini-grid, and off-grid systems are also gaining attention and are relied upon to meet communal energy needs as they also ensure close proximity between the load and source [23]. Solar PV systems have also been identified as positively influencing economic-, societal- and environmental developments [24]. It has been projected by the International Renewable Energy Agency that to achieve the 1.5 °C drop in temperature by 2050 [25], the PV system infrastructure must be enlarged from 1 TW to 15 TW [26]. Various combinations of renewable energy sources are now being mixed and installed to meet energy needs. Sultan *et al.* [27] combined solar PV, wind, and hydropower sources in Egypt's national power system and determined that the clean energy hybrid system reduces the energy exchange with the national grid. The diversification of different renewable energy sources helps to achieve energy security, environmental sustainability, and improve the economy [28,29]. Decentralized standalone mini-grid or off-grid systems provide more reliability, efficiency and require lower infrastructure, installation, and operation costs [30,31]. The decentralized system is economical because energy is generated at the end-user (residential, commercial, and industrial) level. For a renewable energy supply such as solar, a decentralized power generation system that integrates battery storage ensures that energy is generated and stored for use whenever the sun or grid-connected power is not available [32].

The stored energy is generally recommended to be supplied to energy-efficient appliances. Hoseinzadeh *et al.* [33] considered the energy output derivable from solar PV installed on the façade of a high-rise building in Iran. The solar PV installation had to supply about 20% of the energy required for lighting in a critical month (June) of the year. The result of the study showed that solar PV produced the required energy and, in addition, produced up to 51% of the energy required to meet lighting needs in several other months of the year. The clean energy produced from the solar PV installed on the building facades helped to avoid an annual emission of 87 tons of greenhouse gas that would have been released into the atmosphere if conventional energy sources had been considered to supply the required energy.

A hybrid system could consist of a combination of one or several conventional sources and renewable energy sources. The combination aims to reduce the greenhouse gas emissions associated with the sole reliance on fossil fuels and to pilot a gradual transition to clean energy sources [34,35]. Although the focus of this research is on energy efficiency improvement and not a hybrid system, it is worth noting that numerous detailed studies on hybrid systems have been reported in the literature [19,36,37]. Salisu *et al.* [19] investigated a hybrid power generation system by combining solar PV, wind, battery storage, and diesel generators in a remote village in North-central Nigeria. The study found the integration of more renewable energy sources with diesel generators to be economically viable and environmentally friendly. Rehman *et al.* [36] investigated the concepts, technical challenges, and potentials of hybrid systems, while Upadhyay and Sharma [37] reported on hybrid architecture with specific considerations on sizing, control, and configuration.

One of the many challenges in the adoption and deployment of renewable energy for residential electrification by building developers (irrespective of building architecture) has always been the technicalities involved in the design,

124 *Clean energy for low-income communities*

determination, and economic analyses of the renewable source infrastructure components [38,39]. The technicalities are too complex for developers and or building owners with little or no technical skills to make a decision on the adoption and integration of clean energy technology solutions [40,41]. These challenges are being mitigated, as reported by Janssen *et al.* [42], who conducted a study with the sole aim of providing an approach simple enough for building owners to directly estimate losses from their solar PV systems. The model was designed in such a way that building owners are only required to input their daily energy consumption to determine the influence of snow losses on their PV installations. To improve the knowledge of building owners and solar PV developers for mini-grid developments, Jurasz *et al.* [43] reported the influence of three different energy magnitudes, namely the real load, monthly adjusted typical load, and typical daily load, on the overall economics of a solar-battery system with varying levels of reliability. The results showed that using daily energy consumption only underestimates the system cost by 1.2%.

As energy users make a gradual transition to cleaner energy sources for electricity generation in buildings [44], one important factor to consider is the contribution of energy efficiency [45,46]. Chwieduk [47] emphasized the role of energy efficiency and suggested architectural models to improve energy efficiency. This is because increasing energy consumption and associated environmental effects have become major challenges in many countries [48–50]. Energy efficiency provides measures to reduce energy consumption and pollution and thus contributes to the achievement of sustainable development in all energy sectors [51,52]. Coelho *et al.* [53] and Fikru [54] emphasized the importance of energy efficiency in buildings. According to Menezes *et al.* [55] critical consideration should be given to the consumption and demand of electrical power as well as equipment specification and usage in order to understand and make a building more energy-efficient. Opoku *et al.* [56] reported the implementation of energy efficiency measures such as retrofitting of air conditioning units, lighting, and ventilation fans for a case study at the Kwame Nkrumah University of Science and Technology, Kumasi-Ghana reduced monthly total energy consumption by 163.40 MWh when tailored energy-efficient best practices were implemented. Though the decision to integrate energy efficiency best practices in buildings is not that easy [57–59], most building operators and owners would first consider, among other factors, their financial status, experiences from previous energy efficiency implementation, building location, age, and type, as well as energy cost.

These are so important because energy efficiency helps reduce energy consumption, associated costs, and environmental adverse effects [60] as well as the infrastructure needed to setup clean energy generation [61]. Ogedengbe *et al.* [62] conducted investigations toward encouraging the adoption of renewable energy by developing a tool referred to as EnerghxPlus for optimizing renewable energy projects. Abolarin *et al.* [61] in an investigation on the adoption of energy efficiency best practices in lighting use found that significant savings can be obtained and up to a 45% reduction in carbon dioxide emissions can be achieved when conventional light bulbs are retrofitted with energy-efficient light bulbs. In order to achieve sustainable development whereby clean and affordable supplies are available for energy users, there is a need to understand the link between energy demand in the built environment, measures to reduce greenhouse gas emissions, and a change in behavior [63]. So, building owners need to know that as they try to reduce the direct impact of the use of fossil fuels by advocating for renewable energy deployment, energy efficiency and conservation should first be given priority.

Energy end-users who are interested in transitioning from conventional energy supply to alternative and clean energy systems would require an approach that will enable them to understand the dynamics of clean energy systems. This understanding would assist decision-makers in determining the level of clean energy technologies that meet their energy needs, as the present approach or tools available to them are too technical to make decisions about adopting clean energy best practices. This technicality, along with other challenges, impedes the adoption of clean energy technologies and thus requires attention to achieve a smooth transition from fossil fuels to sustainable sources of power generation.

The purpose of this study is to simplify the technicalities associated with clean energy adoption. The study intends to determine the significance of energy efficiency retro-commissioning and renewable energy technologies in providing clean electricity. In this study, a residential duplex in Lagos, Nigeria, has been used for the purpose of this analysis. The study focuses on identifying and comparing a base case (the present scenario) with an alternative case (an energy-efficient and improved energy-efficient case), using the energy consumption pattern in the selected building. The aim of this approach is to provide reliable information to both technical and non-technical individuals willing to switch from grid-connected electricity supply to clean energy alternatives. Also, end-users would be informed on best practices in clean energy technologies and the implications of their adopted energy efficiency measures.

5.2 Data collection

The methodologies followed in this study utilized Level 1 and partly Level 2 of the energy audit procedures, based on the recommendation of the American Society of Heating, Refrigerating and Air Conditioning Engineers (ASHRAE) standard 211-2018 [64]. The procedure involved the collection and analysis of data using an energy audit approach and an analysis tool developed called Electronic Energy Audit and Solar siZing (e-EASZ) tool—a VBA-based excel spreadsheet.

This tool is user-friendly, has data input, collection, analysis, and storage capabilities for conducting an energy audit of a building and appliances, and can model the energy consumption patterns of different appliances. The input data relating to each appliance category (lighting, air conditioning, and other appliances) are stored using Excel's Visual Basic for Application (VBA) and stored in a worksheet created for each appliance. The input data in each worksheet are stored in a unique manner, as the input relating to each appliance's characteristic is stored according to space characteristics on a new row in the respective worksheet of lighting, air conditioning, and other appliances. The VBA makes it possible for the tool to accept the argument corresponding to each parameter in the schematic of the tool's architecture summarized in Figure 5.1. This systematic arrangement enables data aggregation and analysis and presents the results in a predetermined format. Figure 5.1 shows the parameters and the mapping of the energy audit data as well as the solar sizing section of the tool to provide the size, capacity, and quantity of energy-efficient appliances and solar energy components required to meet the client's energy needs.

The building envelope considered for this study is a single-storey residential building constructed in 2014 with a total gross floor area of 660 m^2, located in Lagos, Nigeria. The building consists of three living rooms, two dining spaces, four bedrooms, a master bedroom, a kitchen, laundry, a gym, and a study room. A preliminary visit showed that the sources of electricity supply to the building include (1) the national grid, (2) a 33 kVA soundproof diesel generator, and (3) a string of four batteries each connected to a 5 kVA inverter. The batteries are recharged using either the diesel generator or the national grid.

An energy audit of the facility was conducted by a team of trained and certified energy auditors. The team identified different energy-consuming appliances

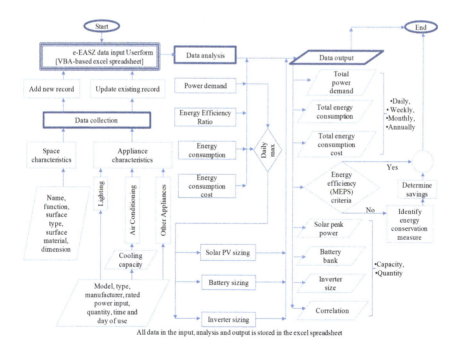

Figure 5.1   Schematic of the e-EASZ tool: a VBA-based excel spreadsheet with capabilities of data input, collection, analysis, and output used for energy audit

installed in the building and categorized them into lighting, air conditioning, and other appliances (entertainment, sports, etc.). The diesel generator and the national grid are used to power all the energy-consuming appliances within the building, while the battery storage is used to power selected loads such as lighting, fans, and entertainment devices. The light bulbs in the building are composed of both energy-efficient and inefficient lights. The light bulbs include an incandescent bulb (being used as a grid-supply indicator), linear fluorescent tubes, compact fluorescent lamps (CFL), LED downlights, and LED spotlights. These lights are installed on both the building's interior and exterior, as well as around the fence.

The cooling devices installed in the building are split air-conditioning units of different cooling capacities and input power ratings. The auditing team found that the air conditioning units are a mix of those that do and do not meet the Nigerian Minimum Energy Performance Standards (MEPS) for air conditioners in terms of their energy efficiency ratio (EER), while some fall within the 1-star and 2-star energy efficiency ratings.

In Nigeria, an air conditioner is said to be energy efficient if its EER falls within the set standard, as summarized in Table 5.1. This standard for Nigeria is the NIS 943_2017 Minimum Energy Performance Standards (MEPS) and Labels for air conditioners. Other appliances in the building include refrigerators, a standalone water dispenser, televisions, and standing fans (rechargeable and non-rechargeable). The relevant variables of this building's appliances, such as room name, room function, appliance type, rated input power, maximum input power, the number of appliances, time of use (times when the appliance was switched ON and OFF) every day of the week (Monday to Friday), room occupancy (number of people within the space), etc., were inputted into the e-EASZ tool.

Once the form has been completed on the first day of the audit, the tool provides the opportunity to update the data from day 1 for the remaining days of the week. The tool also provides an opportunity to analyze the daily and weekly energy consumption of the various appliances. The daily energy consumption is an important variable when designing and sizing a solar energy system. The tool helps to identify appliances that are energy efficient and those that are not, particularly for lighting and air conditioning units. The data collected from the building considered in this study are further processed in the data reduction section.

Table 5.1   Acceptable EER rating of air conditioners for Nigeria [65]

Energy class	EER levels
1-star	$2.80 \leq EER < 3.20$
2-star	$3.20 \leq EER < 3.60$
3-star	$3.60 \leq EER < 4.20$
4-star	$4.20 \leq EER < 5.00$
5-star	$EER \geq 5.00$

128   *Clean energy for low-income communities*

5.3   Data reduction

5.3.1   *Energy consumption analysis*

Power demand is regarded as the rated input power, $P_{ri,i}$, of each appliance and is taken from the appliance nameplate. The quantity, $Q_{ty,i}$, of each appliance was determined by taking a count of appliances, n_i, with the same rated input power and performing the same function, i. The total power demand, P_{TD}, of each appliance in a room and the entire building was determined as a product of the rated power input of each appliance, P_{ri}, and $Q_{ty,i}$ (5.1):

$$P_{TD} = \sum_{n_i=1}^{n_i} \sum_{i=1}^{i} P_{ri,i} Q_{ty,i} \tag{5.1}$$

For lighting, the power demand on each light bulb or lamp is the sum of the lamp power and ballast-rated power input. The light bulbs for which ballast-rated power input was accounted for are fluorescent light bulbs (CFLs, 2 ft and 4 ft fluorescent lamps).

The daily duration of appliance use, $t_{adu,i}$, for each day of the week was determined as the difference between the time the appliance was switched OFF versus the time the appliance was switched ON. This time record is in a 24-h format which means that the magnitude of the point in time the appliance was switched ON is less than the magnitude of the point in time the appliance was switched OFF. The appliance duration of use for a day is expressed as (5.2):

$$t_{adu,i} = \sum \left| t_{OFF,i} - t_{ON,i} \right| \tag{5.2}$$

where i is an index for five of the days of a week, Monday, Tuesday, Wednesday, Thursday, and Friday.

The daily energy consumption, EC_d, of each appliance, was determined as the product of the total power demand P_{TD} and the daily duration of use, $t_{adu,i}$ of each appliance (5.3):

$$EC_d = P_{TD} \times t_{adu,i} \tag{5.3}$$

The weekly energy consumption is obtained as the product of the appliance's total power demand and the sum of daily hours of appliance use for a period of 1 week (5.4):

$$EC_w = P_{TD} \times \sum_{i}^{n} t_{adu,i} \tag{5.4}$$

where n is the number of days in a week on which the energy consumption analysis is based.

The monthly energy consumption, EC_m, was determined as a product of the daily energy consumption, EC_d, and the number of days in a month, N_{dm}, (5.5):

$$EC_m = EC_d \times N_{dm} \tag{5.5}$$

Energy efficiency improvements   129

The annual energy consumption, EC_a, was determined as the product of the daily energy consumption, EC_d, and the number of days, N_{dy}, in a year (5.6):

$$EC_a = EC_d \times N_{dy} \qquad (5.6)$$

The energy consumption cost, ECC, was determined as the product of energy consumption of the period of interest (day, week, month, or year) and the unit cost of energy, ECC_u, measured in cost per kWh. For instance, for the week, the energy consumption cost was determined (5.7):

$$ECC_w = EC_w \times ECC_u \qquad (5.7)$$

For air conditioners, the energy efficiency ratio, EER, is generally used to determine the compliance level with respect to energy efficiency standards. EER was determined as the ratio of the cooling capacity of the air-conditioning unit, P_{CC}, to the manufacturer's rated input power, P_{ri}, printed on the appliance nameplate by the manufacturer (5.8):

$$EER = \frac{P_{CC}}{P_{ri}} \qquad (5.8)$$

Depending on the analysis of interest, a savings calculation could be made for days, weeks, or years.

Since the levels of efficiency for the air conditioning units are determined using the EER, where the EER of an air conditioner was found to be less than the minimum acceptable (as summarized in Table 5.1), an acceptable EER_e value was used to determine the rated input power rating of the energy-efficient alternative. The only assumption is that the cooling capacity of both cases is the same. The rated input power of the energy-efficient case, $P_{ri,e}$, was determined using the EER_p of the base case (energy inefficient case), rated power input, $P_{ri,p}$, of the energy inefficient case, and the EER_e of the energy-efficient case (5.9):

$$P_{ri,e} = \frac{EER_p \times P_{ri,p}}{EER_e} \qquad (5.9)$$

For the EER_e to be acceptable, the value must be greater than or equal to 2.8 (which is the minimum acceptable EER value for any energy-efficient air conditioner in Nigeria). As summarized in Table 5.1, the higher the value EER, the more energy-efficient the air-conditioning unit.

5.3.2   Energy efficiency improvements

During the audit, the team identified opportunities for energy efficiency improvement for appliances such as lighting, air conditioners, energy use behavior, etc. The energy management opportunities (EMOs) were identified for the purposes of reducing energy consumption, as well as investigating the influence of the improvement on the solar photovoltaic system components needed. Two cases were investigated, the first being the base (present) case and the second is the alternative or energy efficient case (where the energy consumption is improved). The two cases were compared in this study in order to determine and showcase the benefits of adopting energy efficiency best practices in energy use.

130   *Clean energy for low-income communities*

Once the energy-efficient appliances used in the first case were identified, a similar analysis was conducted for the energy-efficient scenario using (5.1)–(5.8). Thus, the energy consumption saving, EC_s, for each appliance was determined using the difference in the energy consumption of the present case, EC_p, and the energy-efficient case, EC_e (5.10):

$$[EC]_s = [EC]_p - [EC]_e \tag{5.10}$$

Energy consumption saving is deemed to have been achieved when the value of EC_s is positive.

The percentage of energy consumption saving, $EC_{s\%}$, was determined for each appliance using the ratio of the difference in the energy consumption of the present case, EC_p, and the energy-efficient case, EC_e, to the energy consumption of the present case, EC_p (5.11):

$$[EC]_{s\%} = \left[\frac{[EC]_p - [EC]_e}{[EC]_p} \right] \times 100\% \tag{5.11}$$

5.3.3   Solar sizing analysis

In the solar sizing section of the e-EASZ, the tool was developed to make it easy for users to select and size appliances or sets of appliances for the solar PV system design. This provides an added advantage to both the designer and energy user to determine which appliances are a priority to be powered using solar photovoltaic technology. The entire sizing involves determining the quantity and size of the solar panel array, battery bank capacity, and inverter required to meet the energy end-user's load requirements. The solar PV input parameters are summarized in Table 5.2.

5.3.4   Solar PV sizing

The determination of the total PV array capacity and number of solar panels are generally based on daily energy consumption [66]. During the energy audit of the building, energy consumption-related data were collected for a period of 5 days (Monday to Friday). The maximum daily energy consumption within the 5 days of the data collection has been used as illustrated in Figure 5.1. The e-EASZ searches for the daily energy consumption within the duration of data collection and then selects the maximum value of energy consumption, $EC_{i,max}$. In line with the general rule [66], the maximum daily energy consumption is increased by 30% to account for losses in the system.

The peak solar PV power to meet the daily energy consumption was determined using the ratio of maximum daily energy consumption, $EC_{i,max}$, to the solar sun hours, S_{sh}, of the building location (Eqn. 5.12):

$$P_{sp} = \frac{1.3 \times [EC_i]_{max}}{S_{sh}} \tag{5.12}$$

Energy efficiency improvements   131

Table 5.2   Parameters of the solar PV module, battery, and inverter

Parameters	Symbols	Values	Units
Solar PV module			
Efficiency of solar PV module	η_{sp}	21.5	%
Length of solar PV module	L_{sp}	1.835	m
Solar PV cell temperature in current time step	T_c	24	°C
Solar PV derating factor	f_{sp}	0.33	
Solar radiation of PV array	G_T	4.8	kW/m^2
Solar radiation of PV array at STC	G_{STC}	1	kW/m^2
Solar sun hours	S_{sh}	4.8	h
Temperature coefficient of power	α_p	−0.29	%/°C
Unit peak power of solar PV module	P_{up}	400	W
Width of solar PV module	W_{sp}	1.016	m
Battery			
Battery nominal voltage	V_{bn}	12	V
Days of autonomy	D_{oa}	2	Days
Unit battery capacity	B_{scu}	220	Ah
Inverter			
Inverter efficiency	η_{inv}	90	%
Power factor	$\cos\varphi$	1	

Using the capacity of a unit solar panel as P_{up}, the number of solar panels to be installed to meet the daily energy consumption was determined as the ratio of the peak solar power to the unit power of a panel (Eqn. 5.13):

$$N_{sp} = Even\left[Roundup\left(\frac{P_{sp}}{P_{up}}, 0\right)\right] \tag{5.13}$$

5.3.5   Energy yield of the solar PV array

The expected yield of the peak solar power was determined using correlations from the literature [27,67,68] and then compared with the daily energy requirement of this study. In the correlations of Sultan *et al.* [27] and Ogedengbe *et al.* [62], the yield from the solar PV array was determined as the product of the number of solar panels, N_{sp}, the efficiency of the PV, η_{sp}, unit capacity of a PV module, P_{up}, solar sun hours, S_{sh}, and the ratio of solar radiation at standard test condition, G_{STC}, to the solar radiation on site, G_T (5.14):

$$E_y = N_{sp} \times \eta_{sp} \times P_{up} \times \frac{G_{STC}}{G_T} \times S_{sh} \tag{5.14}$$

Ye *et al.* [67] published that the energy output from the solar array could be determined using (5.15):

$$E_y = P_{sp} \times f_{sp} \times S_{sh} \times \left[\frac{G_T}{G_{STC}}\right] \times \left[1 + \alpha_P(T_C - T_{C,STC})\right] \tag{5.15}$$

132   *Clean energy for low-income communities*

where P_{sp} is the peak power of the PV module, f_{sp} is the PV derating factor $(0 \leq f_{pv} \leq 1)$, G_T is the solar radiation of the PV array, G_{STC} is the solar radiation of the PV array at standard test condition, α_p is the temperature coefficient of the module in the current time step, T_C is the PV cell temperature is the current time step and $T_{C,STC}$ is the cell temperature under standard test condition (25 °C). Ilieva and Iliev [68] reported that solar energy yield could be determined using the product of the solar sun hours (S_{sh}), the efficiency of the PV (η_{sp}), the length (L_u) and breadth (B_u) of the unit module and the number of modules (N_{sp}) required as (5.16):

$$E_y = S_{sh} \times \eta_{sp} \times L_u \times B_u \times N_{sp} \tag{5.16}$$

5.3.6 *Battery sizing*

Since the solar radiation needed to provide energy is only available during the day, there is a need to incorporate a battery storage system to help store the energy for use when the sun is not available. The battery is to be charged using the solar panel. The battery input parameters are summarized in Table 5.2. The capacity of the battery bank, B_{sc}, in ampere-hours, required for the group of appliances to be powered was determined using the maximum daily energy consumption, $EC_{i,max}$, days of autonomy (2 days), D_{oa}, battery life factor (0.85), B_{lf}, permissible depth of discharge (70%), D_{od}, and the nominal voltage of the battery, V_{bn} [66] (5.17):

$$B_{sc} = \frac{[EC_i]_{max} \times D_{oa}}{B_{lf} \times D_{od} \times V_{bn}} \tag{5.17}$$

The quantity of batteries required in the battery bank was determined as the ratio of the battery bank to the unit capacity of a battery (5.18):

$$N_b = Even\left[Roundup\left(\frac{B_{sc}}{B_{scu}}, 0\right)\right] \tag{5.18}$$

5.3.7 *Inverter sizing*

The loads to be powered in the building require alternating current (AC). Thus, there is a need to convert the direct current supply from the solar PV and batteries to alternating current, using an inverter. The inverter input parameters are summarized in Table 5.2. The required size of the inverter is a function of the total power demand P_{TD}, inverter power factor ($\cos\varphi$), and inverter efficiency (η_{inv}). The rated input of the inverter, P_{inv}, required to transform the DC power to AC power was determined (5.19):

$$P_{inv} = \frac{P_{TD}}{\cos\varphi \times \eta_{inv}} \tag{5.19}$$

5.4 Model validation

To ensure that the data reduction procedure is consistent with the literature, the study compared the maximum daily energy requirement for an appliance within the building with the solar PV energy yield correlations in the work of Sultan et al. [27], Ogedengbe et al. [62], Ye et al. [67], and Ilieva and Iliev [68]. The solar PV energy yield must be higher than the maximum daily energy requirement for the energy needs to be met and for the data reduction process used in this study to be accurate. The validation is illustrated for the lighting appliances, and the analysis has been performed using the e-EASZ tool. The solar PV energy yield using correlation from literature and the maximum daily energy requirement of the lighting in the present case are determined and compared as illustrated in Figure 5.2.

Figure 5.2(a) indicates that the daily energy requirement of the lighting installed in the building would be adequately met, as the demand is lower compared with the energy yield obtained from the solar PV system. Figure 5.2(b) indicates the normalized maximum daily energy demand with the solar PV energy yield. When the daily energy requirement of the present lighting case was compared with the

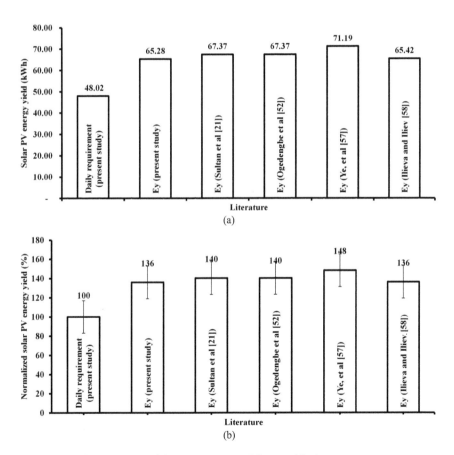

Figure 5.2 Comparison of the present case (a) actual lighting energy requirement (in kWh) and (b) normalized (in percentage) daily energy requirement of the present study with energy output from the literature

134   *Clean energy for low-income communities*

correlations of Sultan *et al.* [27], Ogedengbe *et al.* [62], Ye *et al.* [67], and Ilieva and Iliev [68], the results showed an increased solar PV energy yield of approximately 40%, 40%, 48%, and 36%, respectively compared with the maximum daily energy requirement. This increase in energy yield could be attributed to the 30% compensation considered to account for losses and variability in the solar radiation intermittency in the system. If the intermittent solar radiation had not been accounted for in this analysis, the increase in the solar PV energy yield corresponding to the correlations would have been 7%, 7%, 13%, and 4%, respectively. This verification means that the daily energy requirement would be met by the solar PV array and battery bank capacity designed in this study. This validation shows an excellent agreement with the literature. The data reduction procedure earlier presented, and the results in the subsequent section are hereby validated.

5.5   Results and discussion

The results of the energy consumption, capacities and number of solar PV modules, batteries and inverters required to meet the daily energy requirements for both the present case and the alternative (energy efficient) case are presented in this section. The savings accruable by transitioning from the present case to the alternative case are presented, as well as correlations developed that could be used for the prediction of the solar PV array and battery bank capacities.

5.5.1   Present case

The weekly energy consumption of the appliances installed in the building is summarized in Table 5.3. Table 5.3 summarizes the total daily energy consumption (TDEC) as well as the total appliance energy consumption (TAEC) in a week.

Table 5.3 presents the daily energy consumption values of the three categories of appliances, and Figure 5.3 illustrates the percentage contribution of each day's energy consumption. Figure 5.3 shows a relatively significant difference in terms of daily energy consumption. The results indicate that the highest energy consumption occurred on Friday with approximately 22%, while the lowest energy consumption occurred on Tuesday (18.56% ≈ 19%).

For each appliance category, there was a significant difference in their contribution to the overall weekly energy consumption. As illustrated in Table 5.3, the Air conditioners consumed the most energy, followed by the Other Appliances and

Table 5.3   Summary of the present case weekly appliance energy consumption

Appliances	Monday	Tuesday	Wednesday	Thursday	Friday	TAEC (kWh)
Lighting	45.05	44.00	43.15	43.89	48.02	224.12
Air Conditioners	188.26	167.20	186.99	173.71	213.73	929.89
Other Appliances	50.07	50.99	57.40	50.98	48.97	258.41
TDEC (kWh)	283.39	262.20	287.54	268.58	310.72	1412.42

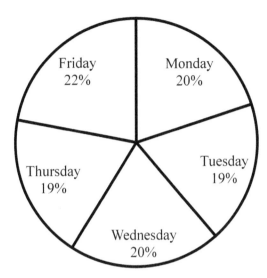

Figure 5.3 Daily energy consumption of all the appliances in the building—present case

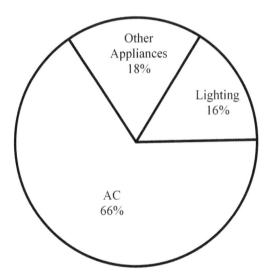

Figure 5.4 Comparison of energy consumption per appliance category

Lighting categories. This is expected as the rated input power of the air conditioners is far higher than that of the other appliances and lighting. From the results (see Figure 5.4), the air conditioners consumed 66% of the weekly energy use, followed by other appliances with 18% of energy use and then lighting with 16% of energy use.

136 *Clean energy for low-income communities*

5.5.1.1 Lighting

In the building under study, there are a total of 165 light bulbs to meet the lighting needs. The quantity of the different light bulbs installed is summarized in Table 5.4. CFL (146), Linear Fluorescent—2 ft (6), Linear Fluorescent—4 ft (1), LED downlights (8), and LED spotlight (3). The power rating of the CFLs is between 15 W and 56 W, the linear fluorescent tubes are rated between 18 W and 38 W, the incandescent light bulb is rated at 100 W and the LED downlights and spotlights are rated at 10 W each. The CFLs comply with the Nigerian Industrial Standard, NIS 747:2012 [69]. In terms of energy consumption and costs, the building under consideration falls within service band C and the non-maximum demand of the Eko Electricity Distribution Plc, with a rate of ₦ 43.01/kWh [70] (equivalent to $0.11/kWh). The CFLs consumed the highest amount of energy at 212.10 kWh, which cost ₦ 9,122.62 ($23.64) weekly. This is due to having the highest quantity installed, high-rated power input, and long hours of use.

Consumers require lighting on a daily basis. Thus, the energy consumption pattern for lighting needs is met, as illustrated in Figure 5.5. The daily energy

Table 5.4 Light bulbs presently installed in the building, their weekly energy consumption, and associated costs

Bulb type	Qty	EC_w [kWh]	ECC_w [₦]	ECC_w [$]
CFL	146	212.10	9,122.62	23.64
LF 2ft	6	3.02	130.06	0.34
LF 4ft	1	2.50	107.35	0.28
Incandescent	1	5.30	227.95	0.59
LED down light	8	0.40	17.20	0.04
LED spotlight	3	0.8	34.19	0.09
Total	165	224.12	9,639.38	24.98

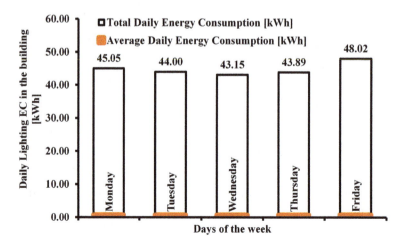

Figure 5.5 Daily lighting energy consumption in the building for the present case

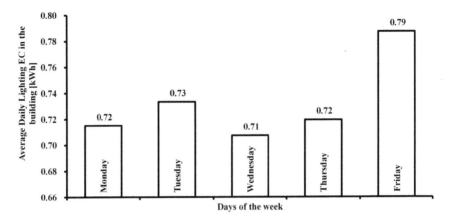

Figure 5.6 Average daily lighting energy consumption as a function of the days of the week

consumption of lighting in each room or space within the building is provided in Figure 5.6, as well as the average daily energy consumption. In Figure 5.5, the daily energy consumption of lighting in all the rooms was determined to be 45.05 kWh, 44.00 kWh, 43.15 kWh, 43.89 kWh, and 48.02 kWh for Monday, Tuesday, Wednesday, Thursday, and Friday, respectively.

Figure 5.6 illustrates the variation in the average daily energy consumption for lighting as a function of the day of the week. The average daily lighting energy consumption was determined by averaging the daily lighting energy consumption for each space of the building. The results indicate that there is minimal variation in the average daily energy consumption. The average daily lighting energy consumption in all the rooms was obtained as approximate values of 0.72, 0.73, 0.71, 0.72, and 0.79 kWh for Monday, Tuesday, Wednesday, Thursday, and Friday, respectively.

When the percentage contributions of the daily energy consumption were compared, the results indicated minimal differences in lighting used in the building for different days of the week. Each day's lighting use is approximately between 19.25% and 21.43%, as shown in Figure 5.7.

5.5.1.2 Air conditioning units

There are fourteen (14) air conditioning units in the building. The units are split non-inverter-type air conditioners and are installed in the living rooms, dining room, and bedrooms. The seven units have cooling capacities ranging between 3,490 and 3,600 W and a rated input power of 1,160–1,170 W. Six units have cooling capacities ranging between 5,130 and 5,280 W with a rated input power of 1,900–2,100 W, while one has a cooling capacity of 5,400 W with a rated input power of 1,680 W. Table 5.3 summarizes the daily energy consumption of air conditioners in the building. The results indicate the daily energy consumption for Monday, Tuesday, Wednesday, Thursday, and Friday to be 188.26, 167.20, 186.99,

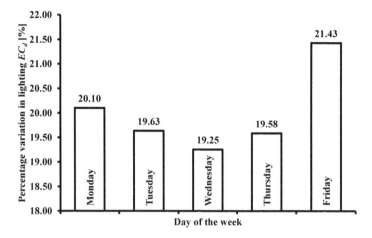

Figure 5.7 Percentage variations in lighting energy consumption per day of the week

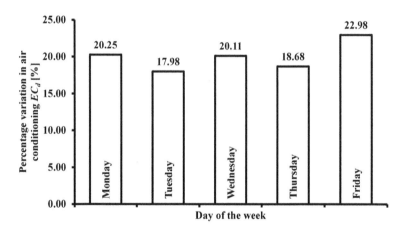

Figure 5.8 Percentage variation in daily energy consumption of air conditioning units

173.71, and 213.73 kWh, respectively. Figure 5.8 illustrates that most of the energy is consumed by the air conditioning units, 22.98% on Friday, followed by Monday and Wednesday with 20.25% and 20.11%, while Tuesday and Thursday account for 17.98% and 18.68%, respectively. The variation in the use of air conditioners could be attributed to a pattern of occupancy behavior and consciousness toward energy savings, ambient temperature leading to high heat gain inside the building, as well as the number of occupants in the rooms where the units are installed.

To ascertain the energy efficiency compliance level of the air conditioning units, the energy efficiency ratio (EER), which outlines the usable output power in the form

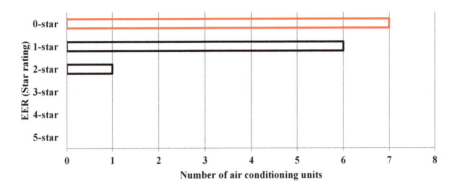

Figure 5.9 Comparison of the number of air-conditioning units under different energy efficiency ratios (EER)

of cooling capacity to the rated power input [71] of each unit, was determined with the e-EASZ tool [65] and compared with the Nigerian Industrial Standard, NIS 943:2017, on the Minimum Energy Performance Standards (MEPS) and Labels for Air-conditioners [65]. The results are shown in Figure 5.9 and indicate that one of the units complied with the two-star standard of an air conditioner with an EER of 3.21, six of the units conformed with the one-star rating standard of air conditioners with an EER ranging between 3.10 and 3.14. The remaining units have an EER of less than 2.8. The air conditioning units that did not comply with the minimum standard are referred to as zero stars in this study, as illustrated in Figure 5.9. This indicates an opportunity for energy efficiency improvement in the air-conditioning units.

5.5.1.3 Other appliances

The electrical loads belonging to the "Other Appliances" category include standing fans, televisions, water dispensers, decoders, a home theatre, and refrigerators. In a tropical country like Nigeria, ambient temperatures usually vary within the range from 20 °C to 35 °C. Energy users resort to using standing fans as an alternative to air conditioners. The reason is the consciousness that the air conditioners consume more energy, which is also associated with a higher cost of energy use. Thus, in situations where it is not absolutely necessary to use air conditioners to modify the room temperature to a more comfortable level, standing fans are used, as was the case with the building under study. There are six standing fans, four of which are rated 30 W, with one rated at 65.5 W and another at 150 W. There are three televisions with power ratings ranging from 158 W to 172 W.

The overall energy consumption associated with "Other Appliances" as presented in Table 5.3, is given as 258.41 kWh. The use of energy-consuming appliances in this category varied daily depending on the need for energy by the end-users in the building. As illustrated in Figure 5.10, Monday's energy consumption was 50.07 kWh, Tuesday's was 50.99 kWh, Wednesday's was 57.40 kWh, Thursday's was 50.98 kWh, and Friday's was 48.97 kWh. The percentage contribution of daily energy consumption is presented in Figure 5.11.

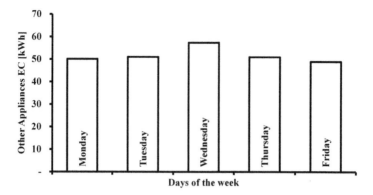

Figure 5.10 Variation of daily energy consumption of the "Other Appliances" category

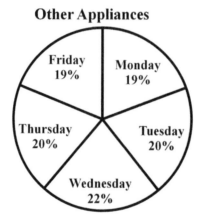

Figure 5.11 Percentage of daily variation of energy consumption in "Other Appliances" category

Figures 5.10 and 5.11 indicate that Wednesday has the highest energy consumption, showing a change in the behavior of occupants on this day. Furthermore, during the energy audit, the water dispensers and refrigerators were found to always be switched ON. This caused the water dispenser to account for 18% of energy use, while the refrigerators accounted for 68% of the energy consumption in the "Other Appliances" category.

5.5.2 Alternative case—energy-efficient case
5.5.2.1 Lighting
The energy auditors found that 87% of the light bulbs were CFLs. Despite being regarded as energy-efficient light bulbs, this provided another opportunity for

energy management namely, to retrofit the CFLs with LEDs. Over the years, lighting technology has improved, and the latest LEDs with an efficacy of up to 165 lm/W are recommended for minimizing lighting loads. Compared to conventional incandescent and compact fluorescent lights, LED bulbs are the best current alternative to meet lighting needs because of their increased energy efficiency [72]. LEDs also have some of the longest lifespans [73–75], the best energy savings and illumination [76], produce the least heat, and are easy to operate. According to a report from the International Electrotechnical Commission (IEC), where LEDs were compared to incandescent lamps in terms of energy efficiency, it was stated that LEDs are in excess of about 90% more energy efficient. Compared with CFLs, LED lamps consume less energy, have longer lifetimes, are robust, and can be easily controlled with fast switching [77].

The lighting improvement measures recommended for this residential building to achieve the standard and required illumination levels of the different spaces include retrofitting (1) all the fluorescent lamps and incandescent bulbs with LEDs, (2) all the CFLs with LED bulbs with a rated power input of 20 W, and (3) the fluorescent tubes with LED tubes. The quantities of the individual lamps are summarized in Table 5.5 and comply with the IEC 62717 standard on the LED modules for general lighting requirements—luminaire performance.

Table 5.5 illustrates that implementation of the suggested energy management opportunities can reduce the weekly energy consumption due to lighting in the building from 224.12 kWh (see Table 5.3) to 128.08 kWh. The weekly cost associated with energy consumption was reduced from ₦9 639.38 ($24.98) to ₦5 509.00 ($14.28), translating to a weekly energy cost savings of ₦4 130.38 ($11.51). The projected annual savings represent a substantial sum of money that could be added to the consumers' distributable income. The lighting-related energy efficiency measures predict cost savings of about 43% due to the improvement of energy efficiency. Further improvements are possible through adopting energy use best practices with regard to occupancy sensors as well as retrofitting with lower-rated power LED bulbs in spaces where the illumination levels provided by the lower-rated power LED bulbs are suitable and adequate.

Table 5.5 *Recommended lighting improvement in the building showing the new weekly energy consumption and cost reductions*

Bulb type	Qty	P_{ri} [W]	P_{TD} [kW]	EC [kWh]	ECC [₦]	ECC [$]
LED bulb	147	20	2.94	120.93	5,201.18	13.48
LED down light	8	10	0.08	3.29	141.53	0.37
LED 2ft	6	8	0.048	1.97	84.92	0.22
LED 4ft	1	16	0.016	0.66	28.31	0.07
LED spotlight	3	10	0.03	1.23	53.07	0.14
Total	165		3.114	128.08	5,509.00	14.28

5.5.2.2 Air conditioning units

The scope of energy efficiency improvement is limited to the seven air conditioning units with *EER* values less than 2.8. *EER* values less than 2.8 are regarded as inefficient with respect to energy use, according to the Nigerian Minimum Energy Performance Standards (MEPS) [65]. Considering the different criteria, including the installation point, thermal comfort, rated power input, cost, etc., air conditioning units with an *EER* within the 2-star rating ($3.20 \leq EER < 3.6$) category are proposed to replace the inefficient ones. The total number of 1-star units would remain at 6, while the 2-star air conditioners would be eight, as illustrated in Figure 5.12. This measure reduces the total power demand of the air conditioning units from 22.16 to 15.87 kW. The decision to consider air conditioning units with higher *EER* and in accordance with the Nigerian MEPS also helps to select between air conditioning units with either a higher cooling capacity or with lower-rated input power. This is very important because adhering to energy efficiency standards and best practices contributes significantly to improved occupancy comfort [78,79]. The proposed air conditioning units are equipped with inverter technology and are energy efficient. These air conditioning units have been found to have higher initial capital costs compared with the non-inverter type. Fortunately, the inherent benefits of energy savings make the inverter-technology air conditioning units cost-effective, as the energy savings accrued over the life of the units can pay off the initial capital investment, thus ensuring a shorter payback period [80–83].

The predicted energy savings due to the implementation of the energy efficiency measures are 26%, as the weekly energy consumption was reduced from 929.89 kWh to 685.59 kWh. This means that the daily energy consumption was reduced to 141.89, 126.85, 140.62, 126.27, and 149.96 kWh from Monday to Friday, respectively.

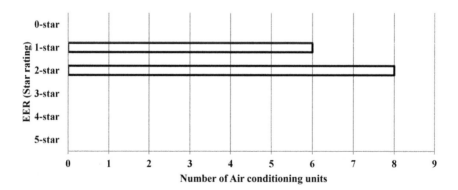

Figure 5.12 Energy efficiency improvements from the 0-star air conditioning units to the 2-star air conditioning units in compliance with the Nigerian MEPS standards

5.5.2.3 Other appliances

While fans typically consume less energy, their levels of efficiency and energy savings depend on usage patterns. Shah *et al.* [84] pointed out that the improvement of fans could lead to 50% energy savings when the best technology (a brushless DC motor instead of a fan with an induction motor) is combined with energy-saving best practices. Making the right decision on the choice of a fan is very important, as this contributes to a reduction in energy consumption. The efficiency of a brushless DC motor is in the order of 90%, while that of an induction motor is about 75% [85,86]. These appliances are often not switched off when not needed. This type of behavior provides little or no energy-saving opportunities. Appliances not in use should be turned off or regulated in such a way that a minimal amount of energy is being consumed.

With respect to behavioral practices observed during the energy audit of the building under study, a 20% reduction in energy could be achieved if these energy-consuming appliances are turned off when not in use. The adoption of this strategy can reduce the weekly energy consumption due to other appliances from the base case of 258.41–206.73 kWh, and the associated energy cost (at the rate of ₦ 43.01/kWh, equivalent to $0.11/kWh) of ₦ 11 114.10 ($28.81) reduces to ₦ 8 891.28 ($23.04). This translates to a weekly energy and cost savings of 51.68 kWh and ₦ 2 222.82 ($5.76).

5.5.2.4 Powering the appliances with renewable energy supply

The abysmal supply of electricity in many developing countries is now promoting the installation of renewable energy technologies, such as solar PV systems. The building under study has already installed four 300 W mono-crystalline PV modules, charging four 220 Ah tubular batteries. These DC power sources are transformed to AC by a 5 kVA inverter with a system voltage of 48 V. Unfortunately, this is not enough to meet the daily energy needs of the building's occupants.

By ensuring that this building gradually transitions from the use of fossil fuel-powered generators to sustainable, modern, and clean energy, the e-EASZ tool developed for the energy audit and solar sizing has been deployed to determine the (1) solar PV array capacity, (2) size of battery banks, and (3) capacity of the inverter that could be required to meet different categories and combinations of electrical loads.

The solar PV sizing was determined by the tool in such a way that the capacity and numbers of the solar energy components required to meet the daily energy needs of (1) lighting alone, (2) air conditioners alone, (3) other appliances alone, (4) a combination of lighting and air conditioners, (5) lighting and other appliances, and (6) air conditioners and other appliances are determined for both the present case and the energy-efficient case. Since the design of solar PV systems is generally based on daily energy use, the tool selects the day with the maximum energy consumption for this analysis. For the batteries, 2 days of autonomy are allowed, and for the inverter, a power factor of unity and efficiency of at least 90% are suggested and considered in the analysis.

Figure 5.13 is a schematic illustration of the system configuration of the solar PV system, battery storage, DC-to-AC inverter, and AC loads. The solar PV

144 *Clean energy for low-income communities*

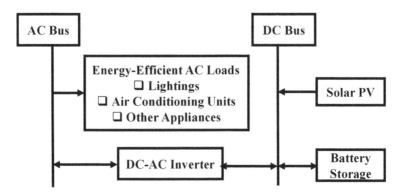

Figure 5.13 Schematic of the clean energy system configuration

Table 5.6 The capacity and number of solar PV modules, batteries, and inverters needed to power different combinations of the present (base case) energy-consuming appliances

Present (base) case	Scenarios	$[EC]_d$ [kWh]	P_{sp} [kW]	B_{sc} [Ah]	$\approx P_{inv}$ [kVA]	N_{sp} [-]	N_b [-]
1DC load	DC load	10.47	3.2	3,080	0	8	14
1L	Lighting	48.02	13.6	13,640	6	34	62
1AC	Air conditioner	213.73	58.4	60,280	25	146	274
1OA	Other appliances	57.40	16	16,280	3.5	40	74
1L_DC	Lighting and DC load	58.49	16	16,720	6	40	76
1L_AC	Lighting and Air conditioner	261.75	71.2	73,480	31	178	334
1L_OA	Lighting and Other appliances	105.42	28.8	29,920	10	72	136
1L_AC_OA	Lighting, Air conditioner and Other appliances	319.15	87.2	89,760	34	218	408
1DC Load_ L_AC_OA	Lighting, DC load, Air conditioner and Other appliances	329.62	89.6	92400	34	224	420

Note: "1" in the first column of this table represents present case

modules and batteries of direct current are connected to the DC Bus where the direct current is converted to alternating current by the inverter, which is connected to the AC Bus, through which the AC electrical loads draw their power.

The results of the analysis for the present or base case scenario are summarized in Table 5.6. Table 5.6 illustrates that, as energy consumption increases, the solar energy system infrastructure also increases. For example, the solar PV array capacity of 13.6 kW (34 units of solar PV modules rated at 400 W each) is required to meet the present case lighting requirement with a daily energy consumption of 48.02 kWh. The size of the battery bank required is 13,640 Ah (62 units of batteries each rated at 220 Ah) with an approximate inverter size of 6 kVA. When the daily energy consumption increases to 319.15 kWh in order to meet the lighting, air conditioning, and other appliances, a solar PV array capacity of 87.2 kW (218 units of solar PV modules rated at 400 W each) and a battery bank capacity of 89,760 Ah (408 units of batteries each rated at 220 Ah) with an approximate inverter size of 34 kVA are required. The number of solar PV modules and batteries was determined using (5.13) and (5.18).

To enable the building owner to predict the solar PV array, battery bank capacities, and size of inverter required to meet desired energy consumption, Figures 5.14 and 5.15 have been generated using the results obtained from this study in order to provide that guidance. These figures illustrate an excellent linear

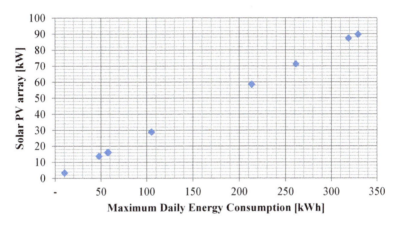

Figure 5.14 Solar PV array as a function of present case maximum daily energy consumption

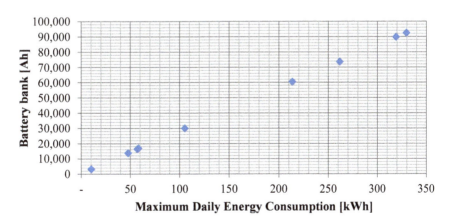

Figure 5.15 Battery bank capacity as a function of present case maximum daily energy consumption

146 *Clean energy for low-income communities*

relationship with an R² value between the maximum daily energy consumption and the size of solar PV modules and batteries.

Similarly, the solar PV array, battery bank capacity, and the size of the inverter are determined for the energy-efficient case scenarios, as summarized in Table 5.7. The solar PV array has been plotted as a function of daily energy consumption, as illustrated in Figure 5.16.

The size of the battery bank obtained for the different consumption scenarios has been plotted as a function of energy consumption, as shown in Figure 5.17. Figures 5.16 and 5.17 illustrate that as energy consumption increases, the solar PV

Table 5.7 The number and capacities of solar PV modules, batteries, and inverters needed to power different combinations of the alternative (energy efficient) case appliances

Case 2	Description of scenarios	$[EC]_d$ [kWh]	P_{sp} [kW]	B_{sc} [Ah]	$\approx P_{inv}$ [kVA]	N_{sp}	N_b
2L	Lighting	27.08	8.00	7,920	3.5	20	36
2AC	Air conditioner	149.96	40.8	42,240	18	102	192
2OA	Other appliances	45.92	12.8	13,200	3	32	60
2L_DC	Lighting and DC load	37.55	10.4	10,560	3.5	26	48
2L_AC	Lighting and air conditioner	177.04	48	49,720	22	120	226
2L_OA	Lighting and other appliances	72.99	20	20,680	6.5	50	94
2L_AC_OA	Lighting, air conditioner, and other appliances	222.95	60.8	62,480	25	152	284
2DC_L_AC_OA	Lighting, DC load, air conditioner, and other appliances	233.43	64	65,560	25	160	298

Note: "2" in the first column of this table represents alternative case

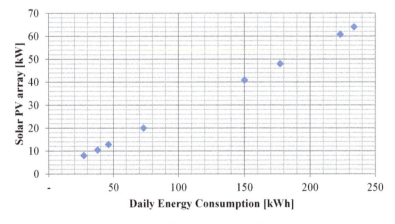

Figure 5.16 Alternative (energy efficient) case—solar PV array as a function of maximum daily energy consumption

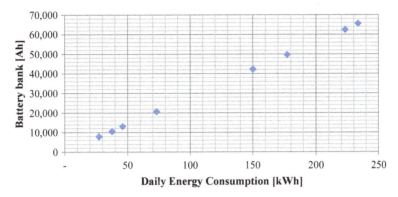

Figure 5.17 Alternative (energy efficient) case—battery bank capacity as a function of maximum daily energy consumption

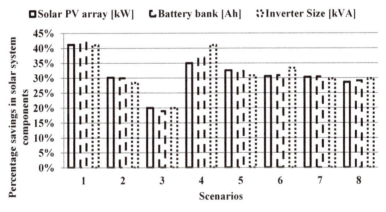

Figure 5.18 Percentage savings in solar PV system infrastructure due to energy efficiency measures

array and battery bank capacity requirements increase. The relationship between the solar PV array, and the daily energy consumption is linear. A similar relationship is obtained when the battery bank capacity is plotted against the daily energy consumption.

Consideration of energy efficiency measures resulted in a reduction in the solar PV array capacity, battery bank, and size of the inverter required to meet the energy needs, as illustrated in Figure 5.18. From Figure 5.18, the implementation of energy efficiency measures across the different scenarios saved between 19% and 42% in infrastructure—capital and operational costs. The integration of energy efficiency measures with renewable energy sources contributes to increasing energy sustainability in the building [87–89].

148 *Clean energy for low-income communities*

5.5.2.5 Correlation to predict the number of solar module and batteries

To enable prediction and decision-making of the solar PV array capacity and battery bank with known daily energy consumption, this study developed correlations to ease the gradual (modular) transition. As illustrated in Figure 5.14, the relationship between the daily energy consumption and the calculated solar PV array is linear. Similarly, in Figure 5.15, the relationship between daily energy consumption and the battery bank capacity required to meet the energy consumption is linear as well.

5.5.2.6 Base case

The correlation to predict the solar PV array with known daily energy consumption for the base case is plotted in Figure 5.14, developed as (5.20), and plotted as illustrated in Figure 5.19(a). This correlation has a *p*-value of $\ll 0.1$. This indicates

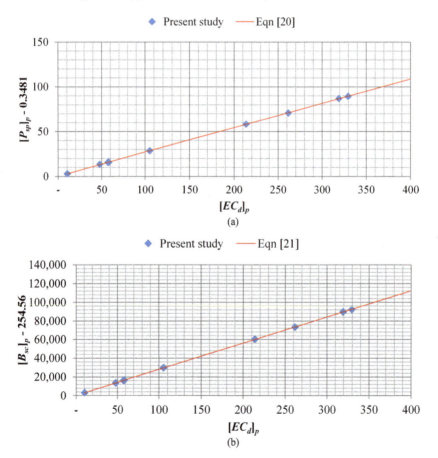

Figure 5.19 Comparison of this present case's (a) solar PV array as a function of daily energy consumption with the correlation developed in (5.20), (b) battery bank capacity as a function of daily energy consumption with the correlation developed in (5.21)

a significant relationship between the solar PV size and the maximum daily energy consumption. The regression analysis conducted by determining the root mean square error (RMSE) and the coefficient of variation of the root mean square error (CV_{RMSE}) at a 95% confidence level indicates that for the base case when the solar PV size was plotted as a function of the maximum daily energy, the RMSE is 0.192, while the CV_{RMSE} is 0.45%. The ASHRAE sets the maximum CV_{RMSE} value to be 25% [90]. This indicates that (5.20) predicts the base case solar PV size with a predictive capacity that is acceptable.

$$[P_{sp}]_p = 0.2713 \times [EC_d]_p + 0.3481 \quad (5.20)$$

Similarly, the correlation to predict the battery bank capacity of the base case plotted in Figure 5.15 was developed as (5.21). The CV_{RMSE} for the base case model, which relates the battery size to the maximum daily energy consumption, is illustrated in Figure 5.19(b) for the different scenarios. In the base case, it is 0.3%. The p-value of the relationship is $<<<0.1$.

$$[B_{sc}]_p = 361.15 \times [EC_d]_p + 193.83 \quad (5.21)$$

The comparison of the correlation in (5.20) with Figure 5.14 shows that the correlation agreed with the calculated data with a maximum deviation of 2%. The correlation to predict the battery bank capacity using daily energy consumption deviated from the calculated data by a maximum of 1%.

The correlation to predict either the solar PV array using the battery bank capacity or vice versa is also developed in this study (Eqn. 5.22):

$$[P_{sp}]_p = 9.69 \times 10^{-4} [B_{sc}]_p + 0.5609 \quad (5.22)$$

Eqn. (5.22) correlates well with the audit data as shown in Figure 5.20 with a maximum deviation of 2%. The solar PV module's characteristics can be

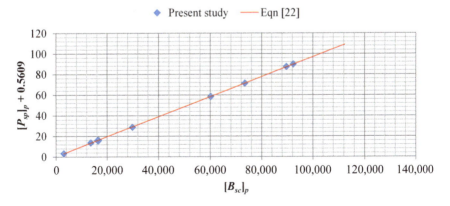

Figure 5.20 Comparison of the calculated solar PV array as a function of battery bank capacity with the correlation developed as (5.22) for the base case scenario

determined using (5.12) and (5.13), while battery characteristics can be determined using Eqns. (5.17) and (5.18). The CV_{RMSE} for the relationship between the PV size and the battery size is 0.5%, while the *p*-value is $\lll 0.1$. These statistical values indicate that the model correlates well with the results of this study.

5.5.2.7 Alternative case

For the alternative case, which is referred to as the energy-efficient case, correlations were developed to enable the easy prediction of the solar PV array and battery bank capacities corresponding to the different appliance scenarios.

The correlation to predict the solar PV array as a function of the daily energy consumption for the energy-efficient case displayed in Figure 5.16 was developed as (5.23). The CV_{RMSE}, which indicates the closeness of the predicted values of the alternative case solar PV size to the energy-efficient consumption mean (Figure 5.21(a)), is 0.85%. This is in line with the ASHRAE standard of a

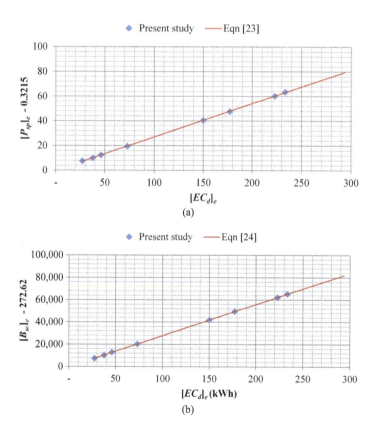

Figure 5.21 Comparison for the alternative case: (a) solar PV array as a function of daily energy consumption with the correlation developed as (5.23), (b) battery bank capacity as a function of daily energy consumption with the correlation developed as (5.24)

maximum of 25%.

$$[P_{sp}]_e = 0.2712 \times [EC_d]_e + 0.3215 \quad (5.23)$$

The correlation to predict the battery banks of the energy-efficient case plotted in Figure 5.17 was developed as a function of daily energy consumption as (5.24). The comparison of the predicted battery size as a function of the energy-efficient consumption as shown in Figure 5.21(b) indicates the CV_{RMSE} to be 0.33%.

$$[B_{sc}]_e = 279.42 \times [EC_d]_e + 272.62 \quad (5.24)$$

The correlations to predict the solar PV array (Eqn. (5.23)) and battery banks (Eqn. (5.24)) were compared with the results obtained in this present study, as illustrated in Figure 5.21. The results indicate that the correlation in (5.23) as illustrated in Figure 5.21(a) deviated by a maximum of 6%, while (5.24) as illustrated in Figure 5.21(b) deviated from the present study by 1%. This strong relationship indicates excellent agreement between the correlations and the results of the present study.

For the same daily energy consumption, the correlation for the alternative case was also developed to predict the solar PV array as a function of battery bank capacity (Eqn. 5.25). The CV_{RMSE} of the model for the relationship between the solar PV size and the battery size of the alternative case is 0.75%.

$$[P_{sp}]_e = 9.706 \times 10^{-4} \, [B_{sc}]_e + 0.0569 \quad (5.25)$$

When compared with the present study, as illustrated in Figure 5.22, the correlation indicates strong agreement with a maximum deviation of 2%.

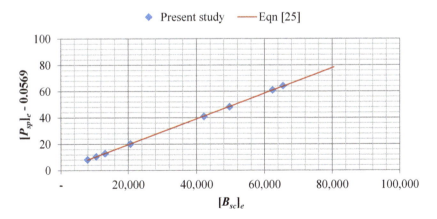

Figure 5.22 Comparison of the present study's solar PV array as a function of battery bank capacity with the correlation developed as (5.25) for the alternative case scenario

152 *Clean energy for low-income communities*

5.5.2.8 Solar PV energy yield

The maximum daily energy demand due to the use of energy-efficient lighting was compared with the solar PV energy yield, as illustrated in Figure 5.23. Figure 5.23(a) shows the actual energy (in kWh), for the lighting fixtures. Figure 5.23(b) illustrates the normalized demand with solar PV yield for all alternative case scenarios in

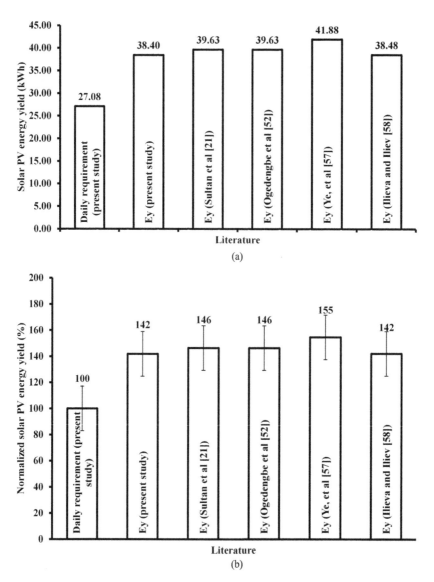

Figure 5.23 Comparison of the alternative case (a) actual (in kWh) and (b) normalized (in percentage) daily energy requirement of the present study with energy output from the literature

Energy efficiency improvements   153

Table 5.7. The normalized comparison is necessary to compare the energy yield of all scenarios with their respective maximum daily energy demand. This is because the daily energy requirements, as well as the energy yields of all the scenarios, are not the same. The results show that the maximum daily energy demand will be adequately met with an excess energy yield of at least 42% when the solar intermittency and losses are taken into consideration, while the yield would have been in excess of 13% when the variation in solar irradiation intermittency is not considered.

This is an indication that the choice of the solar PV components selected to meet the daily energy requirement was suitable and that the daily energy requirement would be met in excess of at least 42% energy yield. This means that the energy requirement would be met by the solar PV array and battery bank capacity designed in this study. Figure 5.23(b) illustrates that the energy yield from the present study compared well with the correlations from the literature.

5.6   Conclusions

A significant reduction in energy and cost of energy, as well as solar PV infrastructure accruable from the installation of energy-efficient appliances, have been demonstrated in this study under two cases—a present (base) case and a proposed energy-efficient case. The cases investigated are carried out for a five-bedroom residential duplex in Lagos, Nigeria. For each case, scenarios comprising different combinations of appliances are analyzed. The scenarios comprised a combination of lighting, DC loads, air conditioning units, and other appliances.

A solar PV system was sized for each scenario to determine the influence of adopting increased energy efficiency measures. In general, the energy efficiency measures considered in the study led to a significant reduction in the solar PV infrastructure components by 19%–42%.

Correlations between the solar PV array capacity and battery bank, as well as daily energy consumption, were developed and presented to enable prediction and inform better decision-making. The regression analysis conducted on the models indicates that the models predicted the actual results of this study closely, with the CV_{RMSE} ranging from 0.3% to 0.85%. The range is within the maximum acceptable CV_{RMSE} value of 25% of the ASHRAE. Therefore, the study illustrates that energy efficiency in residential buildings can be improved by installing energy-efficient appliances and adopting energy-saving behaviors.

The maximum daily energy requirements of the different cases were compared with the solar PV energy yield correlations from the literature. The results showed that the daily energy requirements for the various cases and scenarios would be adequately met. The energy yield showed an excess of at least 42% over the maximum daily energy demand when the solar intermittency is taken into account, or at least 13% excess yield when the variation in solar intermittency and losses are not considered.

154   *Clean energy for low-income communities*

Further work in this study shall investigate economic indicators to support the investigation into the possible benefits of energy efficiency interventions and retrofits in residential buildings under the different load scenarios presented.

Declaration of competing interest

The authors declare that there are no known financial interests or personal relationships that could have appeared to influence the work reported in this paper.

Acknowledgments

The authors extend appreciation for the support and collaboration between the researchers from the University of the Free State, South Africa and National Centre for Energy Efficiency and Conservation, Nigeria.

References

[1] Babatunde, O., Buraimoh, E., Tinuoye, O., Ayegbusi, C., Davidson, I., and Ighravwe, D.E., "Electricity sector assessment in Nigeria: the post-liberation era," *Cogent Engineering*, vol. 10, no. 1, p. 2157536, 2023, doi:10.1080/23311916.2022.2157536.

[2] Somoye, O.A., "Energy crisis and renewable energy potentials in Nigeria: A review," *Renewable and Sustainable Energy Reviews*, vol. 188, p. 113794, 2023, doi:https://doi.org/10.1016/j.rser.2023.113794.

[3] Ikemba, S., Song-hyun, K., Scott, T.O., Ewim, D.R.E., Abolarin, S.M., and Fawole, A.A., "Analysis of solar energy potentials of five selected south-east cities in nigeria using deep learning algorithms," *Sustainable Energy Research*, vol. 11, no. 1, p. 2, 2024, doi:10.1186/s40807-023-00096-7.

[4] Jolaiya, E., Akintola, M., and Nafiu, O., "Mapping access to electricity in urban and rural Nigeria," in *Open Mapping towards Sustainable Development Goals: Voices of YouthMappers on Community Engaged Scholarship*, P. Solís and M. Zeballos, Eds. Cham: Springer International Publishing, 2023, pp. 141–149.

[5] Ndukwu, M.C., Onwude, D.I., Bennamoun, L., *et al.*, "Nigeria's energy deficit: the challenges and eco-friendly approach in reducing the energy gap," *International Journal of Sustainable Engineering*, vol. 14, no. 3, pp. 442–459, 2021, doi:10.1080/19397038.2020.1842546.

[6] Falaiye, H., "Nigerians opt for alternative energy sources amid high fuel costs," *The Punch Newspaper*, Nigeria: The Punch Newspaper, 2024.

[7] Bou-Rabee, M., Baharudin, Z., Sulaiman, S., and Naz, M., "Wind power characteristics and feasibility for electrical energy generation," *International Journal of Strategic Energy and Environmental Planning*, vol. 2, no. 1, pp. 30–46, 2020.

[8] Uhunmwangho, R., Odje, M., and Okedu, K.E., "Comparative analysis of mini hydro turbines for Bumaji Stream, Boki, Cross River State, Nigeria," *Sustainable Energy Technologies and Assessments*, vol. 27, pp. 102–108, 2018, doi:https://doi.org/10.1016/j.seta.2018.04.003.

[9] Mei, B., Barnoon, P., Toghraie, D., Su, C.-H., Nguyen, H.C., and Khan, A., "Energy, exergy, environmental and economic analyzes (4E) and multi-objective optimization of a PEM fuel cell equipped with coolant channels," *Renewable and Sustainable Energy Reviews*, vol. 157, p. 112021, 2022, doi:https://doi.org/10.1016/j.rser.2021.112021.

[10] Khandaker, S., Bashar, M.M., Islam, A., Hossain, M.T., Teo, S.H., and Awual, M.R., "Sustainable energy generation from textile biowaste and its challenges: a comprehensive review," *Renewable and Sustainable Energy Reviews*, vol. 157, p. 112051, 2022, doi:https://doi.org/10.1016/j.rser.2021.112051.

[11] Hansen, K., Breyer, C., and Lund, H., "Status and perspectives on 100% renewable energy systems," *Energy*, vol. 175, pp. 471–480, 2019, doi:https://doi.org/10.1016/j.energy.2019.03.092.

[12] Agyekum, E.B., "Techno-economic comparative analysis of solar photovoltaic power systems with and without storage systems in three different climatic regions, Ghana," *Sustainable Energy Technologies and Assessments*, vol. 43, p. 100906, 2021, doi:https://doi.org/10.1016/j.seta.2020.100906.

[13] Bhatti, A.R., Salam, Z., and Ashique, R.H., "Electric vehicle charging using photovoltaic based microgrid for remote islands," *Energy Procedia*, vol. 103, pp. 213–218, 2016, doi:https://doi.org/10.1016/j.egypro.2016.11.275.

[14] Karmaker, A.K., Rahman, M.M., Hossain, M.A., and Ahmed, M.R., "Exploration and corrective measures of greenhouse gas emission from fossil fuel power stations for Bangladesh," *Journal of Cleaner Production*, vol. 244, p. 118645, 2020, doi:https://doi.org/10.1016/j.jclepro.2019.118645.

[15] Wang, Q., Xue, M., Lin, B.-L., Lei, Z., and Zhang, Z., "Well-to-wheel analysis of energy consumption, greenhouse gas and air pollutants emissions of hydrogen fuel cell vehicle in China," *Journal of Cleaner Production*, vol. 275, p. 123061, 2020, doi:https://doi.org/10.1016/j.jclepro.2020.123061.

[16] Wan Mansor, W.N., Abdullah, S., Che Wan Othman, C.W.M.N., Jarkoni, M. N.K., Chao, H.-R., and Lin, S.-L., "Data on greenhouse gases emission of fuels in power plants in Malaysia during the year of 1990–2017," *Data in Brief*, vol. 30, p. 105440, 2020, doi:https://doi.org/10.1016/j.dib.2020.105440.

[17] Maduekwe, M., Akpan, U., and Isihak, S., "Road transport energy consumption and vehicular emissions in Lagos, Nigeria: an application of the LEAP model," *Transportation Research Interdisciplinary Perspectives*, vol. 6, p. 100172, 2020, doi:https://doi.org/10.1016/j.trip.2020.100172.

[18] Ameyaw, B., Yao, L., Oppong, A., and Agyeman, J.K., "Investigating, forecasting and proposing emission mitigation pathways for CO_2 emissions from fossil fuel combustion only: a case study of selected countries," *Energy Policy*, vol. 130, pp. 7–21, 2019, doi:https://doi.org/10.1016/j.enpol.2019.03.056.

[19] Salisu, S., Mustafa, M.W., Olatomiwa, L., and Mohammed, O.O., "Assessment of technical and economic feasibility for a hybrid PV-wind-diesel-battery energy system in a remote community of north central Nigeria," *Alexandria Engineering Journal*, vol. 58, no. 4, pp. 1103–1118, 2019, doi: https://doi.org/10.1016/j.aej.2019.09.013.

[20] Abolarin, S.M., Shitta, B.M., Aghogho, E.M., Nwosu, P.B., Aninyem, C.M., and Lagrange, L., "An impact of solar PV specifications on module peak power and number of modules: a case study of a five-bedroom residential duplex," *IOP Conference Series: Earth and Environmental Science*, vol. 983, no. 1, p. 012056, 2022, doi:10.1088/1755-1315/983/1/012056.

[21] Zhang, Z., Li, R., Zhao, C., and Li, F., "Cross-characterization of PV and sunshine profiles based on hierarchical classification," *Energy Procedia*, vol. 103, pp. 15–21, 2016, doi:https://doi.org/10.1016/j.egypro.2016.11.242.

[22] Zeng, P. and Wei, X., "Measurement and convergence of transportation industry total factor energy efficiency in China," *Alexandria Engineering Journal*, vol. 60, no. 5, pp. 4267–4274, 2021, doi:https://doi.org/10.1016/j.aej.2021.03.032.

[23] Zeeshan, M., Jamil, M., Islamia, J.M., Marg, M.M.A.J., and Nagar, J., "Active filter based harmonic mitigation technique for islanded microgrids," *Alternative Energy and Distributed Generation Journal*, vol. 2, no. 2, pp. 17–41, 2020.

[24] Grubert, E. and Zacarias, M., "Paradigm shifts for environmental assessment of decarbonizing energy systems: emerging dominance of embodied impacts and design-oriented decision support needs," *Renewable and Sustainable Energy Reviews*, vol. 159, p. 112208, 2022, doi:https://doi.org/10.1016/j.rser.2022.112208.

[25] Aghaei, M., Fairbrother, A., Gok, A., *et al.*, "Review of degradation and failure phenomena in photovoltaic modules," *Renewable and Sustainable Energy Reviews*, vol. 159, p. 112160, 2022, doi:https://doi.org/10.1016/j.rser.2022.112160.

[26] Gielen, D., Gorini, R., Leme, R., *et al.*, *World Energy Transitions Outlook: 1.5 °C Pathway*, Abu Dhabi: International Renewable Energy Agency, 2021.

[27] Sultan, H.M., Zaki Diab, A.A., Kuznetsov Oleg, N., and Zubkova Irina, S., "Design and evaluation of PV-wind hybrid system with hydroelectric pumped storage on the National Power System of Egypt," *Global Energy Interconnection*, vol. 1, no. 3, pp. 301–311, 2018, doi:https://doi.org/10.14171/j.2096-5117.gei.2018.03.001.

[28] Chan, M.K., Lim, J.M.Y., and Kumaran, P., "Harvesting heat energy as alternative renewable energy," *Alternative Energy and Distributed Generation Journal*, vol. 2, no. 2, pp. 8–16, 2020.

[29] Gaur, M.K., Shrivastava, A., and Pandit, R.K., "Performance analysis and sustainability assessment of a building integrated solar PV system," *International Journal of Ambient Energy*, vol. 45, no. 1, p. 2283762, 2024, doi:10.1080/01430750.2023.2283762.

[30] Kuang, Y., Zhang, Y., Zhou, B., *et al.*, "A review of renewable energy utilization in islands," *Renewable and Sustainable Energy Reviews*, vol. 59, pp. 504–513, 2016.

[31] Amin, M. A smart self-healing grid: In pursuit of a more reliable and resilient system [in my view]. *IEEE Power and Energy Magazine*, vol. 12, pp. 112–110, 2014.

[32] Kishek, K. and Zawaydeh, S., "Economic advantages of decentralized solar PV systems," *International Journal of Strategic Energy and Environmental Planning*, vol. 2, no. 2, pp. 41–55, 2020.

[33] Hoseinzadeh, P., Khalaji Assadi, M., Heidari, S., *et al.*, "Energy performance of building integrated photovoltaic high-rise building: case study, Tehran, Iran," *Energy and Buildings*, vol. 235, p. 110707, 2021, doi:https://doi.org/10.1016/j.enbuild.2020.110707.

[34] Bhattacharjee, S. and Nandi, C., "Design of a voting based smart energy management system of the renewable energy based hybrid energy system for a small community," *Energy*, vol. 214, p. 118977, 2021, doi:https://doi.org/10.1016/j.energy.2020.118977.

[35] Das, B.K., Hasan, M., and Rashid, F., "Optimal sizing of a grid-independent PV/diesel/pump-hydro hybrid system: a case study in Bangladesh," *Sustainable Energy Technologies and Assessments*, vol. 44, p. 100997, 2021, doi:https://doi.org/10.1016/j.seta.2021.100997.

[36] Rehman, S., El-Amin, I.M., Ahmad, F., *et al.*, "Feasibility study of hybrid retrofits to an isolated off-grid diesel power plant," *Renewable and Sustainable Energy Reviews*, vol. 11, no. 4, pp. 635–653, 2007, doi:https://doi.org/10.1016/j.rser.2005.05.003.

[37] Upadhyay, S. and Sharma, M.P., "A review on configurations, control and sizing methodologies of hybrid energy systems," *Renewable and Sustainable Energy Reviews*, vol. 38, pp. 47–63, 2014, doi:https://doi.org/10.1016/j.rser.2014.05.057.

[38] Adedeji, P.A., Olatunji, O.O., Madushele, N., and Rensburg, N.J.V., "Techno-economic analysis of solar PV-assisted hydroponic system - a case study in Johannesburg, South Africa," in *ASME 2022 Power Conference*, Pittsburgh, Pennsylvania, USA, July 18–19, 2022.

[39] Olatunji, O.O., Adedeji, P.A., Madushele, N., Rasmeni, Z.Z., and van Rensburg, N.J., "Hybrid standalone microgrid for agricultural last-mile: a techno-economic analysis," *Energy Reports*, vol. 8, pp. 980–990, 2022/11/01/ 2022, doi:https://doi.org/10.1016/j.egyr.2022.10.274.

[40] Kachapulula-Mudenda, P., Makashini, L., Malama, A., and Abanda, H., "Review of renewable energy technologies in Zambian households: capacities and barriers affecting successful deployment," *Buildings*, vol. 8, no. 6, pp. 1–14, 2018.

[41] van Blommestein, K.C. and Daim, T.U., "Residential energy efficient device adoption in South Africa," *Sustainable Energy Technologies and Assessments*, vol. 1, pp. 13–27, 2013, doi:https://doi.org/10.1016/j.seta.2012.12.001.

[42] Janssen, E., Nixon, D., Bruyn, S.D., Amdurski, G., and Hilaire, L.S., "Simple approach to estimating pv system snow losses applied to long-term pv generation datasets for different tilt angles and mounting styles," *Alternative Energy and Distributed Generation Journal*, vol. 2, no. 2, pp. 59–76, 2020.

158 *Clean energy for low-income communities*

[43] Jurasz, J., Guezgouz, M., Campana, P.E., and Kies, A., "On the impact of load profile data on the optimization results of off-grid energy systems," *Renewable and Sustainable Energy Reviews*, vol. 159, p. 112199, 2022, doi:https://doi.org/10.1016/j.rser.2022.112199.

[44] Pearre, N.S. and Swan, L.G., "Renewable electricity and energy storage to permit retirement of coal-fired generators in Nova Scotia," *Sustainable Energy Technologies and Assessments*, vol. 1, pp. 44–53, 2013, doi:https://doi.org/10.1016/j.seta.2013.01.001.

[45] El-Darwish, I. and Gomaa, M., "Retrofitting strategy for building envelopes to achieve energy efficiency," *Alexandria Engineering Journal*, vol. 56, no. 4, pp. 579–589, 2017, doi:https://doi.org/10.1016/j.aej.2017.05.011.

[46] Radwan, A.F., Hanafy, A.A., Elhelw, M., and El-Sayed, A.E.-H.A., "Retrofitting of existing buildings to achieve better energy-efficiency in commercial building case study: hospital in Egypt," *Alexandria Engineering Journal*, vol. 55, no. 4, pp. 3061–3071, 2016, doi:https://doi.org/10.1016/j.aej.2016.08.005.

[47] Chwieduk, D.A., "Towards modern options of energy conservation in buildings," *Renewable Energy*, vol. 101, pp. 1194–1202, 2017, doi:https://doi.org/10.1016/j.renene.2016.09.061.

[48] Li, X. and Ma, D., "Financial agglomeration, technological innovation, and green total factor energy efficiency," *Alexandria Engineering Journal*, vol. 60, no. 4, pp. 4085–4095, 2021, doi:https://doi.org/10.1016/j.aej.2021.03.001.

[49] Wang, Q. and Zhao, C., "Dynamic evolution and influencing factors of industrial green total factor energy efficiency in China," *Alexandria Engineering Journal*, vol. 60, no. 1, pp. 1929–1937, 2021, doi:https://doi.org/10.1016/j.aej.2020.11.040.

[50] Mehic, M., Duliman, M., Selimovic, N., and Voznak, M., "LoRaWAN end nodes: security and energy efficiency analysis," *Alexandria Engineering Journal*, vol. 61, pp. 8997–9009, 2022, doi:https://doi.org/10.1016/j.aej.2022.02.035.

[51] Shi, L., Liu, S., and Bao, M., "Empirical analysis on manufacturing energy efficiency of Yangtze River Basin under environmental constraints and its impactors," *Alexandria Engineering Journal*, vol. 60, no. 6, pp. 5147–5155, 2021, doi:https://doi.org/10.1016/j.aej.2021.04.046.

[52] Papadakis, N. and Katsaprakakis, D.A., "A review of energy efficiency interventions in public buildings," *Energies*, vol. 16, no. 17, p. 6329, 2023. [Online]. Available: https://www.mdpi.com/1996-1073/16/17/6329.

[53] Coelho, S., Russo, M., Oliveira, R., Monteiro, A., Lopes, M., and Borrego, C., "Sustainable energy action plans at city level: a Portuguese experience and perception," *Journal of Cleaner Production*, vol. 176, pp. 1223–1230, 2018, doi:https://doi.org/10.1016/j.jclepro.2017.11.247.

[54] Fikru, M.G., "Electricity bill savings and the role of energy efficiency improvements: a case study of residential solar adopters in the USA," *Renewable and Sustainable Energy Reviews*, vol. 106, pp. 124–132, 2019, doi:https://doi.org/10.1016/j.rser.2019.02.028.

[55] Menezes, A.C., Cripps, A., Buswell, R.A., Wright, J., and Bouchlaghem, D., "Estimating the energy consumption and power demand of small power equipment in office buildings," *Energy and Buildings*, vol. 75, pp. 199–209, 2014, doi:https://doi.org/10.1016/j.enbuild.2014.02.011.

[56] Opoku, R., Adjei, E.A., Ahadzie, D.K., and Agyarko, K.A., "Energy efficiency, solar energy and cost saving opportunities in public tertiary institutions in developing countries: the case of KNUST, Ghana," *Alexandria Engineering Journal*, vol. 59, no. 1, pp. 417–428, 2020, doi:https://doi.org/10.1016/j.aej.2020.01.011.

[57] Dolšak, J., Hrovatin, N., and Zorić, J., "Factors impacting energy-efficient retrofits in the residential sector: the effectiveness of the Slovenian subsidy program," *Energy and Buildings*, vol. 229, p. 110501, 2020, doi:https://doi.org/10.1016/j.enbuild.2020.110501.

[58] Owolabi, A.B., Emmanuel Kigha Nsafon, B., Wook Roh, J., Suh, D., and Huh, J.-S., "Measurement and verification analysis on the energy performance of a retrofit residential building after energy efficiency measures using RETScreen Expert," *Alexandria Engineering Journal*, vol. 59, no. 6, pp. 4643–4657, 2020, doi:https://doi.org/10.1016/j.aej.2020.08.022.

[59] Valizadeh, J., Sadeh, E., Javanmard, H., and Davodi, H., "The effect of energy prices on energy consumption efficiency in the petrochemical industry in Iran," *Alexandria Engineering Journal*, vol. 57, no. 4, pp. 2241–2256, 2018, doi:https://doi.org/10.1016/j.aej.2017.09.002.

[60] Abolarin, S.M., Gbadegesin, A.O., Shitta, B.M., and Adegbenro, O., "Energy (lighting) audit of four University of Lagos halls of residence," *Journal of Engineering Research*, vol. 16, no. 2, pp. 1–10, 2011.

[61] Abolarin, S.M., Gbadegesin, A.O., Shitta, M.B., *et al.*, "A collective approach to reducing carbon dioxide emission: a case study of four University of Lagos Halls of residence," *Energy and Buildings*, vol. 61, pp. 318–322, 2013, doi:https://doi.org/10.1016/j.enbuild.2013.02.041.

[62] Ogedengbe, E.O.B., Aderoju, P.A., Nkwaze, D.C., Aruwajoye, J.B., and Shitta, M.B., "Optimization of energy performance with renewable energy project sizing using multiple objective functions," *Energy Reports*, vol. 5, pp. 898–908, 2019, doi:https://doi.org/10.1016/j.egyr.2019.07.005.

[63] Syed, A., "Engineering policies for the climate change resilient built environment," *International Journal of Strategic Energy and Environmental Planning*, vol. 2, no. 1, pp. 22–29, 2020.

[64] *Standard for Commercial Building Energy Audits*, ANSI, ASHRAE, and ACCA, Atlanta, 2018. [Online]. Available: https://www.ashrae.org/File%20Library/Technical%20Resources/Bookstore/previews_2016437_pre.pdf

[65] *Minimum Energy Performance Standards (MEPS) and Labels for Air Conditioners*, NIS_943_2017, Nigeria, 2017.

[66] *Leonics*. How to design solar PV system [Online] Available: http://www.leonics.com/support/article2_12j/articles2_12j_en.php

[67] Ye, B., Jiang, J., Miao, L., Yang, P., Li, J., and Shen, B., "Feasibility study of a solar-powered electric vehicle charging station model," *Energies*, vol. 8, pp. 13265–13283, 2015.

160 *Clean energy for low-income communities*

[68] Ilieva, L.M. and Iliev, S.P., "Feasibility assessment of a solar-powered charging station for electric vehicles in the North Central region of Bulgaria," *Renewable Energy and Environmental Sustainability*, vol. 1, no. 12, pp. 1–5, 2016.

[69] *Self-Ballasted Lamps for General Lighting Services—Performance Requirements*, NIS_747:2012, 2012.

[70] *Nigerian Electricity Regulatory Commission Multi Year Rariff Order* 2020. (2020). [Online] Available: https://nerc.gov.ng/index.php/library/documents/MYTO-2020/MYTO-2020-for-EKEDC/

[71] Luo, L., Chen, J., Tang, M., Tian, H., and Lu, L., "Energy efficiency and environmental assessment of an integrated ground source heat pump and anaerobic digestion system," *Journal of Building Engineering*, vol. 54, p. 104613, 2022, doi:https://doi.org/10.1016/j.jobe.2022.104613.

[72] Chen, H.-Y., Whang, A.J.-W., Chen, Y.-Y., and Chou, C.-H., "The hybrid lighting system with natural light and LED for tunnel lighting," *Optik*, vol. 203, p. 163958, 2020. 02/01/2020, doi:https://doi.org/10.1016/j.ijleo.2019.163958.

[73] Li, G., Tan, Z.-K., Di, D., *et al.*, "Efficient light-emitting diodes based on nanocrystalline perovskite in a dielectric polymer matrix," *Nano Letters*, vol. 15, no. 4, pp. 2640–2644, 2015, doi:10.1021/acs.nanolett.5b00235.

[74] Xu, Y., Sun, C., Skibniewski, M.J., Chan, A.P.C., Yeung, J.F.Y., and Cheng, H., "System dynamics (SD) -based concession pricing model for PPP highway projects," *International Journal of Project Management*, vol. 30, no. 2, pp. 240–251, 2012, doi:https://doi.org/10.1016/j.ijproman.2011.06.001.

[75] Wu, B.-S., Hitti, Y., MacPherson, S., Orsat, V., and Lefsrud, M.G., "Comparison and perspective of conventional and LED lighting for photobiology and industry applications," *Environmental and Experimental Botany*, vol. 171, p. 103953, 2020, 03/01/2020, doi:https://doi.org/10.1016/j.envexpbot.2019.103953.

[76] Li, S.G., Tu, G., and Zhou, Q., "An optimal design model for tunnel lighting systems," *Optik*, vol. 226, p. 165660, 2021, doi:https://doi.org/10.1016/j.ijleo.2020.165660.

[77] IECQ_LED:2015, "IECQ scheme for LED lighting: a valuable qualification and supply chain management tool for LED manufacturers," *International Electrotechnical Commission*, Sydney, Australia, 2015. *[Online]*. Available: https://www.iecq.org/about/brochures/pdf/IEC_IECQ_LED_brochure_LR.pdf

[78] Taib, N.S.M., Ahmad Zaki, S., Rijal, H.B., *et al.*, "Associating thermal comfort and preference in Malaysian universities' air-conditioned office rooms under various set-point temperatures," *Journal of Building Engineering*, vol. 54, p. 104575, 2022, doi:https://doi.org/10.1016/j.jobe.2022.104575.

[79] Li, Y., Bonyadi, N., and Lee, B., "A parallel decomposition approach for building design optimization," *Journal of Building Engineering*, vol. 54, p. 104574, 2022, doi:https://doi.org/10.1016/j.jobe.2022.104574.

[80] Siriwardhana, M. and Namal, D.D.A., "Comparison of energy consumption between a standard air conditioner and an inverter-type air conditioner operating in an office building," *SLEMA Journal*, vol. 20, no. 1–2, pp. 1–6, 2017, doi:http://doi.org/10.4038/slemaj.v20i1-2.5.

[81] Sukri, M. and Jamali, M., "Economics analysis of an inverter and non-inverter type split unit air-conditioners for household application," *Journal of Engineering and Applied Sciences*, vol. 13, 06/01 2018.

[82] Opoku, R., Edwin, I.A., and Agyarko, K.A., "Energy efficiency and cost saving opportunities in public and commercial buildings in developing countries – the case of air-conditioners in Ghana," *Journal of Cleaner Production*, vol. 230, pp. 937–944, 2019, doi:https://doi.org/10.1016/j.jclepro.2019.05.067.

[83] Panasonic, A. Residential Air Conditioner—Wall Type—CS-YS12-UKA [Online] Available: https://www.panasonic.com/africa/consumer/air-solutions/residential-air-conditioners/wall-type/cs-ys12-uka.html

[84] Shah, N., Sathaye, N., Phadke, A., and Letschert, V., "Efficiency improvement opportunities for ceiling fans," *Energy Efficiency*, vol. 8, no. 1, pp. 37–50, 2015, doi:10.1007/s12053-014-9274-6.

[85] Desroches, L.B. and Garbesi, K., *Max tech and Beyond: Maximizing Appliance and Equipment Efficiency by Design,* Berkeley, CA: Lawrence Berkeley National Laboratory, 2011. [Online]. Available: https://digital.library.unt.edu/ark:/67531/metadc828559/m2/1/high_res_d/1047752.pdf

[86] Sathaye, N., Phadke, A., Shah, N., and Letschert, V., *Potential Global Benefits of Improved Ceiling Fan Energy Efficiency*, Berkeley, CA: Ernest Orlando Lawrence Berkeley National Laboratory, 2013.

[87] Belussi, L., Barozzi, B., Bellazzi, A., *et al.*, "A review of performance of zero energy buildings and energy efficiency solutions," *Journal of Building Engineering*, vol. 25, p. 100772, 2019, doi:https://doi.org/10.1016/j.jobe.2019.100772.

[88] Gholinejad, H.R., Loni, A., Adabi, J., and Marzband, M., "A hierarchical energy management system for multiple home energy hubs in neighborhood grids," *Journal of Building Engineering*, vol. 28, p. 101028, 2020, doi:https://doi.org/10.1016/j.jobe.2019.101028.

[89] Mariano-Hernández, D., Hernández-Callejo, L., Zorita-Lamadrid, A., Duque-Pérez, O., and Santos García, F., "A review of strategies for building energy management system: model predictive control, demand side management, optimization, and fault detect & diagnosis," *Journal of Building Engineering*, vol. 33, p. 101692, 2021, doi:https://doi.org/10.1016/j.jobe.2020.101692.

[90] *Measurement of Energy, Demand, and Water Savings*, Atlanta, GA, United States: ASHRAE, 2014.

Chapter 6

Sustainable energy solutions for rural electrification in a low-income community

Reza Babaei[1], David S-K. Ting[1] and Rupp Carriveau[1]

Addressing the simultaneous provision of electricity, heat, and water to rural areas is a pervasive global challenge. This study focuses on optimizing a poly-generation hybrid system that integrates PV, wind turbine, Combined Heat and Power (CHP) unit, battery, and brackish water reverse osmosis desalination, designed for warm climates, to meet the essential energy needs of Sar Goli village and a health clinic in Khuzestan province, Iran. Unlike previous studies, this research conducts sensitivity analyses considering diverse economic and climate conditions, evaluating the grid breakeven distance, environmental impact, and technical performance. The proposed 51.2 kW PV/10 kW WT/10 kW CHP/96 kWh BT/23.8 kW CNV system with reverse osmosis desalination demonstrates a cost of electricity (COE) and net present cost (NPC) of \$0.161/kWh and \$107,203, respectively. The study highlights that increased solar irradiation and wind speed contribute to cost efficiencies in renewable energy, resulting in lower NPC and COE values. However, rising diesel prices pose economic challenges for diesel-dependent systems, emphasizing the importance of strategic planning for resilient energy solutions. Additionally, improving boiler efficiency significantly reduces fuel consumption and CO_2 emissions, emphasizing the interconnected nature of thermal load levels and environmental impact, guiding the path toward enhanced sustainability.

Keywords: Hybrid energy system; Renewable energy; Desalination; Polygeneration

Nomenclature

BT	Battery
CHP	Combined Heat and Power
CNV	Converter

(Continues)

[1]Turbulence & Energy Laboratory, University of Windsor, Canada

164 *Clean energy for low-income communities*

(*Continued*)

COE	Cost of Electricity
CRF	Capital Recovery Factor
FPA	Flower Pollination Algorithm
HES	Hybrid Energy System
IC	Initial Cost
kWh	Kilowatt-hour
LA	Lead-Acid
Li-Ion	Lithium-Ion
LPSP	Loss of Power Supply Probability
Ni–Fe	Nickel–Iron
NPC	Net Present Cost
PV	Photovoltaic
RF	Renewable Fraction
RO	Reverse Osmosis
TAC	Total Annual Cost
TNPV	Total Net Present Value
WT	Wind Turbine

6.1 Introduction

A critical and far-reaching global challenge is delivering electricity to rural, low-income, and thinly populated communities without easy access to the national electricity grid. This pressing issue underscores the intricate web of economic, social, and infrastructural disparities that persist on a global scale. According to the 2019 report from the International Energy Agency (IEA), a staggering 15% of the worldwide rural population encounters formidable obstacles in securing reliable access to electricity.

Here are two main approaches to meeting energy demands: continuing to rely on fossil fuels or transitioning towards a combination of renewable energy sources. However, the use of fossil fuels has several negative impacts, such as environmental degradation and the fact that these resources are finite. Therefore, combining different renewable energy sources seems more attractive for some countries. This is a hybrid renewable energy system that integrates renewable energy resources, conventional energy sources, and energy storage devices. However, since renewable energy sources are intermittent, using conventional energy systems and storage devices as backups can reduce reliability and lead to excess electricity in hybrid renewable energy systems [1].

Providing freshwater for drinking and other purposes constitutes a fundamental human requirement. However, this necessity encounters considerable challenges due to the rapid expansion of the global population [2]. Conversely, only 1% of the world's total water resources are available as freshwater, with the remainder existing as saline water (97%) and frozen water (2%) [3].

In addressing this issue, desalination emerges as a viable solution. Desalination involves a process where salt is separated from saline water using specialized methods, enabling the utilization of freshwater for various needs such as drinking and agriculture. Several methods exist for salt separation, with thermal and membrane processes being the two primary accepted categories. Despite various technologies falling under these classes, reverse osmosis (RO), with a 65% installed capacity, is widely adopted [4].

The combined heat and power (CHP) system is a beneficial framework for concurrently generating heat and power from a singular fuel source, utilizing the recovered waste heat to generate additional energy. Noteworthy attributes of the CHP system include increased efficiency, reduced fuel consumption, and diminished greenhouse gas emissions [5].

Sinha and Chandel [6], in their evaluation of software tools for optimizing renewable energy systems, affirmed that HOMER exhibits remarkable precision in delineating the techno-economic attributes of an energy system. Numerous studies have been conducted to explore the application of various hybrid energy configurations and HOMER optimization tools in supplying electricity, water, and heat and their combinations. The subsequent section introduces several of these research endeavors. In some previous investigations, the focus has been on employing hybrid energy systems exclusively for water supply. For example, Ibrahim *et al.* [7] conducted a numerical study exploring two-hybrid systems: PV/Wind Turbine (WT)/Diesel Generator (DG)/Battery (Bat) and PV/DG/Hydrokinetic Turbine (HKT)/BT. These systems were designed to cater to the electricity needs of a desalination seawater RO unit with a capacity of 1 m^3/h and a power demand of 4.38 kW in an Egyptian city. Utilizing optimization results from HOMER software, the optimal configuration for the first scenario involved 8.3 kW of PV, one wind turbine, 4.9 kW of diesel generator, 15 units of battery, and a 5.29 kW converter. The corresponding Reliability Factor (RF), Cost of Energy (COE), and water cost were 52.5%, 0.2252 $/kWh, and 1.10 $/$m^3$, respectively. Conversely, the second optimized configuration comprised 2.82 kW of PV, three hydrokinetic turbines, 4.9 kW of diesel generator, 15 units of battery, and a 0.984 kW converter, achieving a 98.2% RF, 0.1216 $/kWh COE, and a water cost of 0.56 $/$m^3$.

Kumar Nag and Sarkar [8] presented a modeled hybrid energy system for a rural community in India by incorporating solar, wind, hydrokinetic, and bioenergy components. The analysis, encompassing optimization and sensitivity evaluations, was conducted using the HOMER software. The primary focus was on enhancing system efficiency while maintaining cost-effectiveness. The study revealed that a more diverse array of renewable energy sources is more effective in meeting future energy demands, as a logistic growth model projected. Over the periods (2012–2021) to (2042–2051), the optimized cost of electricity (COE) is minimized, demonstrating increased electric generation while minimizing environmental impact. Notably, despite the potential for alternative combinations to generate more power, the need for excess power storage resulted in increased COE.

166   *Clean energy for low-income communities*

Atallah *et al.* [9] conducted a HOMER software simulation to fulfill the electricity requirements of a Reverse Osmosis (RO) unit, producing 557.22 kWh/day, with a daily output of 100 m^3, in Nakhl, North Sinai, Egypt. Their study encompassed the evaluation of 11 distinct hybrid systems to ascertain the optimized configuration. According to the findings, the optimal system comprised 160 kW of PV, 50 kW of DG, 39.3 kW of the converter, and 190 units of lead-acid batteries, achieving a Reliability Factor (RF) of 93.1%. The corresponding Cost of Energy (COE) and Net Present Cost (NPC) were proportional to 0.107 $/kWh and 502,662 $, respectively. Additionally, the carbon dioxide emissions associated with the optimized scenario amounted to 122,897 kg/year.

Some other research studies used MATLAB® to investigate the feasibility of renewable HES in remote or rural areas. For instance, Kumar *et al.* [10] focused on HES using three distinct battery technologies—lithium-ion (Li-Ion), nickel–iron (Ni–Fe), and lead-acid (LA)—in conjunction with a diesel generator to ensure an uninterrupted power supply to grid-disconnected villages in India. Samy *et al.* [11] research focused on conducting a techno-economic feasibility study for off-grid solar photovoltaic fuel cell (PV/FC) hybrid systems to provide electricity to remote areas and isolated urban regions in Egypt. The study's objective function is formulated based on the total annual cost (TAC), with optimization achieved through the implementation of the Flower Pollination Algorithm (FPA), a recently efficient metaheuristic method. The FPA was employed to estimate the optimal number of both PV panels and the FC/electrolyzer/H$_2$ storage tanks set, ensuring the attainment of the lowest total net present value (TNPV); FPA Algorithm exhibited the least fulfillment time and optimal performance compared to the other algorithms.

The other literature review findings, including additional studies and their respective optimal configurations, are summarized in Table 6.1. This table provides a comprehensive overview of the various hybrid energy systems applied to water desalination, showcasing system components, optimal configurations, and key performance metrics across different studies.

After analyzing the research above, it is evident that HOMER Pro software has not been used to investigate the techno-economic optimization of a hybrid energy system (HES) that simultaneously provides electricity, heat, and freshwater in a low-income community. Reference [17] indicated that multi-energy systems have several benefits, such as reducing environmental impact, improving system efficiency, and lowering costs. However, renewable energy sources are not widely used in urban heating systems, so incorporating Combined Heat and Power (CHP) co-generation units with hybrid renewable energy systems could be a step towards a more sustainable and environmentally friendly energy planning approach.

In this investigation, a configuration comprising a PV module, wind turbine, CHP unit, battery, converter, and a CHP package system, in conjunction with a brackish water reverse osmosis desalination system, is proposed for supplying electricity, heat, and water to a remote area and health clinic in Iran. Following the optimization of the hybrid renewable energy system, an in-depth examination of

Sustainable energy solutions for rural electrification 167

Table 6.1 Electrical and thermal load profiles during a year

Reference	System components	Optimal configuration details	Performance metrics	Key findings
Karimi *et al.* [12]	PV/WT/ DG/BT	RO: 0.5 kW DG, 2.75 kW PV, 12 battery units, 1.5 kW converter, COE $0.161/kWh, NPC $7,337, EDR: 1.375 kW PV, 9 battery units, 0.6 kW converter, COE $0.174/kWh, NPC $3,636	COE, NPC	Hybrid energy system applied to EDR and RO desalination units using PV/WT/DG/Bat; optimization based on experimental results and simulation using HOMER software.
Setiawan *et al.* [13]	PV/WT/ DG/BT	WT/DG: COE $0.437/kWh, IC $185,900, NPC $632,159	COE, IC, NPC	Hybrid energy system (PV/WT/DG/Bat) supplying electricity and water for RO desalination system in the Maldives; post-tsunami scenario simulation with opti- mized configuration using HOMER software.
Mehrjerdi [14]	PV/WT/ DG/BT for RO; PV/ WT/DG/ Bat/ Boiler for MSF and MED	RO: Electric energy 3 kWh/m^3, NPC approx. $69M MSF and MED: Electric energy 3 kWh/m^3, Thermal energy supplied by boiler	NPC	HOMER is used for covering electric load and water (4,800 m^3/ day) with three desalination technologies; compari- son of output results and cost analysis.
Waqar *et al.* [15]	PV/DG/ BT/Boiler (on/off-grid)	Gilgit city: Optimal with 2 PV systems, 2 diesel generators, waste heat recovery; COE $0.049/kWh, NPC $5.79M Lahore: Opti- mal for minimization of annual greenhouse gas emissions.	NPC, COE	CHP set optimization (PV/DG/Bat/Boiler) for heat and electricity supply in six cities of Pakistan; different sce- narios for minimizing NPC, COE, and greenhouse gas emissions.
Yuan *et al.* [16]	PV/BT with wood- syngas CHP	Optimal: 0.5 kW PV, 0.65 kW wood gas generator, 3 battery units, 0.5 kW converter; COE $0.351/kWh, NPC $3,572, CO$_2$ emissions 6,490 kg/year	COE, NPC, CO$_2$ emissions	Off-grid hybrid system in China for electric and thermal load reduction; wood- syngas CHP with PV/ Bat; comparison of feasible cases.

EDR: energy dissipative reverse osmosis; MSF: multi-stage flash; and MED: multi-effect distillation.

the CHP performance in the optimal scenario is conducted. Additionally, sensitivity analyses are performed on various influential parameters, including renewable energy resources, boiler efficiency, fuel price, and available resources, aiming to generalize the findings for application in other rural communities.

The articles examined above highlight a growing interest in the global shift to clean energy, particularly with the unique features of hybrid energy systems. Nevertheless, there are notable gaps in current research: (i) the majority of analyzed hybrid energy systems have been primarily designed to meet a single load requirement (electricity, heat, or water) in diverse locations such as islands, rural areas, or urban settings, with limited exploration of multi-generation systems that harness surplus energy and simultaneously recover waste heat; (ii) there is a lack of studies investigating the impact of boiler efficiency on the economic performance of hybrid energy systems; and (iii) insufficient attention has been given to the measurement of Loss of Power Supply Probability (LPSP). The technique of this study can be applied to evaluate HES performance in any location by adjusting renewable resources and load profiles.

6.2 Methods and materials

6.2.1 Study area

In this study, Sar Goli village in Khuzestan province, Iran, situated at 32°25′26″N 49°36′23″E, as illustrated in Figure 6.1, was selected as the research site. Sar Goli is a low-income community grappling with the absence of direct access to essential utilities such as electricity, heat, and freshwater. The selected research site

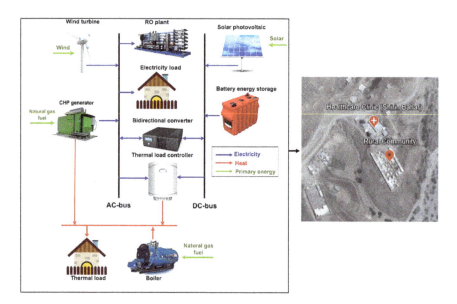

Figure 6.1 Location of Sar Goli village and nearby healthcare clinic

underscores the significance of implementing sustainable solutions, and the off-grid hybrid energy system depicted in Figure 6.1 aims to address these challenges by providing an alternative and reliable source of electricity, heat, and purified water for the community. As per the most recent research conducted by the Statistical Center of Iran in 2016 [3], the village is home to approximately 140 inhabitants, distributed across 25 families.

6.3 System overview

The proposed design is shaped by either renewable or fossil fuel-driven electricity sources, primarily focusing on ensuring a balanced power distribution throughout the year (8,760 hours). Consequently, each equipment component undergoes mathematical representation, and the computation involves determining the quantity of power generated and stored by each element. Figure 6.1 illustrates the arrangement of the assumed HES designed to supply electricity, freshwater, and thermal loads for 100 residential households in the specified regions. To provide uniform and stable electricity, the grid-isolated power supply comprises eight main components: PV panels, wind turbines (WT), CHP units, CHP plants, boilers, battery packs, and converters. The mathematical description of the configuration is represented in the following subsection.

6.3.1 Renewable resources

Figure 6.2(a) and (b) illustrates the hourly interval data on irradiation and wind speed in Sar Goli village, offering insights into the variation of these crucial meteorological factors throughout the year. The data is collected from NASA's meteorological resource data center (NASA) [18]. In terms of irradiation, the values steadily increase from January (2.97 kWh/m^2) to June (7.74 kWh/m^2), indicating a progressive rise in solar energy availability during the first half of the year. Subsequently, there is a slight decrease in irradiation during July (7.31 kWh/m^2) and August (6.85 kWh/m^2), likely influenced by seasonal weather patterns. Wind speed, however, shows a more consistent trend, with relatively lower values in the earlier months and a gradual increase from June (3.34 m/s) to August (3.26 m/s).

6.3.2 Desalinaion unit

The proposed reverse osmosis (RO) desalination plant in these targeted areas aims to provide essential freshwater to rural users in Sar Goli village. The brackish water is first stored in a tank before being pumped through high-pressure pumps to the pretreatment section. Next, it goes to the membrane modules, where two streams are created: permeate and brine. The product water, which is derived from the RO plant, is stored in a separate tank. Post-treatment involves adjusting the pH and disinfecting the permeated water through a dosing system.

Figure 6.3 presents a schematic diagram that outlines the various components of the RO plant. These include the headworks, which consist of well pumps, holding tanks, and feedwater pumps, as well as the pretreatment units, which

170 *Clean energy for low-income communities*

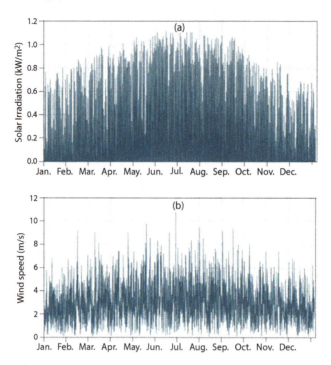

Figure 6.2 Yearly profiles of (a) solar irradiation and (b) wind speed in Sar Goli village

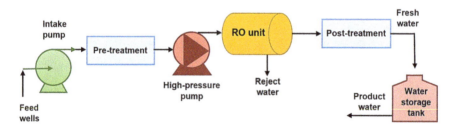

Figure 6.3 Schematic figure of the intended RO desalination plan

comprise multi-media and cartridge filters. The RO modules, high-pressure pumps, and post-treatment adjustments for permeate water flow (pH and chlorination) are also illustrated. The rejected brine is safely injected into distant injection wells located away from the well pumps and the RO plant.

6.3.3 Load profile

A health clinic was established to improve healthcare accessibility in rural areas. Sar Goli village's population is limited, resulting in relatively low electricity

Sustainable energy solutions for rural electrification 171

consumption. Table 6.2 presents the yearly profile of the electrical clinic, rural load, and thermal load. The daily clinic load is estimated to be 9.66 kWh/day with a peak load of 2.05 kW. The rural part of the village has a daily load of 169.03 kWh/day and a peak load of 35.90 kW. Deferrable loads are electric loads that are often used in desalination plants to control the pumps on a specific schedule. These deferrable load bar charts show that deferrable loads are aligned with other load

Table 6.2   Electrical and thermal load profiles during a year

Load category	Load profile
Clinic electrical load Average load (kWh) = 7.04 Peak load (kWh) = 35.90 Load ratio (%) = 20	
Rural electrical load Average load (kWh) = 0.40 Peak load (kWh) = 2.05 Load ratio (%) = 20	
RO thermal load Average load (kWh) = 0.46 Peak load (kWh) = 2.31 Load ratio (%) = 20	
RO deferrable electrical load demand Average load (kWh) = 4.16 Peak load (kWh) = 10 Load ratio (%) = 42	
Overall load allocation	

172 *Clean energy for low-income communities*

categories, with high demand in summer and lower demand in winter. Overall load allocation reveals that the highest and lowest load goes to meeting rural electrical load at 58% and clinic electrical load at 3%, respectively.

6.3.4 CHP plant

To ensure a reliable source of power while also prioritizing sustainability, an HES system can be augmented by integrating backup generation to support the energy needs of the renewable components; by utilizing a CHP, initial, operating, and maintenance costs can also be reduced. CHP has been chosen to support the energy needs of renewable components. The specification of the CHP unit is found in Table 6.3.

The following equation defines the CHP efficiency:

$$\eta_g = \frac{3600 \quad P_e}{\rho_f \left(P_{gen}F_0 + P_eF_1\right)} \tag{6.1}$$

where η_g is the CHP unit efficiency (%), P_e is the output power (kW), ρ_f is the fuel density (kg/m^3), P_{gen} is the rated CHP power (kW), F_0 is the CHP unit fuel curve intercept co-efficient (L/h/rated kW or m^3/h/rated kW), and F_1 is the fuel curve slope (L/h/ output kW or, m^3/h/output kW). The following relation calculates the real CHP units' fuel consumption in L/h:

$$m_{\text{fuel}} = F_0 Y_{gen} + F_1 P_{gen} \tag{6.2}$$

where Y_{gen} is the rated capacity of the CHP plant (kW).

6.3.5 PV module

The PV module generates DC electricity in proportion to the solar radiation it receives. However, the actual power output of the module is lower than its theoretical output due to various factors like dust accumulation, shading, snow cover, wiring losses, and aging. The derating factor represents this reduction in output. To calculate the power output of a solar panel, you can use the following equation [20]:

$$p_{PV} = W_{PV}f_{PV}\frac{G_T}{G_S}\left[1 + \alpha_p(T_C - T_S)\right] \tag{6.3}$$

Table 6.3 Technical and economic data of the CHP plant [19]

Parameter	Value
Power output (kW)	10
Initial cost ($)	5,000
Replacement cost ($)	5,000
O&M cost ($/op. hour)	0.3
Fuel type	Diesel
Fuel price ($/m^3)	0.1
Heat recovery ratio (%)	30
Lifetime (hours)	15,000

Sustainable energy solutions for rural electrification 173

In this equation, W_{pv} represents the peak power output of the PV array in kilowatts, f_{pv} is the PV derating factor in %, G_T is the solar radiation incident during the current hour measured in kWh/m², G_S is the incident radiation at standard test conditions, which is 1 kW/m². a_p is the temperature coefficient in %/°C, T_C is the PV module temperature in the current time step measured in °C and T_S the PV module temperature in standard test conditions is 25 °C. The number of solar PV panels can vary, and for each 10 kW flat plate, specifications are presented in Table 6.4.

6.3.6 Wind turbine

This area has been selected for a 10 kW AC voltage wind turbine based on factors such as cut-in and cut-out wind speed values, hub height, and wind turbine cost. To calculate the power output of the wind turbine, we can use the following equation [1]:

$$P = \frac{1}{2}\rho A V^3 C_{PC} \tag{6.4}$$

where ρ is the air density (1.225 kg/m³), V is the wind speed (m/s), A is the rotor swept area (m²), and C_{pc} is the maximum power coefficient. The technical key data of the selected wind turbine is prepared in Table 6.5.

6.3.7 Battery storage

Batteries function as energy storage devices by converting and storing electricity in chemical form. This stored energy can be retrieved, recharged, and reused to ensure

Table 6.4 Technical and economic data of considered PV module [21]

Parameter	Value
Output (kW)	10
Initial cost ($)	9,000
Replacement cost ($)	8,500
O&M cost ($/year)	100
Lifetime (years)	20

Table 6.5 Wind turbine specifications

Parameter	Value
Output (kW)	10
Initial cost ($)	9,500
Replacement cost ($)	9,000
O&M cost ($/year)	30
Lifetime (years)	20

174 *Clean energy for low-income communities*

an uninterrupted power supply. To ensure the battery bank's durability and optimal performance, maintaining a battery charge of at least 20% is crucial. Table 6.6 shows the general specification of the battery technology considered for this study.

The following equation shows how values of battery energy can be estimated [22].

$$Q_{battery} = Q_{battery,0} + \int_0^\tau V_{battery} I_{battery} dt \tag{6.5}$$

where $Q_{battery,0}$ (kWh) is the initial battery charge, $V_{battery}$ (V) is the battery voltage, and $I_{battery}$ (A) is the battery current.

The state of battery charge is expressed by (6.12).

$$B_{soc} = \frac{Q_{battery}}{Q_{battery,\ max}} \times 100(\%) \tag{6.6}$$

6.3.8 Converter

The converter aims to facilitate energy exchange between direct current (DC) and alternating current (AC), similar to the functions performed by an inverter or rectifier. It involves converting DC power from the PV module and battery output into AC. Conversely, when surplus wind energy is generated, a rectifier converts AC power into DC for storage in the battery storage system. Table 6.7 provides

Table 6.6 Battery specifications

Parameter	Value
Nominal voltage (V)	12
Nominal capacity (kWh)	1
Capital cost ($)	300
Replacement ($)	300
O&M ($/year)	10
Lifetime (years)	15

Table 6.7 Technical and economic data of considered converter [23]

Parameter	Value
Capital cost ($/battery)	300
Replacement cost ($)	300
O&M cost ($/year)	10
Lifetime (years)	15

Sustainable energy solutions for rural electrification 175

comprehensive data on the chosen power converter. The power rating of these converters can be calculated using the equation provided below [24]:

$$P_{inv} = \frac{P_{peak}}{\eta_{inv}} \tag{6.7}$$

where P_{peak} is the peak load demand, and η_{inv} is inverter efficiency.

The system's life cycle total cost can be characterized by NPC, which involves the initial O&M, replacement, and resource-related costs such as fuel cost over the project lifetime. The total NPC ($) is measured by the following equation [25,26]:

$$C_{npc,tot} = \frac{C_{ann,tot}}{CRF(i, R_{proj})} \tag{6.8}$$

Here, $C_{ann,tot}$ is the total annualized cost ($/year), i is the annual real interest rate (%), and $CRF(i, n)$ is the capital recovery factor, which is calculated by (6.9) [27]:

$$CRF(i, n) = \frac{i(1 + i)^n}{(1 + i)^n - 1} \tag{6.9}$$

which n is the lifetime of the project (year), and i is the annual real interest rate, which is determined by (6.6)

$$i = \frac{i - f}{1 + f} \tag{6.10}$$

where i is the nominal interest rate (%), and f is the annual inflation rate (%). To measure the LCOE ($/kWh), the following equation is used [28]:

$$LCOE = \frac{C_{ann,tot}}{L_{ann,load}} \tag{6.11}$$

Here, $L_{ann,load}$ is the total electricity consumption per year (kWh/year), and $C_{ann,tot}$ is the total annualized cost ($/year).

Moreover, one of the principal objectives inherent to the optimization paradigm is the assurance of system reliability. In the context of this research, the evaluation of system reliability is undertaken through the metric of Loss of Power Supply Probability (LPSP). The LPSP index is delineated as the quotient of the aggregate power supply failure duration over the total reporting time. This particular metric, extensively documented in pertinent literature, is established as a standard for the comprehensive assessment of system reliability. Significantly, LPSP is consistently quantified in relation to the percentage of insufficient power time.

$$LPSP = \sum_{t=1}^{8760} \left(\frac{P_{pv}(t) + P_{WT}(t) + P_{CHP}(t) + P_{batt}(t) < P_{load}(t)}{8760} \right) \tag{6.12}$$

LPSP is assigned numerical values of 0 or 1. An LPSP value of zero indicates the absence of any unmet load throughout the entirety of the simulation period.

176   *Clean energy for low-income communities*

Conversely, an LPSP value of 1 signifies a persistent presence of unmet loads, contingent upon the constraint outlined below:

$$LPSP \leq LPSP_{max} \tag{6.13}$$

The term $LPSP_{max}$ denotes the maximum acceptable loss of load over the entire simulation period, as defined by the controller. This constraint is applicable exclusively to instances where LPSP is non-zero. For example, a 5% LPSP corresponds to 438 h, equivalent to 1.2 h per day, while a 1% LPSP translates to 0.24 h of unmet load each day.

6.4   Results and discussion

This section provides a detailed analysis of the optimization outcomes, specifically focusing on the performance of the CHP unit. A sensitivity analysis is conducted to investigate the impact of various parameters on the system, which serves as a valuable reference for decision-makers in the energy sector, guiding strategic planning and investment efforts towards sustainable and economically viable energy solutions. Table 6.8 outlines four optimal hybrid energy system configurations, each with specific capacities. The winning solution with the lowest NPC is Option I where the system comprises 51.2 kW of photovoltaic (PV), 10 kW of wind turbine (WT), 10 kW of combined heat and power (CHP), 96 kW of battery storage (BT), and 23.8 kW of conversion (CNV). This configuration leads to an NPC of $207,203 and a COE of $0.161/$. Option II emphasizes 136 kW of PV, 529 kW of BT, and 38.1 kW of CNV, resulting in an NPC of $490,590 and a COE of $0.382/$. Option III integrates 148 kW of PV, 10 kW of WT, 483 kW of BT, and 36.1 kW of CNV, yielding an NPC of $495,572 and a COE of $0.386/$. Finally, Option IV emphasizes 490 kW of WT, 10 kW of CHP, 712 kW of BT, and 39.8 kW of CNV, resulting in an NPC of $1,012,129 and a COE of $0.789/$. In summary, Option I, combining both PV and WT elements, proves to be the most economically efficient option. On the other hand, Option IV, predominantly utilizes WT without

Table 6.8   Techno-economic optimization result of the proposed hybrid energy systems

Option	HES	PV (kW)	WT (kW)	CHP (kW)	BT (kW)	CNV (kW)	NPC ($)	COE ($/kWh)
I	PV-WT-CHP-BLR-BT-CNV	51.2	10	10	96	23.8	207,203	0.161
II	PV-BLR-BT-CNV	136	–	–	529	38.1	490,590	0.382
III	PV-WT-BLR-BT-CNV	148	10	–	483	36.1	495,572	0.386
IV	WT-CHP-BLR-BT-CNV	–	490	10	712	39.8	1,012,129	0.789

integrating PV, appears to be the least financially beneficial. This underscores that configurations that prioritize WT over PV might not yield equivalent economic advantages in regions rich in solar energy potential.

Table 6.9 depicts the techno-economic optimization results for the proposed hybrid energy systems. Option III mirrors Option II, featuring high electrical production of 249,223 kWh, a surplus of 135,607 kWh, and a 100% renewable electricity fraction. Option III's LPSP is 0.0005, marking it as a compelling choice. On the other hand, Option IV demonstrates a unique combination, producing 51,234 kWh of electricity with 64,154 kWh of excess electricity. The thermal production is 35,464 kWh with 32,770 kWh of excess thermal energy; the electricity renewable fraction is 70%. The LPSP for Option IV is 0.0010. Overall, these detailed metrics enable a comprehensive evaluation of the performance and system reliability associated with each option in the energy system.

Figure 6.4 displays the cost distribution among system components. The CHP plant claims the highest cost share at 34%, followed by PV panels at 26% and batteries at 29%. Converters and wind turbines evenly share the remaining 10%. This breakdown explains the financial prioritization among system components, guiding resource allocation.

Figure 6.5 provides the monthly electricity production share for the output of a hybrid energy system comprising PV, WT, and a 10 kW CHP unit. In January, the PV system generated 6,075 kW, the WT system produced 71 kW, and the 10 kW CHP unit contributed 3,984 kW. These values are provided for each month,

Table 6.9 Techno-economic optimization result of the proposed hybrid energy systems

Option	Electrical production (kWh)	Excess electricity (kWh)	Thermal production (kWh)	Excess thermal energy (kWh)	Electricity renewable fraction (%)	LPSP
I	140,705	34,627	37,421	33,316	64	0.0005
II	229,336	112,934	4,105	0	100	0.0006
III	249,223	135,607	4,105	0	100	0.0005
IV	51,234	64,154	35,464	32,770	70	0.0010

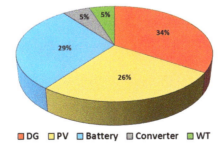

Figure 6.4   Cost allocation for the energy components

178 Clean energy for low-income communities

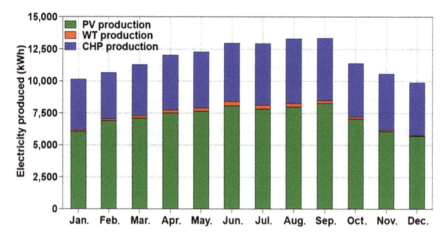

Figure 6.5 Contribution of electricity production from each energy component

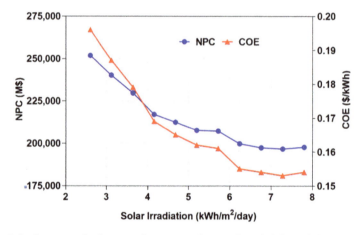

Figure 6.6 Impact of solar irradiation on the NPC and COE of the winning solution

offering insights into the performance and variations of the hybrid energy system over a year. The data reveals the monthly fluctuations in output for each component, providing valuable information for energy planning, system optimization, and performance analysis.

Figures 6.6–6.8 provide insightful perspectives on the economic dynamics of the winning HES under varying conditions. Figure 6.6 reveals an inverse relationship between irradiation levels and both NPC and COE, demonstrating that higher solar irradiation contributes to reduced costs. Figure 6.7 highlights a consistent decline in NPC and COE values as wind speed increases, indicating cost efficiencies associated with higher wind speeds in wind energy generation.

Sustainable energy solutions for rural electrification   179

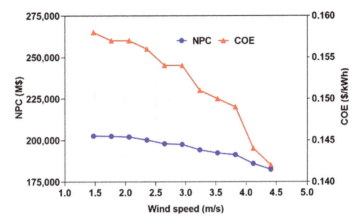

Figure 6.7 Impact of wind speed on the NPC and COE of the winning solution

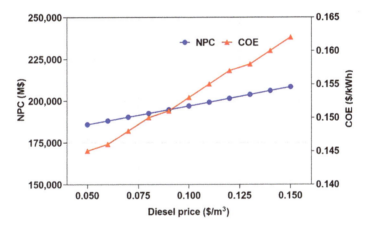

Figure 6.8 Impact of the diesel price of the CHP unit on the NPC and COE of the winning solution

Conversely, Figure 6.8 demonstrates a direct correlation between rising diesel prices and increasing NPC and COE values, emphasizing the economic challenges of elevated fuel costs in diesel-dependent energy systems.

Figures 6.9 and 6.10 offer valuable insights into distinct systems' energy efficiency and environmental impact. Figure 6.1 demonstrates a clear inverse relationship between efficiency, fuel usage, and carbon dioxide (CO_2) emissions. As the boiler efficiency improves from 30% to 90%, there is a consistent reduction in fuel consumption (509–170 liters per year) and CO_2 emissions (1,018–340 kg/year), emphasizing the significance of enhancing boiler performance for lower environmental impact. In the second scenario concerning thermal load, the data showcases a proportional increase in fuel usage (89.9–270 liters per year) and CO_2

180 Clean energy for low-income communities

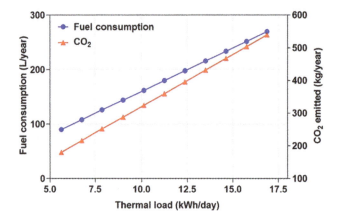

Figure 6.9 Impact of thermal load on the fuel consumption and CO_2 emissions of the winning solution

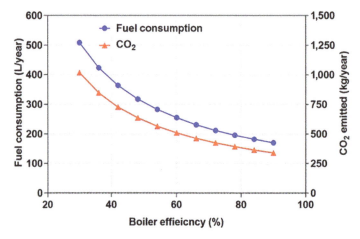

Figure 6.10 Impact of boiler efficiency on the fuel consumption and CO_2 emissions of the winning solution

emissions (180–540 kg/year) with rising thermal load levels from 5.625 to 16.875 kWh per day. This underscores the direct correlation between thermal load, fuel consumption, and environmental consequences, providing crucial insights into the energy efficiency and sustainability considerations associated with varying thermal load levels.

6.5 Conclusion

Addressing the challenge of providing electricity, heat, and water simultaneously in rural areas is a major global issue. This study aimed to optimize a poly-generation hybrid energy system (HES) combining different energy sources and a desalination process. The HES is designed for warm climates and is intended to meet the basic energy needs of Sar Goli village and a health clinic in Khuzestan province, Iran. The primary findings of the research include:

(i) Option I, which integrates both PV and WT components, emerges as the most cost-effective choice. In contrast, Option IV, which relies primarily on WT without incorporating PV, stands as the least economically advantageous. This highlights that configurations emphasizing WT over PV in areas abundant in solar energy potential may not achieve comparable economic benefits.

(ii) Option II (PV-BLR-BT-CNV) has the highest electrical production, complete reliance on renewable sources, and a relatively higher LPSP, suggesting its robustness and potential cost-effectiveness. Options I (PV-WT-CHP-BLR-BT-CNV) and III (PV-WT-BLR-BT-CNV) also exhibit high electrical production and renewable fractions. Option IV, while unique in its combination, shows comparatively lower electrical production and a higher LPSP, indicating potential efficiency considerations.

(iii) Higher solar irradiation and increased wind speed contribute to cost efficiencies in renewable energy, reflected in lower NPC and COE values. Conversely, rising diesel prices pose economic challenges for diesel-dependent systems. These findings stress the need for resource-specific optimization in renewables and strategic planning for resilient energy solutions.

(iv) Enhancing boiler efficiency is crucial for lowering environmental impact, as seen in the consistent reduction of fuel consumption and CO_2 emissions. The proportional relationship between rising thermal load levels, increased fuel usage, and higher CO_2 emissions underscores the interconnected nature of these factors. These findings highlight the importance of strategic system performance improvements to reduce environmental consequences and enhance overall sustainability.

References

[1] T. Chen, M. Wang, R. Babaei, M. E. Safa, and A. A. Shojaei, "Technoeconomic analysis and optimization of hybrid solar-wind-hydrodiesel renewable energy systems using two dispatch strategies," *Int. J. Photoenergy*, vol. 2023, pp. 1–20, 2023, doi:10.1155/2023/3101876.

[2] M. A. Eltawil, Z. Zhengming, and L. Yuan, "A review of renewable energy technologies integrated with desalination systems," *Renew. Sustain. Energy Rev.*, vol. 13, no. 9, pp. 2245–2262, 2009, doi:10.1016/j.rser.2009.06.011.

[3] A. Shahsavari, M. A. Vaziri Rad, F. Pourfayaz, and A. Kasaeian, "Optimal sizing of an integrated CHP and desalination system as a polygeneration plant for supplying rural demands," *Energy*, vol. 258, p. 124820, 2022, doi:10.1016/j.energy.2022.124820.

[4] M. A. M. Khan, S. Rehman, and F. A. Al-Sulaiman, "A hybrid renewable energy system as a potential energy source for water desalination using

reverse osmosis: A review," *Renew. Sustain. Energy Rev.*, vol. 97, pp. 456–477, 2018, doi:10.1016/j.rser.2018.08.049.

[5] M. A. Bagherian and K. Mehranzamir, "A comprehensive review on renewable energy integration for combined heat and power production," *Energy Convers. Manag.*, vol. 224, p. 113454, 2020, doi:10.1016/j.enconman.2020.113454.

[6] S. Sinha and S. S. Chandel, "Review of software tools for hybrid renewable energy systems," *Renew. Sustain. Energy Rev.*, vol. 32, pp. 192–205, 2014, doi:10.1016/j.rser.2014.01.035.

[7] M. M. Ibrahim, N. H. Mostafa, A. H. Osman, and A. Hesham, "Performance analysis of a stand-alone hybrid energy system for desalination unit in Egypt," *Energy Convers. Manag.*, vol. 215, p. 112941, 2020, doi:10.1016/j.enconman.2020.112941.

[8] A. K. Nag and S. Sarkar, "Modeling of hybrid energy system for futuristic energy demand of an Indian rural area and their optimal and sensitivity analysis," *Renew. Energy*, vol. 118, pp. 477–488, 2018, doi:10.1016/j.renene.2017.11.047.

[9] M. O. Atallah, M. A. Farahat, M. E. Lotfy, and T. Senjyu, "Operation of conventional and unconventional energy sources to drive a reverse osmosis desalination plant in Sinai Peninsula, Egypt," *Renew. Energy*, vol. 145, pp. 141–152, 2020, doi:10.1016/j.renene.2019.05.138.

[10] P. P. Kumar, R. S. S. Nuvvula, Md. A. Hossain, *et al.*, "Optimal operation of an integrated hybrid renewable energy system with demand-side management in a rural context," *Energies*, vol. 15, no. 14, pp. 1–50, 2022, doi:10.3390/en15145176.

[11] M. M. Samy, S. Barakat, and H. S. Ramadan, "A flower pollination optimization algorithm for an off-grid PV-fuel cell hybrid renewable system," *Int. J. Hydrogen Energy*, vol. 44, no. 4, pp. 2141–2152, 2019, doi:10.1016/j.ijhydene.2018.05.127.

[12] L. Karimi, L. Abkar, M. Aghajani, and A. Ghassemi, "Technical feasibility comparison of off-grid PV-EDR and PV-RO desalination systems via their energy consumption," *Sep. Purif. Technol.*, vol. 151, pp. 82–94, 2015, doi:10.1016/j.seppur.2015.07.023.

[13] A. A. Setiawan, Y. Zhao, and C. V. Nayar, "Design, economic analysis and environmental considerations of mini-grid hybrid power system with reverse osmosis desalination plant for remote areas," *Renew. Energy*, vol. 34, no. 2, pp. 374–383, 2009, doi:10.1016/j.renene.2008.05.014.

[14] H. Mehrjerdi, "Modeling and optimization of an island water-energy nexus powered by a hybrid solar-wind renewable system," *Energy*, vol. 197, p. 117217, 2020, doi:10.1016/j.energy.2020.117217.

[15] A. Waqar, M. Shahbaz Tanveer, J. Ahmad, M. Aamir, M. Yaqoob, and F. Anwar, "Multi-objective analysis of a CHP plant integrated microgrid in Pakistan," *Energies*, vol. 10, no. 10, p. 1625, 2017, doi:10.3390/en10101625.

[16] J. Yuan, J. Xu, and Y. Wang, "Techno-economic study of a distributed hybrid renewable energy system supplying electrical power and heat for a rural house in China," *IOP Conf. Ser. Earth Environ. Sci.*, vol. 127, p. 012001, 2018, doi:10.1088/1755-1315/127/1/012001.

[17] E. Guelpa, A. Bischi, V. Verda, M. Chertkov, and H. Lund, "Towards future infrastructures for sustainable multi-energy systems: A review," *Energy*, vol. 184, pp. 2–21, 2019, doi:10.1016/j.energy.2019.05.057.

[18] "NASA Prediction Of Worldwide Energy Resources." [Online]. Available: https://power.larc.nasa.gov/

[19] R. Babaei, D. S. Ting, and R. Carriveau, "Feasibility and optimal sizing analysis of stand-alone hybrid energy systems coupled with various battery technologies: A case study of Pelee Island," *Energy Reports*, vol. 8, pp. 4747–4762, 2022, doi:10.1016/j.egyr.2022.03.133.

[20] M. Bagheri, S. Hamid, M. Pakzadmanesh, and C. A. Kennedy, "City-integrated renewable energy design for low-carbon and climate-resilient communities," *Appl. Energy*, vol. 239, pp. 1212–1225, 2019, doi:10.1016/j.apenergy.2019.02.031.

[21] B. K. Das and F. Zaman, "Performance analysis of a PV/Diesel hybrid system for a remote area in Bangladesh: Effects of dispatch strategies, batteries, and generator selection," *Energy*, vol. 169, pp. 263–276, 2019, doi:10.1016/j.energy.2018.12.014.

[22] C. Li, X. Ge, Y. Zheng, *et al.*, "Techno-economic feasibility study of autonomous hybrid wind/PV/battery power system for a household in Urumqi, China," *Energy*, vol. 55, pp. 263–272, 2013, doi:10.1016/j.energy.2013.03.084.

[23] J. Li, P. Liu, and Z. Li, "Optimal design and techno-economic analysis of a hybrid renewable energy system for off-grid power supply and hydrogen production: A case study of West China," *Chem. Eng. Res. Des.*, vol. 177, pp. 604–614, 2022, doi:10.1016/j.cherd.2021.11.014.

[24] R. Babaei, D. S. Ting, and R. Carriveau, "Optimization of hydrogen-producing sustainable island microgrids," *Int. J. Hydrogen Energy*, vol. 47, no. 32, pp. 14375–14392, 2022, doi:10.1016/j.ijhydene.2022.02.187.

[25] M. Hossain, S. Mekhilef, and L. Olatomiwa, "Performance evaluation of a stand-alone PV-wind-diesel-battery hybrid system feasible for a large resort center in South China Sea, Malaysia," *Sustain. Cities Soc.*, vol. 28, pp. 358–366, 2017, doi:10.1016/j.scs.2016.10.008.

[26] A. Cano, P. Arévalo, and F. Jurado, "Energy analysis and techno-economic assessment of a hybrid PV/HKT/BAT system using biomass gasifier: Cuenca-Ecuador case study," *Energy*, vol. 202, 2020, doi:10.1016/j.energy.2020.117727.

[27] M. H. Jahangir, S. A. Mousavi, and M. A. Vaziri Rad, "A techno-economic comparison of a photovoltaic/thermal organic Rankine cycle with several renewable hybrid systems for a residential area in Rayen, Iran," *Energy Convers. Manag.*, vol. 195, pp. 244–261, 2019, doi:10.1016/j.enconman.2019.05.010.

[28] S. K. Singal, Varun, and R. P. Singh, "Rural electrification of a remote island by renewable energy sources," *Renew. Energy*, vol. 32, no. 15, pp. 2491–2501, 2007, doi:10.1016/j.renene.2006.12.013.

Chapter 7

An introduction to the electrification in remote communities located in ecologically sensitive areas: from planning to implementation experience

Cresencio-Silvio Segura-Salas[1]

The universalisation of electricity access is pivotal for human development and aligns with Sustainable Development Goal 7 (SDG 7) of the United Nations. Despite its importance, delivering electricity to remote areas faces numerous challenges, such as sparse populations, limited conventional energy sources, deficient infrastructure, and accessibility issues. Financial institutions like the World Bank have been instrumental in providing energy access through initiatives like rural electrification using domestic photovoltaic systems. Renewable energy technologies offer promise but vary in feasibility across nations due to factors like production and implementation costs. This chapter addresses challenges in achieving universal electricity access in expansive continental regions located in ecologically sensitive areas, proposing planning strategies including conventional distribution system expansion, microgrid establishment, and stand-alone systems. The chapter also discusses difficulties in implementing and maintaining off-grid generation systems, which are crucial for remote area electrification. Recommendations drawn from experiences in Brazil aim to enhance electrification program planning and implementation worldwide.

Keywords: Electrification in remote communities; Distribution system expansion; Photovoltaic systems; Maintenance planning

The universalization of electricity access stands as a significant global concern, serving as a fundamental pillar for human development and aligning with the Sustainable Development Goals (SDGs) outlined by the United Nations. Pursuing these objectives is essential, particularly in alignment with SDG 7, which advocates for clean and affordable energy access for all [1]. However, delivering access to electricity, particularly in remote areas, presents escalating complexities due to sparse human settlements, limitations on conventional energy sources, deficient

---

[1]Power System Division, Lactec Institute – Paraná, Brazil

186 *Clean energy for low-income communities*

urban infrastructure, challenging accessibility, and distance from major consumption centers.

In developing countries, assistance programs from financial institutions like the World Bank have been actively involved in providing energy access to these households [2]. World Bank and the Global Environment Facility have facilitated rural electrification through domestic photovoltaic systems across different countries globally [3]. These initiatives were pursued where extending the distribution systems was deemed economically unviable. As in China [4], Sri Lanka, and Indonesia [5] most projects were integral components of larger endeavors encompassing electric sector reform, rural electrification, and rural development efforts.

The advancement of renewable energy technologies, along with supportive policies promoting energy generation from these sources, yields varied outcomes regarding the feasibility of renewable generation systems across different nations. This variation is conspicuous, with developed Asian countries achieving economic viability as early as the 2000s, benefiting from more accessible production costs of photovoltaic components, for example. Conversely, in regions like Latin America, achieving viability necessitated an additional 15 years of development for the same technology.

Within this context, this chapter delves into the persisting challenges of achieving universal access to electricity in expansive regions situated within continental zones, particularly in ecologically fragile areas. Leveraging study experiences and derived insights, it deliberates on planning and implementation strategies. Concerning planning, the chapter scrutinizes a multi-criteria heuristic approach aimed at catering to these regions, encompassing conventional distribution system expansion, microgrid establishment, and stand-alone systems. These deliberations unfold within environmentally delicate and logistically intricate domains, where a significant portion of the population has limited economic means.

In the second part of the chapter, some difficulties associated with implementing, operating, and maintaining off-grid generation systems are discussed. These obstacles pose long-term challenges to electrification programs in remote areas, which are critical for consumer satisfaction and financial stability.

Finally, readers will find significant recommendations based on experience in conducting electrification studies in a vast, ecologically sensitive remote region of approximately 90,000 km^2 in Brazil. Readers are expected to identify opportunities for enhancing the planning and implementation of their electrification program.

7.1 Socio-environmental concerns in remote electrification programs

This is one of the most significant criteria for electrification planning in the world's last frontier, not only due to the appeal found in SDG 7 but also because in ecologically sensitive remote regions, meeting sustainability requirements is a matter of survival. It is essential that the impact on natural resources be minimal and reversible, and remain distant from the site of electricity generation. This statement certainly advocates for electricity generation in these locations to be clean and not impact local flora and fauna, as well as for the correct disposal of waste generation. Certainly, this does not imply that the production of electricity generation equipment and inputs remains unsustainable; on the contrary, the aim is to procure equipment where the environmental costs of production are minimal and energy efficiency is maximized. This leads to the choice of electricity generation and distribution technology alternatives, which should be made as holistic as possible, considering the entire life cycle of the technology and the environment [6–10]. An example is photovoltaic electricity generation, which is considered a low environmental impact source, but in an area of native forest, it may be the least recommended due to shading or the need for vegetation suppression [11,12].

On the other hand, supplying electricity to remote areas should consider the analysis of socio-environmental impacts and restrictions, considering the environmental scenario in which these areas are situated. Often, they are characterized as extensive areas and/or various legal preservation instruments. Georeferenced analysis of physical, biotic, and socioeconomic aspects is essential for identifying potential impacts and restrictions on electrification alternatives. Unlike energy potential assessment studies, which typically rely on pre-existing data, planning electricity supply for remote regions necessitates field research.

7.1.1 Socio-environmental characterization [13]

First and foremost, the foundation of any socio-environmental analysis lies in the data that comprise the study area, which requires multidisciplinary knowledge about the region of interest. This involves gathering representative data from secondary surveys and direct field observations guided by a detailed work plan specifying activities, methodology, and expected outcomes. A typical example of secondary data collection corresponds to the survey of the physical environment to study the abiotic factors of the study area, with the characterization of the environment in its components of land, water, and air. Some areas to be considered in the study of the physical environment include geology, geomorphology, pedology, mineral resources, seismicity, speleology, hydrogeology, hydrology, water quality, air quality, climatology, acoustics, and landscape [13]. Among the data that must be directly collected are lifestyle, energy needs, future population demand, and complementary measurements to secondary data, such as current water, air, and acoustic quality, for example.

Additionally, according to [14], socio-environmental characterization involves the survey of geospatial data, preferably from official sources, to compose a georeferenced database. The objective of this phase of the study is the identification of relief forms, water bodies, and human settlements, among others, as well as delimiting the area of interest. However, in remote regions, mainly in developing countries, it is common for geospatial data not to be available in the region of interest. In these cases, primary data collection is necessary, often implying the acquisition of satellite images and the vectorization of features of interest. Knowledge of the exact location of access roads, watercourses, and orographic barriers is a determining factor in the choice of technology alternatives for remote electrification. These data can also indicate possible consumer units, whether dispersed or grouped, serving as the main basis for primary data collection.

188 *Clean energy for low-income communities*

Also, in this type of study, it must be considered first that the area of influence may not correspond to the study area. The areas of influence consider the spatial scope of the repercussions or effects of the actions necessary for the implementation and operation of the activity, which will cause modifications in the various physical, biotic, and anthropic factors that characterize its reference environment.

The survey of the biotic environment involves the study of the biotic factors of the study area, characterizing the environment in terms of flora and fauna, excluding the human component. It should incorporate, besides flora and fauna, conservation units and priority conservation areas. In this survey, existing fauna species and vegetation types occurring in the study region should be characterized. It is important to consider the region's seasonality, with dry and rainy seasons.

On the other hand, the anthropic environment represents humans and their relationships with the environment. It encompasses social, economic, and cultural heritage themes. Censuses and social and economic surveys provide information on demographics, occupation, income, education, and various other indicators, but most of the time they serve only as historical data, so they must be complemented by field research. In the socio economic diagnosis, it is imperative to include the historical context of territorial occupation, socioeconomic conditions, traditional communities, agrarian reform settlements, conflicts, health standards, production dynamics, transportation modes, tourism activities, and other relevant insights about the study area.

Cultural heritage encompasses material and immaterial assets, taken individually or collectively, bearing reference to the identity, action, and memory of the different groups forming society. Material assets refer to scientific, artistic, and technological creations; to works, objects, documents, buildings, and other spaces destined for artistic and cultural manifestations, as well as urban complexes and sites of historical, landscape, artistic, archaeological, paleontological, ecological, and scientific value. Immaterial assets refer to forms of expression and ways of creating, making, and living.

7.1.2 Socio-environmental analysis

In this context, in addition to the possible environmental impacts, the identification of the main conflicts to be addressed is considered, focusing on energy needs and perspectives regarding applicable electrification alternatives. Socio-environmental assessments should inform the choice of the alternative to be presented, in a clear and participatory manner, with the involvement of stakeholders in the region (Public Prosecutor's Office, environmental agencies, non-governmental organizations, and local population), considering the compatibility of social interest in electricity supply with environmental conservation in the region.

In an electrification study, the socio-environmental evaluation of technological alternatives is necessary, i.e., the study and assessment of the socio-environmental impacts of each generation source are considered. According to [14] a practical and common method is the definition of socio-environmental indices, following a process of condensing information starting from primary and secondary data, followed by the analysis of these data to produce indicators, and finally the indices.

Indices are understood as the result of combining a set of indicators associated with each other through a pre-established relationship that gives rise to a new and unique value. The use of indices as a representation of synthesized information is widely used by the scientific community, for example, in quality-of-life assessments [15]. In this association, relative values are assigned to each indicator that makes up the index, and the relationship can be established through statistics, analytical formulation, or mathematical ratio calculation.

In extensive and heterogeneous territories, the best way to define environmental indicators is through geospatial data. Spatial analysis of environmental data is based on the discretization, segmentation, and stratification of the region of interest into homogeneous territorial units. The most commonly used tools for environmental analysis are Geographic Information Systems (GIS), which are computer programs that allow the storage, manipulation, analysis, and display of spatially referenced data and are the basis of digital cartography. There are several methods for geospatial analysis applied to environmental issues, classified as qualitative and quantitative. Both qualitative and quantitative methods (based on mathematical or statistical analyses) use mapping that generates boundaries drawn by an artificial polygonal line in space, almost always imprecise, which does not represent the gradual variations found in the field. An alternative to minimize this effect is to try to represent transitional zones when the scale allows it. This does not mean that the integration product is incorrect, but its result should always be interpreted keeping this limitation in mind. For extensive areas, the map algebra method [16] is of interest, since the multivariate analysis method needs to have a physically well-defined territory, with a thematic base of consistent data, a GIS, and suitable statistical software, information that is not always available.

7.2   Multicriteria electrification approaches

A typical case of a remote, vast region with complex logistics and ecological sensitivity is the Pantanal in the Brazilian state of Mato Grosso do Sul, located on the west-central border with Bolivia and Paraguay, as illustrated in Figure 7.1. The region is recognized as a National Heritage by the Brazilian government and by UNESCO (United Nations Educational, Scientific and Cultural Organization) as a Biosphere Reserve and Natural Heritage of Humanity [13]. Its main features include a vast territorial extension of approximately 90,000 km^2 in Brazilian territory, with seasonal and permanent flooded areas. Due to these characteristics, it will be used to describe a methodology for determining the best electrification alternatives for consumers located in these types of remote regions.

An example of a heuristic methodology is presented in Figure 7.2. A methodology of this nature ought to commence by undertaking the characterization of the cohort of consumers devoid of access to electricity. This characterization frequently involves on-site visits and field studies. For this purpose, it is recommended to first create a cartographic database from the vectorization of orbital images. Orbital orthoimages SPOT6&7 with a spatial resolution of 6 m can be acquired for mapping buildings and existing road systems. This mapping is essential as it serves as a basis for the logistical planning of field teams, and these maps can be embedded in mobile applications for the navigation of research teams.

190 Clean energy for low-income communities

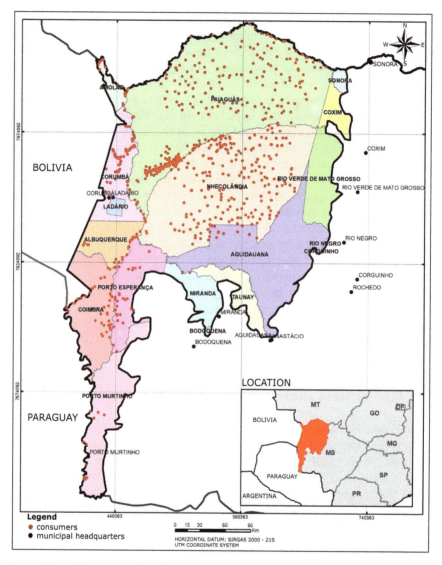

Figure 7.1 The geographical location of the Pantanal within the Brazilian state of Mato Grosso do Sul is of significant academic and scientific interest. In addition to the satellite municipalities within the Pantanal region, it is plausible to scrutinize the consumers who were surveyed during research interviews in the year 2016. During that period, logistical limitations hindered access for all consumers [17].

Field research should include interviews covering the following topics: social, housing, logistical, seasonality, economic activities and production, ownership and habits, energy demand surveys, energy resources, and photographic reports. It is recommended that these topics be surveyed through questionnaires in mobile

An introduction to the electrification in remote communities 191

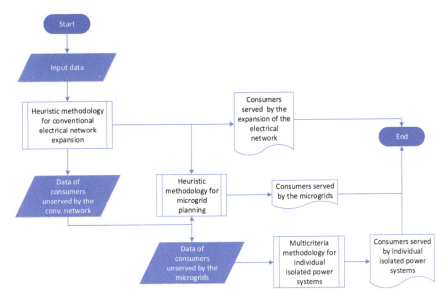

Figure 7.2 Multicriteria heuristic methodology for the selection of alternative electrification in the Pantanal of Mato Grosso do Sul state, Brazil, employing a sequential approach encompassing conventional network expansion, microgrid formation, and the integration of stand-alone systems

applications, which, in addition to collecting questionnaire data, automatically capture the geographic position of the survey, as well as record the routes taken by the teams. The application can be developed to transmit field data to a WEB system when the cell or satellite signal is available, as exemplified in the record presented in Figure 7.3.

A database is thus established, comprising potential users without access to electricity, also containing detailed data on possible roads and pathways to be followed, as well as the respective modes of transportation for people and equipment. Based on the field input data, a distribution system expansion algorithm should determine which consumers can be served by extending boundary feeders, provided they meet environmental, economic, and electrical constraints. Customers ineligible for feeder extension may undergo microgrid formation analysis to serve clusters while remaining isolated customers can be assessed using spatial multicriteria methodology to determine the best alternative energy source. It is observed that this heuristic approach leverages spatial multicriteria methodology to size the microgrid power source. It is noteworthy that this heuristic is generic and can apply to any remote region of vast extension.

For a prospective consumer requiring electrification, the following options may be available: conventional grid, microgrid, and stand-alone systems. Microgrid and standalone system sources, such as solar, wind, diesel, and hybrid configurations, vary based on the area's energy potential. Subsequently, the discussion delves into the examination of socio-environmental variables, which are

192 *Clean energy for low-income communities*

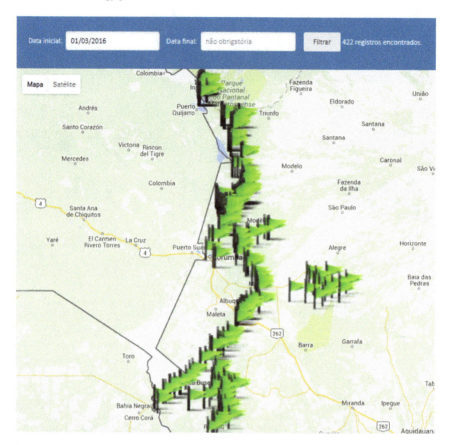

Figure 7.3 Recording of interviews conducted via a web application in the Pantanal municipalities of Corumbá, to be processed in the city of Curitiba, Paraná/Brazil

common across methodologies, followed by their tailored application in specific sections dedicated to each methodology.

7.2.1 Input data—socio-environmental criteria characterization

Incorporating environmental viability requires the consideration of socio-environmental indices (SEIs) and environmental restriction indicators (ERIs). The ERI layer could be created through spatial analysis and overlaying thematic maps in the study area. The addressed themes may include protected areas, priority areas for conservation, ecological-economic zoning, and ecologically sensitive areas. Within the theme of protected areas, layers such as conservation units, buffer zones, ecological corridors, and indigenous lands can be included. So, the ERI map must be based on land use information and the results of fauna and flora field campaigns [18]. Each element in the layer possesses a value or weight, representing

An introduction to the electrification in remote communities   193

an assignment based on specific studies, with weight standardization using scales 1, 3, 5, 7, and 9, like the Saaty scale [19].

The thematic map of ERIs could be developed using a qualitative methodology in geoprocessing tools such as ArcGIS or an open-source QGIS platform. To perform the combination of layers, assigned with different weights, the Combinatorial And operation is usually applied. After all layer intersections, a new range of layers utilized the Standard Deviation method to represent the ERI map. A map, as depicted in Figure 7.4 for the Pantanal case study, could be disseminated into five classes: very low restriction, low, medium, high, and very high [17].

Small-scale solar systems, categorized as standalone systems, are advisable for all environmental restriction classes, while large-scale solar sources are discouraged from the medium restriction onwards, representing forested or flooded areas. Wind systems are not recommended in areas with high environmental restrictions, particularly where conservation units are located. These areas are protected by law and were created to safeguard local flora and fauna. Genset systems are not advisable in areas with medium environmental restrictions, especially in flooded areas, which are more sensitive to this technology due to the use of fossil fuels.

Conversely, the SEI index, comprising environmental indicators related to electric power source alternatives, particularly emphasizes land use, necessitating spatial evaluation. In this application, these indicators derive from two matrices: the first correlating indicators with each power source alternative, and the second linking land use classes with the power source alternative. As a result, Figure 7.5 presents the geospatial SEI index, correlated with solar and wind sources, considering all environmental and land use indicators, which serves for the analysis of power source alternatives in multicriteria methodologies. For instance, in the case of photovoltaic and wind power sources, the environmental impact assessment considers the quantity of panels or turbines deployed. If more panels or turbines occupy a larger area, increasing potential environmental impacts [20,21], then Figure 7.5 shows the delineation between low and high SEI indices based on the scale of renewable generation capacity. Genset systems' environmental impact assessment hinges on annual fuel consumption [22].

7.2.2   Heuristic approach for the expansion of the conventional distribution system in remote areas

Expanding rural distribution systems in areas with difficult access and socio-environmental restrictions poses complex challenges. Logistical and access hurdles escalate service costs, and environmental aspects must be integrated into analysis and modeling. Also, electrical constraints, such as voltage level and short-circuit level in feeders with considerable lengths, can significantly limit the desired extensions. On the other hand, excessive extension can also bring operational challenges and dissatisfaction with power quality among consumers [23], leading to penalties and financial losses for utilities in highly regulated environments.

194 *Clean energy for low-income communities*

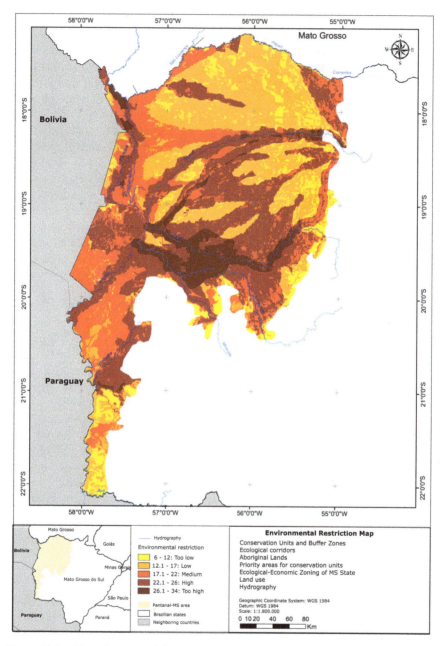

Figure 7.4 *Classification of the environmental constraints index ranges from very low to very high in the Pantanal region of the Brazilian state of Mato Grosso do Sul [17]*

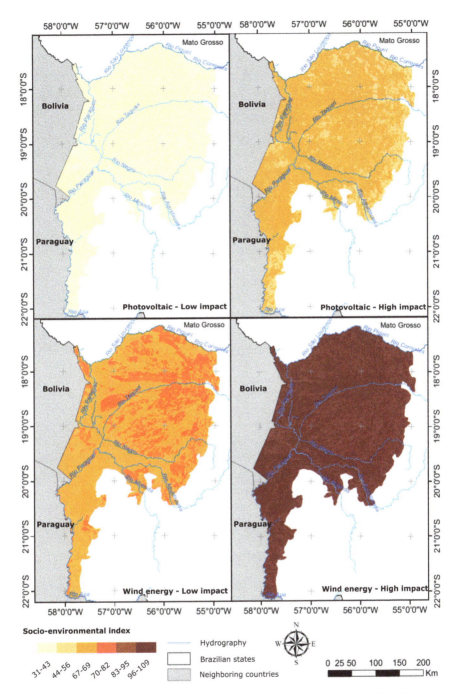

Figure 7.5 Classification of the socio-environmental index ranges from low to high values, for photovoltaic and wind technology in the Pantanal region of the Brazilian state of Mato Grosso do Sul. A low SEI index corresponds to an allocation of 42 m^2 for the PV area and a 1 kW turbine power capacity. Conversely, a high SEI index indicates an allocation range, spanning from 196 to 420 m^2 for PV area and 8 to 10 kW for turbine power capacity [14].

196   *Clean energy for low-income communities*

Typically, a distribution system expansion planning problem entails finding the least costly path while adhering to electrical constraints to serve a consumer from a notable point* in the existing distribution system [24]. In contrast, when the expansion of the distribution system occurs in remote environments, the inclusion of socio-environmental variables becomes necessary. So, owing to the intricacy associated with modeling multidisciplinary and spatial variables, an effective solution to this challenge lies in the heuristic approach. The methodology devised by [17] is heuristic and is delineated into four components: (i) generation of input data, (ii) formation of the semi-connected graph, (iii) ranking of properties, and (iv) evaluation of electrical network expansion.

The methodology's initial phase necessitates geospatial coordinates, monthly energy demands, distances to existing grid points, Euclidean distances between consumers, SEIs, and ERIs along paths to each customer. Additionally, georeferenced data on existing electrical system points and electrical data from the utility's distribution system database are required.

In the second part, the recommendation is to model the problem using graph theory, where vertices represent the geographical coordinates of consumers and notable grid points. Edges denote connections between consumers and grid points and among consumers themselves. At this stage, edges with environmentally restrictive aspects (SEI and ERI) are removed, resulting in a semi-connected graph feasible for traversal.

The third part prioritizes ranking consumers in descending order, from most favorable to least suitable for grid connection, using the Analytic Hierarchy Process method (AHP) [19] or other appropriate multicriteria techniques [25–27]. The ranking considers edge distances, energy consumption, SEIs, and ERIs of connecting edges, along with their respective percentages.

Finally, hypothetical networks (looped feeders) are built for all consumers and edges of the semi-connected graph. These networks are validated through power flow analysis, short-circuit analysis, and extension cost calculations. The algorithm also addresses potential cable reconductoring during electrical studies. Also, an optimization process is conducted to create a minimum-cost expansion, resulting in a hypothetical network optimized for cost, technical feasibility, and socio-environmental factors, to achieve a radial distribution system.

7.2.3 *Microgrid formation methodology*

This kind of methodology aims to establish the largest number of microgrids for consumers not covered by the distribution system expansion approach. The development of the microgrid formation methodology may be rooted in the distribution system expansion problem, as microgrid formation follows the same principle of feeder extension for electricity distribution among users.

According to [17], the microgrid formation planning problem can be subdivided into four optimization subproblems. First, it deals with defining the set of consumers with suitable proximity conditions to potentially form a microgrid, meeting proximity criteria, demand capacity, and logistical constraints. Clustering

---
*Series of structures supporting power distribution lines and equipment.

An introduction to the electrification in remote communities 197

techniques such as K-means [28] and evolutive algorithms [29] can be employed. As a result of this process, besides identifying consumers belonging to potential microgrids, their demand is also known.

Next, the microgrid power source location and size are selected. The site is chosen based on the consumer location with the highest average monthly consumption compared to others in its set, reducing distribution system losses by situating the highest demand near the generation source. The greater the total demand for the microgrid, the greater the capacity of the generation center should be to meet this demand. However, larger generation center capacity results in a larger system size to be installed, often hindering equipment access and transportation, rendering the microgrid unfeasible.

The constructive establishment of the microgrid seeks to determine how consumers within each group will be interconnected and linked to the generation center, along with the optimal electrical configuration for service provision. The configuration of the connection between power sources and loads depends on the number of potential connections between consumers and the generation center. For simplicity, connections can initially be established linearly. This results in the creation of a graph illustrating all conceivable connection possibilities. To design the internal network of the microgrid, readers can employ the distribution system extension methodology outlined in the previous section.

Finally, the selection of generation technology for viable microgrids must be achieved by applying the spatial multicriteria algorithm (described in the next subsection). It is noteworthy that in cases where there are multiple generation centers in a microgrid, the spatial multicriteria algorithm must operate to define the technology of each generation center. However, operational issues must be assessed and valued to ensure that these configurations are advantageous.

7.2.4 Stand-alone systems

Multicriteria analysis technique, which includes decision variables such as social, environmental, financial, logistical, technical, and regulatory aspects, must be included given past instances of widespread use of such systems that returned unsatisfactory results [5,30]. It aims to determine the most appropriate generation technology for individual consumers or groups within microgrids. Notably, it integrates environmental impact variables and socio-environmental constraints as spatial variables.

Typically, a multicriteria model adopts a hierarchical structure [19], as illustrated in Figure 7.6 where criteria, sub-criteria, and alternatives are positioned at various levels.

These levels are divisions within the problem, aiding in the breakdown of a complex issue into manageable parts. Within this framework, all criteria are assessed equally, adhering to the principles of the Analytical Hierarchy Process (AHP) and utilizing Saaty's scale [31] for each alternative under consideration. This assessment can take various forms, including meetings with research teams and experts involved in the project.

Among the five criteria in the model—SEI, ERI, Levelized Cost of Energy (LCOE), regulatory (RGL), and technological (TCN)—the first two must be geospatial, contingent on the geographic coordinates (location) of the electrification site. Figure 7.7 illustrates the data flow for the collection and processing of the five input criteria within the methodology.

198 Clean energy for low-income communities

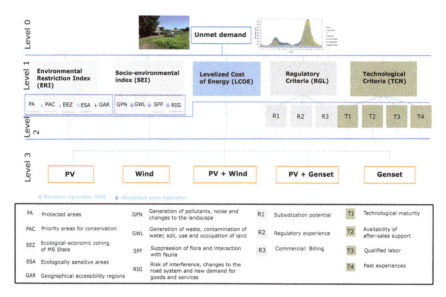

Figure 7.6 *Hierarchical decision-making framework of the spatial multicriteria methodology for determining the suitable generation source alternative for stand-alone and microgrid systems. The criteria ERI and SEI are geospatial. Map algebra can be executed using GIS software platforms such as QGIS and ArcGIS [2].*

7.2.4.1 Cost calculation of stand-alone power generation projects

Capital cost (CAPEX)

Investing in capital assets for stand-alone and individual generation systems can follow various models: system construction and deployment, replacement of components at the end of their lifecycle, retrofitting, or equipment upgrades. The breakdown of construction and deployment costs includes engineering design and project execution, civil works, primary electrical equipment (frequency inverters, charge controllers, photovoltaic modules, micro wind turbines, diesel generators, batteries, among others), additional materials (distribution boxes, wiring, grounding, circuit breakers, metal support profiles, towers, and guy wires for wind turbines), specialized services for transportation, mechanical and electrical assembly, as well as commissioning. Taxes and fees on products and services must also be considered. Capital expenditures depend on the project size, while service costs are contingent upon understanding the logistical and environmental conditions of the installation site, significantly burdening the total CAPEX cost, as this type of project requires specialized firms in the sector.

Operational cost (OPEX)

Operational expenditures denote the financial resources required for project operations. These costs typically encompass operational and maintenance

An introduction to the electrification in remote communities 199

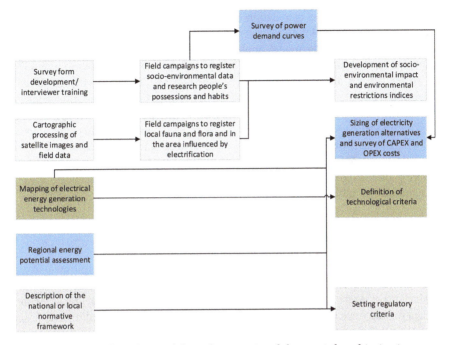

Figure 7.7 Data flow for modeling the criteria of the spatial multicriteria methodology. It is noted that there is a quantitative contribution from the technological and regulatory criteria in the establishment of CAPEX and OPEX, consequently influencing the LCOE, then both criteria were only qualitatively modeled at levels 2 and 1, as shown in Figure 7.6.

activities, including scheduled preventive maintenance actions and their corresponding frequencies, as well as corrective maintenance and fuel expenses, when relevant. The calculation of operation and maintenance (O&M) costs involves aggregating the following elements:

- Operating costs of systems, including fuel expenses for systems with Gensets.
- Cost of preventive maintenance, following specialized guidelines for reference projects, including maintenance actions, periodicity, and the requisite time and cost of skilled labor.
- Costs of corrective maintenance, which can be estimated using a probability distribution model (Weibull) if there is a history of similar data. This is a critical aspect due to the limited availability of data in the literature concerning the failure rates of comparable applications.
- Logistic multiplier factor, contingent upon the installation location of individual systems relative to a base city outside the remote area, and the costs of necessary transportation modes for consumer access.

200   *Clean energy for low-income communities*

The operational cost calculation model should account for projected maintenance actions tabulated with their respective values in man-hours, materials, and frequency. Moreover, the model could integrate a method to address corrective maintenance, enabling probabilistic modeling such as the Weibull distribution [32]. So, corrective maintenance costs can be adjusted by the probability of failure, utilizing the parameters specified for each maintenance table entry. Unfortunately, historical failure data for these systems are scarce and poorly documented.

This model should consider the addition of applicable equipment for standalone systems, as well as their costs and maintenance practices, according to manufacturers' recommendations. The net present value, $NPV_{O\&M}$, is calculated by summing the maintenance costs for each month, based on the frequency stated in the maintenance table and their respective costs, as per (7.1).

$$NPV_{O\&M} = \sum_{m=1}^{2} \frac{\sum_{t=1}^{N} C_m^t}{(1 + d_n)^N} \tag{7.1}$$

The total maintenance costs for each type of maintenance ($m = 1$ preventive, $m = 2$ corrective) are computed for all years t up to the study horizon N and adjusted by the nominal discount rate d_n. The maintenance cost C_m^t is calculated for each year as the sum of each maintenance event occurring in that year, based on its declared interval (fixed interval or according to the Weibull distribution interval for corrective maintenance).

The expense associated with transporting the maintenance team and their equipment is a substantial element within each maintenance cost. The integration of data gathered from field campaigns relies on a modal factor (k), which aggregates the costs of the transportation modes used to travel from the reference location to the target electrification site, the frequency of operational activities performed per year (α), and the distance covered (x), with the project's base city as the origin. Consequently, (7.2) presents the adjusted net present value to encompass the logistical costs of maintenance operations [13].

$$NPV_{O\&M} = \sum_{m=1}^{MT} \frac{\sum_{t=1}^{N} \left[C_m^t + \alpha k x (1 + e)^t \right]}{(1 + d_n)^N} \tag{7.2}$$

It is advisable to include the cost associated with installation logistics as part of the decision-making process for selecting the generation system for each consumer unit. This consideration should account for the distance from the base city and the transportation modes (4 × 4 vehicle, tractor, boat, motorcycle, self-propelled, and airplane) used to reach the consumer and return to the base city. When determining transportation methods, access types can be segregated based on seasonality.

7.2.4.2   Levelized cost of energy

The levelized cost of energy (LCOE) is calculated by dividing the total annualized costs by the annualized energy generated by the system, E_{CU}. Alternatively, it can be expressed as the net present value relationship between generation and costs over the lifespan of the projects. Total costs encompass both the capital costs (CAPEX) and the

An introduction to the electrification in remote communities 201

operational costs (OPEX) in the one-time frame. The LCOE (R$[†]/kWh) offers a concise assessment of the energy generation project's cost performance, enabling comparisons across different projects throughout the entire analysis period, rather than solely focusing on the initial construction and deployment expenses.

Determining the LCOE involves computing the installation cost of the generation systems, encompassing expenses associated with primary equipment, installation materials, civil works, assembly and commissioning services, and transportation costs for installation. Additionally, it requires estimating the replacement CAPEX value, which factors in parts reaching the end of their useful life and replacing defective components according to the corrective action model and the equipment failure estimation model. Depending on local regulations, this expenditure may be classified as OPEX.

Simplistically, the energy accounted for in the LCOE calculation comprises the annualized regulatory consumption of each consumer. This premise presupposes that the size of the generation systems must consistently satisfy the demand. Therefore, the computation of the LCOE for a particular consumer unit, considering the annualized values of CAPEX and OPEX, is expressed by (7.3).

$$LCOE_{CU} = \frac{A_C APEX_{total} + A_O PEX}{E_{CU}} \tag{7.3}$$

7.2.4.3 Regulatory criteria

Existing regulations significantly impact electrification programs in remote regions, as technical and commercial guidelines are often developed based on the electrification of consumers through conventional distribution systems operated by utilities, as in Brazil. Therefore, refining these guidelines to consider all the peculiarities involved in universal service in remote regions is necessary. Before modeling the regulatory criteria, certain aspects need to be analyzed:

- Subsidies for universalization programs (such as the "Light for All" program in Brazil): These funds are typically subsidized by the federal government to finance necessary investments, aiming to avoid affecting tariff moderation in the concession area and to fulfill its social function. However, in the case of the Pantanal region in Mato Grosso do Sul, there is doubt regarding whether most connections in the region could be facilitated through this program. This uncertainty arises because many landowners possess large rural properties with their financial capacity, which contrasts with the financial capacity of potential program consumers, such as rural workers, fishermen, and farmers. Consequently, there is a need to reconsider and enhance universalization subsidies to mitigate any impact on tariff moderation.
- Definition of remote regions for standalone systems: In Brazil, the regulation lacks clarity regarding this definition, suggesting that a remote region must be

---

[†]When Brazilian real (R$) figures are provided in this chapter, they pertain to the year 2016; a factor of 1.75 will be applied to adjust to 2024 R$. 1US$ = 4.96 R$.

202   *Clean energy for low-income communities*

within an existing standalone system. However, this criterion does not apply to the Pantanal region in Mato Grosso do Sul, where the entire surrounding electrical grid is connected to the national interconnected system. Presently, the provision of standalone systems is permitted only in remote regions characterized by consumer dispersion and a lack of economies of scale or in cases involving technical or environmental restrictions preventing conventional network provision [33].

- Monthly guaranteed energy availability: It is imperative to evaluate whether the monthly energy availability stipulated by regulation adequately meets the needs of consumers in the Pantanal region of Mato Grosso do Sul, considering the region's energy potential, generation equipment technology, and consumer consumption profile.
- Deadlines related to compliance with reliability and voltage quality indicators: The established limits and timelines must be evaluated for their feasibility of compliance, considering the unique characteristics of the region, such as transportation challenges, particularly during dry and rainy seasons.
- Inspection, connection, measurement, reading, billing, payment, and reconnection procedures: Evaluate whether the identified flexibilities in service provision through microgrids or stand-alone systems, in comparison to conventional service, are adequate for operating these systems in the Pantanal region of Mato Grosso do Sul. According to [34], 99.1% of the 40,200 stand-alone systems and rural microgrids installed in Brazil from 2009 to 2020 do not employ measurement systems for billing purposes.
- Riverside consumers: The Pantanal region in Mato Grosso do Sul encompasses a notable number of households situated along its riversides, many of which may lack consistent land tenure status, thus rendering them ineligible for service under prevailing regulations.
- Point of coupling: Regulations dictate that the utility must cater to only one delivery point per consumer unit. Nevertheless, the Pantanal region of Mato Grosso do Sul comprises numerous extensive rural estates with multiple worker residences. As per existing regulations, these residences would not receive service from the utility if the connection points were designated at the administrative headquarters.
- Case Studies in Brazil: Most observed cases in Brazil involve the adoption of stand-alone photovoltaic systems. Few applications have been made with microgrids. Auctions for service contracting to supply isolated systems to remote regions still lack relevant information for assessing their application and results. Therefore, the case of the Pantanal, depending on the type of solution to be adopted, will not have many references in other cases in Brazil, necessitating greater attention to regulatory adjustments to enable service provision to the region.

Considering the aforementioned points, it is imperative to establish and formulate regulatory criteria to ensure the feasibility of generation alternatives while minimizing regulatory adjustments and societal impacts. Three matrices have been outlined as regulatory criteria: the potential for regulatory subsidies, the regulatory

An introduction to the electrification in remote communities 203

Table 7.1 Assessment of the sub-criterion technological maturity (T1) for the photovoltaic alternative

Electricity generation alternative	Energy consumption range (kWh/month)						
	<60	80	160	300	600	1,000	>5,000
PVBAT[‡]	no	no	no	no	no	no	no
EOLBAT[§]		no	no	no	no	no	no
Genset	rec	rec	rec	nec	nec	nec	nec
PVBAT + Genset				nec	nec	nec	nec

experience of reference projects in Brazil, and the necessity for measurement systems for billing. These sub-criteria are qualitative, with the potential for regulatory subsidies assessed across projects with "low," "medium," and "high" probabilities of receiving subsidies. Regulatory experience in Brazil is graded as "low" (no regulatory application in Brazil), "medium" (a limited number of projects with regulatory application in Brazil), and "high" (a notable number of projects with regulatory application in Brazil). The requirement for measurement for billing is evaluated as "no" (projects that do not necessitate measurement), "rec" (projects for which consumption measurement is recommended), and "nec" (projects that require consumption measurement). For example, Table 7.1 provides a qualitative assessment of the sub-criteria measurement systems for billing according to the power consumption range in the context of the Pantanal.

Renewable-based systems are less dependent on billing systems. However, the battery degradation cost is a function of energy consumption (cycles of charge and discharge), because in general, more cycles or deep discharges impact the lifetime of the batteries, and therefore more analysis is needed to understand the billing system necessity for these alternatives.

7.2.4.4 Technological criteria

The technological criterion seeks to evaluate the technical characteristics of alternative electricity generation options under review. It is crucial to consider the adoption of mature technologies due to the pressing need for electrification, along with the exploration of innovative solutions through new technologies. This criterion may be subdivided into four or more sub-criteria, depending on the required level of detail. For instance, factors such as technological maturity, availability of post-sales support, skilled labor force, and past experiences can be considered.

The aim is to assess the viability of the project's commercial operation, considering both international and national competitors in the market. The potential for product obsolescence must also be factored in. These sub-criteria aim to ensure that

---

[‡]PVBAT means stand-alone photovoltaic system with a battery energy storage system.
[§]EOLBAT means stand-alone wind power system with a battery energy storage system.

204　*Clean energy for low-income communities*

Table 7.2　Assessment of the sub-criterion technological maturity (T1) for the photovoltaic alternative

Indicators					
Commercial operation	**Competitors in the market (national and international)**	**Possibility of product obsolescence**		**Weight**	**Weight Grade according to the Saaty scale [19] (Indicator × Weight)**
More than 30 years	– High	– Low	– 3		0
Between 15 and 30 years	1 Medium	– Medium	1 2		4
Less than 15 years	– Low	1 High	– 1		1
Rating on the Saaty scale for the sub-criterion (T1):					5

the project benefits from authorized technical support and commercial representation within the national territory, providing swift assistance and support, including spare part delivery. Equally critical is evaluating the capability to train professionals to operate these systems and the experience of industry professionals with such technologies. This criterion should be appraised by consultants or specialists in operation and maintenance within the organization. Table 7.2 offers an example of assessing the technological maturity sub-criterion regarding the photovoltaic system.

In Table 7.2, it is possible to verify that the weights for composing the scores aim to qualify requirements at three levels of fulfillment. If the alternative meets the higher levels, its final weight will be a maximum of 9 (with this method, the minimum score is 3 and the maximum is 9).

7.2.5　Application example in the case of the Pantanal, Mato Grosso do Sul, Brazil

The simulations presented herein were carried out for the Pantanal region in Mato Grosso do Sul, situated within the Brazilian state of Mato Grosso do Sul, as depicted in Figure 7.4. The objective is to provide electricity to 1,824 consumers residing near the conventional grid, estimated via satellite imagery, and to 1,975 consumers within the interior of the Pantanal, primarily identified through on-site visits. The interior area of the Pantanal is demarcated by the dashed border, positioned 5 km outside of the conventional distribution system. The drive to electrify this remote region is rooted in the mandate of National Law 10.438/2002, which ensures universal access to electricity through subsidized programs or supported rate options.

The typologies for the reference projects, considered alternatives in the multicriteria algorithm, include photovoltaic systems with batteries, wind turbines with batteries, combinations of photovoltaic and wind turbines with batteries, and optimized hybrid photovoltaic systems with batteries and Genset.

An introduction to the electrification in remote communities 205

Table 7.3 Minimum design requirements for stand-alone systems according to Brazilian regulatory resolution ANEEL 493/2012 [38]

Monthly Availability (kWh/month)	Reference consumption (Wh/day)	Minimum autonomy (hours)	Minimum Power Demand (W)
13	435	48	250
20	670		250
30	1,000		500
45	1,500		700
60	2,000		1,000
80	2,650		1,250

These typologies were determined based on research into the energy potential of the Pantanal region in Mato Grosso do Sul, utilizing data from the National Environmental Data Organization (SONDA) project under the Solar and Wind Energy Resource Assessment (SWERA) initiative [35], the Brazilian Solar Energy Atlas/Brazilian Wind Potential Atlas (CRESESB) [36], and the Brazilian Bioenergy Atlas [37]. Through this data, the limited potential of biomass and hydro resources in the Pantanal was confirmed, prompting the consideration of technologies with immediate application potential. Solar and wind potentials were examined from both spatial and temporal perspectives, influencing the design of the alternative projects.

For the sizing of the candidate projects, a classification of consumer units was performed according to the monthly availability given in Table 7.3 and defined in Brazilian regulations in 2012 [9][**].

Characterizing the consumption profile of consumer units that have never had access to electricity poses a significant challenge, encompassing daily, seasonal, and annual patterns. Indeed, inherent uncertainties exist regarding medium- and long-term demand growth. The availability of electricity is anticipated to spur the emergence of new applications aimed at enhancing the quality of life for the population. It is important to emphasize that electrification programs should consistently be coupled with energy efficiency initiatives in usage and distribution.

To tackle these challenges, Brazilian regulations stipulate that consumers served by microgrids and stand-alone systems can request load increases after at least 1 year has passed since the initial connection or since the last load increase. Moreover, the design of these systems must include a planning horizon of 5 years, although the project can be contractually dimensioned considering an asset life of 25 years.

For sizing the generation systems, a load profile was established using typical load curves for each consumer type, derived from the ownership and habits survey (OHS) conducted with rural consumers during the third cycle of tariff revisions

---

[**]In 2021 [33], the Brazilian regulatory framework underwent a recent revision, resulting in the establishment of a lower limit of guaranteed monthly availability of 45 kWh per consumer unit.

206 *Clean energy for low-income communities*

Table 7.4 Classification by consumption range following field research of 1,975 interviewed consumers in the interior of the Pantanal, Mato Grosso do Sul/Brazil [18]

Energy consumption range (kWh/month)	Consumer units	Total monthly consumption (kWh/month)	Average monthly consumption (kWh/month)
0–80	266	10,157	38
81–160	690	87,811	127
161–300	569	113,941	200
301–600	390	162,466	417
601–1,000	43	31,856	741
1,001–5,000	17	24,382	1,434

near the Pantanal area. Monthly consumption was determined based on data obtained from the OHSs collected during field visits, considering the current and future needs of the interviewees.

The distribution of the 1,975 consumers in the Pantanal interior based on their consumption range is detailed in Table 7.4. After analyzing the results of the energy consumption calculations, it became evident that it was necessary to consider availability ranges of 160, 300, 600, 1,000, 5,000, and 10,000 kWh/month for the design of the reference projects.

The reference projects underwent simulation using the well-known Homer Energy Software [39]. Cost data for equipment were obtained from manufacturers, specialized retailers, and integrators. The aim of studying the reference projects was to ensure that the selected project could effectively meet the monthly and daily energy availability requirements of the load in the most economically viable manner.

7.2.5.1 Distribution system expansion

The feasibility of expanding the rural distribution system, consisting of 19 feeders, in the Pantanal region of Mato Grosso do Sul was assessed. Figure 7.8 illustrates the Pantanal polygon under study, showing the feeder layout along with land use data that delineate the environmental constraints (ERI).

The simulations only considered the CAPEX cost of building new branches, covering the infrastructure expenses of the network and distribution transformers. The CAPEX limit for feeder extension was established at R$ 38,000 per extended branch, reflecting the CAPEX cost of an 80 kWh/month standard stand-alone PV system with battery. This reference value of 80 kWh/month represented the maximum standard value for meeting isolated system requirements, as outlined in [38].

Another constraint for feeder extension is the voltage level at the point of service, defined as a minimum voltage limit for service at 0.93 p.u., defined in [40]. The scope of the conventional distribution system extension methodology was established for a maximum extension radius of 25 km from the feeders, encompassing all 1,824 consumers near the conventional distribution system and only

An introduction to the electrification in remote communities 207

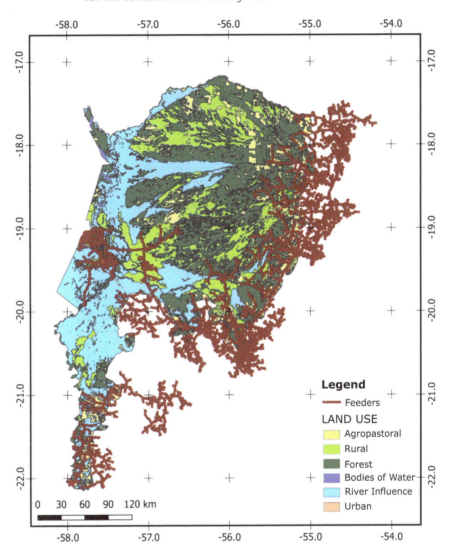

Figure 7.8 Pantanal polygon under study, showing the feeder layout along with land use data that delineate the environmental constraints (ERI) for expanding the conventional distribution system [18]

181 interior consumers planned for distribution system service. The remaining interior consumers proceed directly to the microgrid formation methodology.

As a result of the simulation, a service coverage percentage of 69.43% of the projected consumers was obtained. All belong to the region near the distribution system. Table 7.5 provides a breakdown of consumer services per feeder.

Figure 7.9 depicts the construction expenses for each service branch (feeder extension) relative to the distance of the served consumer from the existing

Table 7.5 The outcome of distribution system expansion in the Pantanal demonstrated an average coverage of 66% [18]. There are no extensions of the feeders beyond the borders of the current distribution system.

Feeder code	Near to the grid – projected	Near to the grid – served	Inland – projected	Inland – served	Served [%]
AQU02	35	35	0	0	100
AQU03	5	5	0	0	100
AQU04	93	45	19	0	40
AQU05	77	55	0	0	71
BON51	60	47	0	0	78
COR01	363	352	2	0	96
COR03	9	4	11	0	20
COR52	441	321	75	0	62
COX03	43	19	3	0	41
MIR01	32	27	0	0	84
MIR51	252	199	29	0	71
MIR52	25	24	0	0	96
PGO01	16	7	1	0	41
PMU01	80	33	5	0	39
PMU02	37	23	3	0	58
RVE01	13	10	0	0	77
RVE51	189	157	17	0	76
RVE52	42	19	12	0	35
SON01	12	10	4	0	63
TOTAL	1,824	1,392	181	0	69

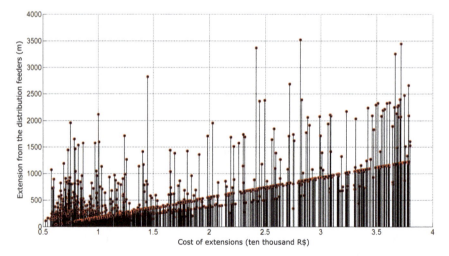

Figure 7.9 Construction expenses for each service branch (feeder extension) relative to the distance of the served consumer from the existing distribution systems. There was not much success in the spatial expansion of the feeders [18].

distribution system feeders. The figure indicates adherence to the R$ 38,000.00 CAPEX limit. The series of points forming a linear pattern along the figure represent the initial service points established, reaching a maximum distance from the distribution systems of approximately 1.25 km for the CAPEX limit. However, some served consumers are situated farther from the existing distribution systems. In such cases, while the initial service point incurs higher costs, subsequent services benefit from lower investment expenses due to the proximity of other consumers, thereby reducing the overall service cost for this group.

Therefore, the solution incurred a total cost of R$ 21,112,946.46 to cater to the 1,392 prospective consumers, amounting to a monthly consumption of 264,400 kWh/month (189 kWh/month/unit). This equates to a relative cost of R$ 79.85 per kWh, resulting in an average service cost of R$ 15,167.35 per consumer.

Figure 7.10 shows the results of the distribution system expansion methodology across all feeders. It illustrates the construction of new branches highlighted in orange, including a detailed view of the expansion in an agro-pastoral region. Despite serving numerous consumers, there was no substantial progress in expanding the feeders within the Pantanal, primarily due to environmental, technical, and economic constraints, in this order. With the completion of this expansion, the 432 consumers near the conventional distribution systems and the 1,975 interior consumers transition to the microgrid formation methodology stage.

7.2.5.2 Microgrids formation

Like the methodology used for expanding the conventional distribution system, if a new microgrid branch traverses an area with significant environmental impact, it is excluded from consideration. However, this exclusion may lead to the fragmentation of a prospective microgrid. For example, if a microgrid is intended to serve 10 consumers, its division may vary based on the environmental features of the area. It could be split into one microgrid serving four consumers and another serving six consumers, or it could be subdivided into a larger number of microgrids.

At this stage, 118 microgrids were established, with 109 serving consumers located in the interior and nine serving consumers near the distribution system. This resulted in a construction cost, excluding the generation source, of R$ 1,118,408 to serve 372 consumers, as outlined in Table 7.6, with a total consumption of 79,320 kWh/month (213 kWh/month/unit). Finally, the generation technologies for the microgrid sources need to be defined. This is achieved using a spatial multi-criteria methodology for stand-alone systems. Essentially, temporal demand data are aggregated, and electrical losses are computed.

In the spatial multicriteria methodology, decision-makers need to assess the weighting of the criteria under analysis. As illustrated in Figure 7.6, the method encompasses five criteria (level 2), and the allocation of weights for these criteria is outlined in Table 7.7. It should be noted that nearly 50% of the weight corresponds to environmental criteria.

The sub-criteria of the technological and regulatory criteria (level 3) are also determined using the pairwise comparison matrix, following the AHP methodology [19].

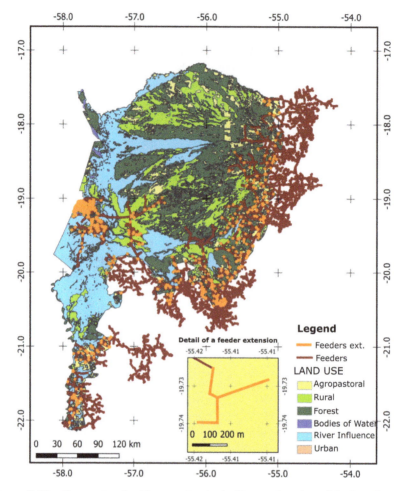

Figure 7.10 Conventional grid expansion in the Pantanal region of the Brazilian state of Mato Grosso do Sul. Results indicate an extension of feeders by less than 4 km. The Ladário area, located in the municipality of Corumbá to the west, experienced significant benefits due to the proximity of existing feeders to unattended consumers without encountering technical or environmental constraints.

Table 7.6 Aggregation of consumers for the establishment of microgrids in the Pantanal region of Mato Grosso do Sul. The results demonstrate the service provision to 372 consumers through 118 microgrids [18].

Microgrids	Projected	Served	[%] Served
Inland	1,975	349	17.7
Near to the distribution systems	432	23	5.3
Total	2407	372	23.0

An introduction to the electrification in remote communities   211

Table 7.7   Weights of the ranking criteria for alternative electricity generation technologies through pairwise comparison [18]

Criteria	Environmental Restriction Index (ERI)	Socio-environmental index (SEI)	Levelized cost of Energy (LCOE)	Regulatory criteria (RGL)	Technological criteria (TCN)
Weight (%)	19	28	35	13	5

Table 7.8   LCOE (R$/kWh) analysis of microgrids with one GC [18]

LCOE (95% percentile)	1 GC	2GC	3GC	1GC[††]
PVBAT	26.1	13.2	10.4	15.4
Genset	9.8	7.4	–	9.8
PVBAT + Genset	13.5	8.5	–	–

The nine microgrids near the conventional grid have only one generation center, comprising seven PV-Battery and two Gensets. In contrast, the microgrids in the interior consist of 98 PV-Battery, seven Genset, and four PV-Battery + Genset hybrids.

The algorithm calculates the LCOE for each microgrid based on the number of generators in the system. Consequently, Table 7.8 provides statistical LCOE values for one generation center (GC), two GC, and three GC for consumers within the interior of the Pantanal.

Figure 7.11 illustrates the spatial allocation of the reference projects for each generation center. It is worth noting that Genset solutions were exclusively selected at the peripheries of the Pantanal.

Ultimately, the 1,626 consumers from the interior and 409 consumers near the distribution systems, who are still without service from the distribution systems or microgrid, need to be assessed using the spatial multicriteria methodology for isolated electricity generation systems.

7.2.5.3   Stand-alone systems

Table 7.9 presents the number of consumers still awaiting coverage according to the guaranteed monthly availability. It is noteworthy that electricity demands equal to or exceeding 1,000 kWh/month signify individuals with favorable economic conditions. According to existing regulations in Brazil, they may opt to finance a portion of the required investment.

The spatial multicriteria simulation yielded 1,621 PV-battery systems, two Genset systems, and three hybrid systems (PV + Battery + Genset) for the 1,626 interior consumers. Similarly, for the 409 consumers near the distribution

---

[††]Microgrid near the conventional distribution system.

212 *Clean energy for low-income communities*

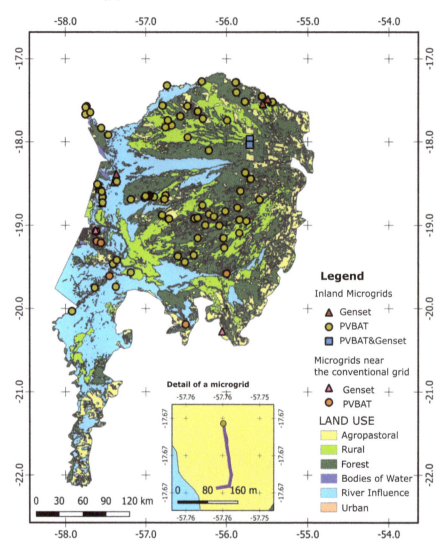

Figure 7.11 The map of the Pantanal with the spatial distribution of microgrids. The presence of small-scale photovoltaic systems with batteries appears to be less pronounced near the distribution system frontier [18].

systems, the simulation resulted in 402 PV battery systems, three Genset systems, and four hybrid systems.

In the multicriteria approach, as discussed earlier, the LCOE criterion is computed for all reference projects to be assessed in the pairwise comparison matrix. Thus, each reference project solution has an associated LCOE value. Tables 7.10 and 7.11 present the statistical LCOE values of the technologies of the reference projects identified as the best option for interior and near-grid consumers, respectively.

An introduction to the electrification in remote communities 213

Table 7.9 Distribution of the remaining unserved consumers based on monthly regulatory consumption after the feeder extension and microgrid formation [18]

	Number of consumers	
Monthly availability (kWh/month)	**Within the interior**	**Next to the distribution systems**
13	59	0
20	15	0
30	19	0
45	38	0
60	29	0
80	52	246
160	558	0
300	481	149
600	324	11
1,000	36	3
5,000	15	0
Total	1,626	409

Table 7.10 Examination of the Levelized Cost of Electricity (LCOE) (R$/kWh) of stand-alone systems for consumers situated in the interior of the Pantanal [18]

LCOE	**Minimum**	**Mean**	**Maximum**	**95th Percentile**
PVBAT	8.47	35.24	5.628.30[‡‡]	61
Genset	8.07	9.02	9.97	10
PVBAT +Genset	9.89	13.39	17.26	17

Table 7.11 Examination of the Levelized Cost of Electricity (LCOE) (R$/kWh) of stand-alone systems for consumers situated near the conventional distribution network [18]

LCOE	**Minimum**	**Mean**	**Maximum**	**95th Percentile**
PVBAT	6.87	11.84	16.35	16
Genset	6.15	6.56	6.89	7
PVBAT +Genset	7.90	8.87	9.63	10

[‡‡]The maximum value refers to a few stand-alone systems with a generation capacity of 10,000 kWh/month located in the interior of the Pantanal with high logistic CAPEX and OPEX.

214 *Clean energy for low-income communities*

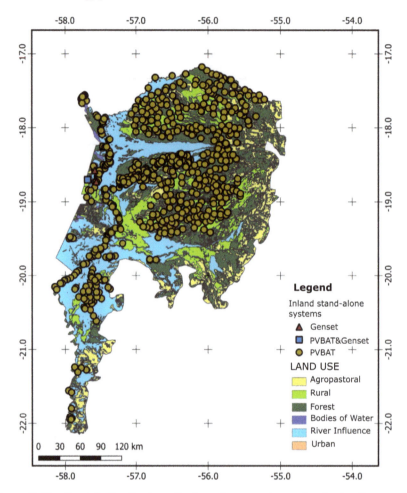

Figure 7.12 Spatial distribution of selected generation technologies for stand-alone systems catering to consumers located within the interior part of the Pantanal. The prevalence of small-scale photovoltaic systems with accompanying batteries is evident [18].

Figure 7.12 describes the spatial distribution of the 1,626 stand-alone systems resulting from the spatial multicriteria method in the interior of the Pantanal, whereas Figure 7.13 illustrates the same for the 409 isolated systems near the distribution systems.

Although the levelized cost of energy for fossil-based systems may be more advantageous (Tables 7.10 and 7.11), the results predominantly favor photovoltaic systems with battery storage. This preference stems from the weighted criteria matrix, which assigns a 28% weight to the SEI index, as illustrated in Table 7.7. Furthermore, fossil-based systems, reliant on a constant supply of fuel, particularly in areas where logistics pose challenges for transportation, are less favored due to the

An introduction to the electrification in remote communities 215

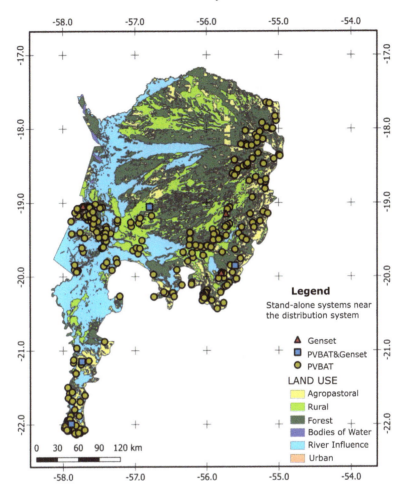

Figure 7.13 *Spatial distribution of chosen generation technologies for stand-alone systems servicing consumers adjacent to the existing distribution systems. The prevalence of small-scale photovoltaic systems with accompanying batteries is readily apparent [18].*

ERI index (19% of weight). From a regulatory perspective, the advantage lies in the absence of a metering system for billing purposes. While the technological criterion holds less weight in decision-making, there is a tendency to favor more established technologies in the country, such as small-scale diesel generators and PV systems.

It is observed that the heuristic strategy, combined with multicriteria techniques, possesses the capacity to provide a solution to a complex multidisciplinary problem, thereby enabling the involvement of various scientific domains. One advantage of this approach is the ability to assess and analyze the partial results of each stage, which will undoubtedly influence subsequent steps. Thus, readers can incorporate and model variables of their interest within the multicriteria

216 *Clean energy for low-income communities*

framework, emphasizing the inclusion of multidisciplinary teams in the modeling and assessment of these criteria. It is noteworthy that in the approach employed, the goal of achieving the global optimal solution is less emphasized compared to finding solutions of excellent quality that meet participants' expectations. Hence, the method outlined in this chapter segment may apply to electrification contexts in vast, remote regions worldwide.

7.3 Electrification model and experiences in prototype implementation

In this section, we outline technical recommendations for the effective deployment of electrification using individual generation systems powered by photovoltaic sources and electrochemical batteries. This guidance has been derived from R&D programs conducted in Brazil, particularly in the Pantanal regions, over a 4-year period from 2017 to 2021. It is particularly relevant in regions where photovoltaic technology is the predominant choice for electricity provision across vast areas.

Battery technology definition: Conducting field and laboratory studies is crucial to determining the most suitable battery technology for the generation system. In this context, Lithium Iron Phosphate (LiFePO4 or LFP) battery technology has emerged as advantageous due to its superior performance throughout its lifespan compared to OPzS[‡‡] and PbC[***] batteries. Moreover, it offers ease of transportation, requires less frequent maintenance, and meets safety, reliability, compatibility, and environmental adaptability requirements.

Implementation of O&M optimization algorithm: This algorithm should tackle the resource sizing challenge of O&M for maintaining isolated PVBAT systems in remote areas. It aims to define all O&M activities, encompassing preventive maintenance, required teams, inventory, and fleets for service execution, along with estimating associated costs and predicting corrective maintenance.

Battery bank lifespan management algorithm: Developing a battery bank lifespan management algorithm is advisable. The objective of this algorithm is to identify the most appropriate operational parameters, which should be programmed into charge controllers at maintenance intervals. This process aims to optimize battery lifespan and maintain efficiency over time.

These recommendations are intended to ensure the effectiveness, safety, and sustainability of photovoltaic generation systems with batteries in remote regions, such as the Pantanal in the state of Mato Grosso do Sul. The insights presented here stem from the R&D project [17], which involved real-world experiments with 23 technologically diverse prototypes across the Pantanal, thus leaving behind a wealth of knowledge and data from the Pantanal I R&D project [18].

[‡‡]OPzS: Vented stationary lead-acid Battery for stationary applications.
[***]PbC: Refers to an advanced lead-carbon battery characterized by the inclusion of carbon nanoparticles in the negative plate composition.

7.3.1 Experiences in electrochemical battery application

Within stand-alone photovoltaic generation systems incorporating batteries, the battery bank stands out as one of the most critical components, potentially constituting over 60% of the total system cost throughout its operational lifespan. Deep battery discharges exert a substantial impact on their durability, frequently resulting in premature replacements [13]. Battery lifespan is influenced by operational variables like state-of-charge limits, electrical current, and standby durations, alongside environmental factors, with temperature being particularly crucial.

Given the intricate nature of these interactions, employing mathematical models is imperative to forecast battery degradation rates based on construction and operational data. In [17], it was noted that precise adjustments to voltage and state-of-charge limits can extend battery life by up to 20% to 30%, contingent upon the technology used. The study encompassed seven battery manufacturers, including three for lead-acid, two for lead-carbon, and two for LiFePO4 batteries. Notably, international production was notable for lead-carbon and lithium-ion battery technologies, underscoring the array of options available in the market and the importance of selecting the most suitable technology for each specific application.

The LiFePO4 batteries exhibited operational characteristics that did not compromise safety, either during handling or operation. Nevertheless, it is imperative to acknowledge that batteries of any technology have the potential to release gases over their lifecycle, which, if ignited by sparks, can lead to fires. Therefore, it is advisable to devise separate enclosures for both batteries and power conditioning components. The PVBAT systems were devised with grounding mechanisms following the stipulations outlined in the Brazilian low-voltage installation standard NBR 5410, which aligns with the international benchmark IEC 364—Low-voltage electrical installations. This meticulous attention to grounding significantly enhances the safety of electrical setups while mitigating the hazards of electric shock and fire incidents.

The design of battery cabinets for small-scale off-grid systems deliberately omitted the installation of forced ventilation systems. Instead, these cabinets were insulated thermally using materials like aluminum fiber. They feature a single opening section located at the bottom to facilitate access to the collection tray, thus minimizing the risk of potential leaks. During operational periods, it was noted that the LiFePO4 batteries performed optimally within the designated "normal" temperature ranges. For example, the operational temperature spectrum of the Pylontech battery spans from 0°C to 50°C, while that of the Unicoba battery ranges from 0°C to 45°C. Figure 7.14 illustrates the temperature distributions of both the environment and the batteries over 87 days, from April to August 2020, documented within a PVBAT system situated in a Pantanal locale. Throughout the morning hours, the cabinet effectively maintained the internal battery temperature at a relatively stable level, with the median approaching 25°C, surpassing the ambient temperature. As the day progressed into the afternoon, the disparity between the temperatures diminished, signifying a slower escalation of the internal cabinet temperature compared to the ambient conditions. Nonetheless, the internal battery temperature consistently remained higher than the ambient temperature.

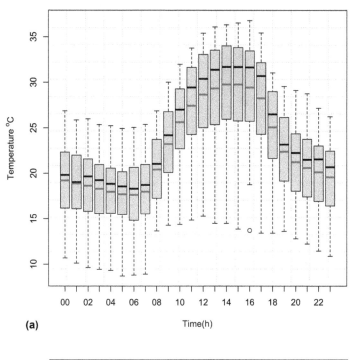

(a)

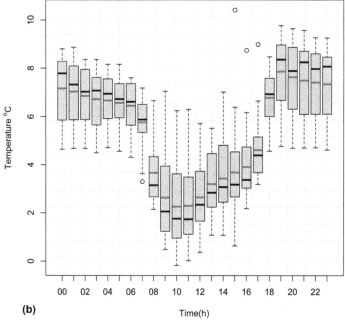

(b)

Figure 7.14 Hourly temperature distribution throughout April to August 2020, documented within a PVBAT system at Cascavel Farm in the Pantanal. (a) Ambient temperature. (b) Average minimum and maximum temperature within the battery cores relative to ambient temperature. The mean temperature is black, and the median one is red [41].

An introduction to the electrification in remote communities 219

These observations imply that thermal insulation proved effective in maintaining the batteries within an acceptable temperature threshold, even amidst elevated ambient temperatures. This aspect is paramount to guaranteeing the efficiency and longevity of batteries in off-grid systems.

Information from various sources indicates that the optimal temperature range for lithium-ion batteries falls between 15°C and 35°C. Maintaining batteries within this range ensures heightened performance with moderate degradation, as corroborated by studies from [42,43]. Furthermore, experimental findings by [44] suggest that the charging characteristics of lithium-ion batteries remain largely unaffected by charging temperatures, spanning from 20°C to 40°C. However, the charge/discharge efficiency of the battery experiences significant deterioration when temperatures dip below 20°C. Thus, the temperatures documented in the aforementioned studies are deemed conducive to the efficient operation of lithium-ion batteries.

The research detailed in [17] underscores the advantages of maintaining lithium-ion batteries at a maximum state of charge below 100%, ideally around 85%. Moreover, a discharge depth below 20% is found to be conducive to prolonging battery life. Consequently, controllers operating in voltage control mode should account for voltage curves concerning the state of charge and discharge of the batteries.

Regarding carbon lead batteries, the study observes that the maximum state of charge typically hovers around 80%, with a minimum threshold of 20%. However, these batteries necessitate periodic full recharges, approximately every 15 days. For lead-acid batteries, the third technology under investigation, two distinct strategies have emerged. In settings with continuous monitoring, lifespan optimization is achievable through a management approach involving a partial state of charge and periodic full recharges. Conversely, in locales with limited monitoring, such as the Pantanal, it is prudent to adopt a more conservative strategy, operating close to full charge, avoiding overcharges, and implementing periodic equalizations, preferably every 6 months or following instances where the batteries are deeply discharged.

7.3.2 On the implementation stage

The deployment of photovoltaic systems in regions with critical environmental characteristics, such as the Pantanal, demands particular attention owing to logistical challenges and the imperative to uphold equipment integrity.

Exploring national and international technical standards for stand-alone systems and adhering to equipment manufacturers' recommendations are paramount. National standards not only ensure equipment standardization but also establish operational procedures that accommodate loads in both direct current and alternating current. However, the absence of specific national standards for individual generation systems and isolated microgrids, both at the construction and equipment levels, has impeded the validation of equipment technical specifications. Relying solely on general standards for low-voltage electrical systems fails to address the unique characteristics of small, isolated systems, thus proving inadequate. Consequently, adaptation based on experience and the advice of designers becomes necessary, potentially resulting in inconsistencies in protection and safety measures.

220   *Clean energy for low-income communities*

In the civil engineering domain, conducting soil sampling studies is imperative to prevent potential failures in the structural foundations of PVBAT systems. It is advisable to thoroughly investigate the soil through sample collection and referencing soil type maps of the region, which enables a stratified analysis of results. During installations in the Pantanal, moist, sandy, or clayey soil types were frequently encountered, posing challenges to the installation of anchor-type foundation structures.

Due to the potential risk of explosion or gas emissions associated with the examined battery technologies, it is advisable not to co-locate them with power conditioning equipment within cabinets. Throughout a 12-month operational period, only one instance of swelling was observed in a LiFePO4 battery pack. However, it is important to note that cell swelling may occur due to excessive deep discharge and might not necessarily indicate a manufacturing flaw [45].

Special consideration must be given to the technical specifications of inverters and charge controllers about the designated grounding system, irrespective of whether they operate on direct or alternating current. This is essential for ensuring electrical protection protocols and the safety of individuals within the photovoltaic installation, commencing with this initial definition.

The challenge of implementing effective planning primarily stems from climatic factors and the difficulty of accessing remote areas. These elements also hinder the transportation of equipment, underscoring the importance of devising strategies to optimize transportation and system assembly. Given the poor road conditions in remote regions, as illustrated in Figure 7.15, transporting panels and batteries requires careful handling to prevent damage to photovoltaic panels and fluid leaks. Such precautions must adhere to manufacturer recommendations and national regulations governing product and personnel transportation. Equipment dimensions and weight should be meticulously evaluated, particularly during river transit. For example, PbC and OPzS battery banks pose greater transportation difficulties compared to lithium battery banks, largely due to their bulk and weight.

Moreover, transportation is impeded by a significant number of gates, deep shallows, and frequent groundings that have resulted in vehicle damage. In addition, consider the extensive distances between consumers, averaging 30 km, and nocturnal wildlife. Owing to these challenges and with the objective of prioritizing team safety, transportation should ideally occur in the morning whenever feasible.

Difficulties also emerge in procuring the services of engineering and construction firms equipped to operate in remote regions. Extreme climates may compromise the adaptability of skilled personnel, notwithstanding their residency in major metropolitan areas. Furthermore, the restricted availability of market equipment for PVBATs and the absence of specific standardization for individual generation systems at the onset of technological introduction necessitate attention [41].

In terms of customer service, regulatory constraints on electrical energy may fail to meet the expectations of all consumers. In the Pantanal, for example, it was common to encounter various capacities of old Diesel generators, which were intermittently used, especially during the night. Ranchers possessed Genset units with capacities ranging from 10 to 50 kVA, used seasonally. For these individuals, PVBAT systems were undoubtedly underestimated, but they constituted less than

An introduction to the electrification in remote communities   221

Figure 7.15 (a) stuck pickup truck in the northern Nhecolândia region of Pantanal during equipment transport mission. (b) 60 kWh/month PVBAT system commissioning at the border of the southern Paraguay river [17].

222 *Clean energy for low-income communities*

3% of the Pantanal population. These consumers may opt to install systems that surpass regulatory standards at their own expense. Nevertheless, for nearly 72% of the population earning less than three minimum wages (R\$ 2,640.00 in 2015 reais and R\$ 4,236 in 2024), such systems are likely to suffice. This assertion is supported by tested PVBATs with a capacity of 160 kWh/month, compliant with current regulations [18].

7.3.3 *Regarding the operations and maintenance plan*

Drawing from past experiences, the operation and maintenance of individual systems or microgrids situated in remote regions are pivotal for the success of electrification initiatives in these areas. Community engagement in such regions must be meticulously upheld to secure active support for field activities. However, numerous projects have neglected this aspect, as evidenced in Peru, Chile, and Ecuador in South America [46], leading to outcomes that failed to meet desired objectives in previous instances.

Undoubtedly, the principal obstacle resides in accessing these remote sites to carry out preventive and corrective maintenance, owing to logistical limitations. In the context of the Pantanal, for example, the development of a methodology was imperative to plan and manage the envisaged total of installations, approaching 2,000 units, to quantify and refine operational expenditures. Such quantification plays a pivotal role in ascertaining the requisite financial resources and the mechanisms by which these resources will be secured, whether through subsidies or via the concession's tariff structure.

Given the uncertainties prevailing in a region characterized by limited infrastructure, notably prone to seasonal droughts and floods, the utilization of multispectral and radar orbital imagery facilitated the identification of flood areas for each month of the year within the Pantanal region. The Normalized Difference Water Index (NDWI) served as a fundamental tool for this purpose. Moreover, comprehensive maps delineating primary and secondary road networks, trails, and waterways were compiled, and tailored to the distinct seasonal patterns of the area. Owing to the widespread distribution of consumers across a 90,000 km^2 expanse, diverse categorizations were essential at the municipal level, enabling the strategic coordination of sequential temporal and spatial installation and maintenance endeavors. Figure 7.16 illustrates the depiction of drought and flood dynamics alongside the implemented categorizations.

A planning algorithm for O&M resources can be developed by considering consumer data, including consumer IDs, associated group IDs, IDs of nearby cities with primary bases, installation dates, and guaranteed energy delivery. Technical equipment specifications and logistical factors, such as seasonal flooding patterns indicating monthly flood occurrences per group, along with routes between nearby cities targeting the most accessible consumer within each group and routes among consumers within the same group, can be incorporated into the algorithm, forming a comprehensive framework.

Of equal significance, the algorithm must accommodate O&M requirements, which vary based on guaranteed energy and generator equipment class. These

An introduction to the electrification in remote communities 223

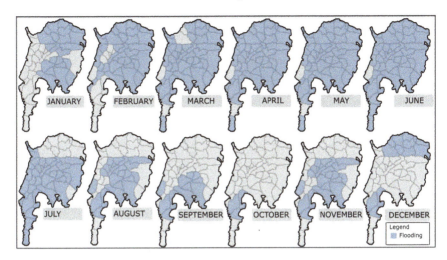

Figure 7.16 Seasonal dynamics of flooding in the Pantanal. Typically, flooding initiates in December from the northern area, reaching its zenith in April and May. The recession phase commences in July, culminating in the driest period in October when natural wildfires are prevalent in the region [17].

requirements involve task lists, task frequencies, the associated workforce, replacement materials, and personal protective equipment. The algorithm can be programmed to conduct calculations annually, enabling field data to inform future equipment wear data, such as actual failure rates. Due to the absence of such data, the author relies on manufacturer-provided data.

Therefore, with such data constraints, an optimization algorithm can select the most appropriate time and resources to realize O&M planning activities, avoiding, for example, flooded groups, but minimizing the interval between preventive maintenance. As an example of this result, the map in Figure 7.17 depicts the outcome of the optimization algorithm, which minimizes travel time variation and mode of transport, selecting the optimal route (shown as the green dashed line) among all available roads (grey line) and waterways (blue line). In addition, the algorithm could estimate the necessary vehicle fleet (ferries, speedboats, tractors, pickup trucks, and quad bikes) and determine team sizes for field and administrative roles, and inventory levels, based on planned schedule O&M activities. Various scenarios, reflecting local labor contributions from stand-alone system users or different battery lifetimes or maintenance timelines, can be employed to assess sensitivities in the total cost.

As highlighted in [47], scant literature exists regarding O&M planning in remote region electrification programs. Reference [48] introduced a methodology for devising maintenance systems and estimating costs for photovoltaic rural electrification, employing a mixed integer linear programming model and a rule-based expert system. The authors calibrated and validated the model within the

224 *Clean energy for low-income communities*

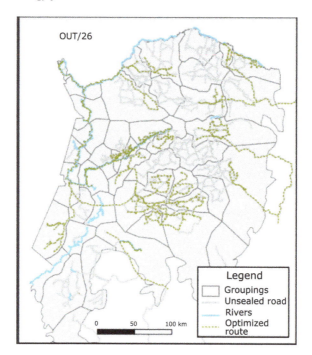

Figure 7.17 Map of modal and routes planning in October/2026. It can be observed that no O&M activities were planned in the southern part of Pantanal for that month, according to the seasonal dynamics of flooding in the Pantanal [47].

Morocco electrification program; however, it lacks restructuring and applicability to vast and ecologically sensitive regions due to its sole reliance on operational costs. Reference [49] exemplifies the development of a comprehensive O&M planning algorithm suitable for any region characterized by a high dispersion of consumers in remote and ecologically sensitive areas.

Also, readers can find valuable O&M recommendations for general PV applications in references [50–52]. We highlight the potential of infrared thermographic diagnosis to detect and classify PV module failures that can be implemented with the aid of drone vehicles.

7.4 Summary

In response to the challenges of servicing consumers situated in remote areas, the initial segment of this chapter delineated the development and implementation of a multidisciplinary decision-making framework aimed at evaluating and selecting the optimal electrification alternatives. This framework is predicated upon the utilization of field data and geospatial analysis of critical decision variables, including socio-environmental factors. Drawing upon insights gleaned from the Pantanal case

An introduction to the electrification in remote communities   225

study, several key observations emerged: (1) the efficacy of individual photovoltaic generation systems over diesel-powered or hybrid alternatives, attributed to the adverse environmental and logistical ramifications associated with diesel generators; (2) the inherent limitations in expanding the distribution systems, primarily due to environmental constraints within the region and, secondarily, owing to economic and technical-electrical constraints; (3) the establishment of microgrids, with exceptions for naturally clustered consumer populations, such as riverside communities, wherein environmental constraints and the widespread dispersal of consumers present additional challenges.

Also, the utilization of geospatial data to discern consumer locations is underscored as crucial for integrating into the analysis the factors that impact operational and investment costs, along with the operational feasibility of the proposed solutions. Additionally, the difficulties associated with gathering field data, stemming from accessibility issues for consumers in the region, are emphasized.

The subsequent section of the chapter furnishes recommendations on the implementation, operation, and maintenance of electrification initiatives, with a particular focus on battery technologies and O&M planning. The text underscores that the outcomes were derived from adherence to general Brazilian standards and the consideration of distinct climatic and environmental factors. Consequently, for the adoption of certain methodologies delineated herein in alternative locales, a thorough evaluation of the region's idiosyncrasies and the technical regulations of the host country, along with environmental and transportation considerations, is imperative. Moreover, the pursuit of solution standardization is advocated to realize economies of scale and bolster operational efficiency.

It is strongly advocated that electrification programs incorporate O&M planning methodologies akin to those observed in the Pantanal case study. This planning framework facilitates the determination of the requisite resources essential for the sustained operation of the electrification program over the medium and long term. Furthermore, it offers valuable insights to inform decision-making processes for local distributors and governmental or private entities overseeing incentives.

As Brazil's intensive remote electrification programs experience a significant resurgence, largely propelled by the decreasing costs of PV and battery storage technologies, it becomes crucial to underscore the importance of establishing comprehensive databases on equipment and system-level performance. Such databases are essential for enhancing the accuracy of resource planning outcomes. As an example, the authors notice no failure rate history currently available for batteries (even for lead-acid) in stand-alone system applications.

Given the critical nature of these systems, which lack backup sources, it is imperative for regulators to set achievable targets for utilities and operators, aligning with current technological advancements. Encouraging the adoption of low-cost monitoring technologies, alongside data-driven AI solutions, can optimize O&M costs and improve consumer satisfaction. Consequently, in numerous remote locations worldwide where satellite signal costs are prohibitive, low power wide area networks (LPWANs), specifically long-term evolution for machines (LTE-M) are essential.

References

[1] Bonsuk Koo, B., Lewis, J., Portale, E., *et al.*: *Measuring Energy Access A Guide to Collecting Data Using "The Core Questions on Household Energy Use,"* LSMS GUIDEBOOK Prepared by The World Bank and the World Health Organization (WHO) (2021).

[2] Hafner, M., Luciani, G.: *The Palgrave Handbook of International Energy Economics*, Palgrave Macmillan, London, 2022, 1st edn.

[3] Martinot, E., Cabraal, A., and Mathur, S.: "World Bank/GEF solar home system projects: experiences and lessons learned 1993–2000," *Renewable and Sustainable Energy Reviews*, 2001, 5(1), pp. 38–57.

[4] Do Gouvello, C., and Song, Y.: "Renewable energy development in China," In *Energy Study*. World Bank (2021). https://doi.org/doi:10.1596/36666.

[5] Sovacool, B.K.: "Success and failure in the political economy of solar electrification: lessons from World Bank Solar Home System (SHS) projects in Sri Lanka and Indonesia," *Energy Policy*, 2018, 123, pp. 482–493.

[6] Lassio, J.G., Magrini, A., and Branco, D.C.: "Life cycle-based sustainability indicators for electricity generation: a systematic review and a proposal for assessments in Brazil," *Journal of Cleaner Production*, 2021, 311, pp. 127568.

[7] Parascanu, M.M., Kaltschmitt, M., Rödl, A., Soreanu, G., and Sánchez-Silva, L.: "Life cycle assessment of electricity generation from combustion and gasification of biomass in Mexico," *Sustainable Production and Consumption*, 2021, 27, 72–85.

[8] Ozturk, M., Dincer, I.: "Life cycle assessment of hydrogen-based electricity generation in place of conventional fuels for residential buildings," *International Journal of Hydrogen Energy*, 2020, 45(50), 26536–26544.

[9] Mahmud, M.A.P., Huda, N., Farjana, S.H., and Lang, C.: "Life-cycle impact assessment of renewable electricity generation systems in the United States,"*Renewable Energy*, 2020, 151, 1028–1045.

[10] Wang, Y., Pan, Z., Zhang, W., Borhani, T.N., Li, R., and Zhang, Z.: "Life cycle assessment of combustion-based electricity generation technologies integrated with carbon capture and storage: a review," *Environmental Research*, 2022, 207, 112219.

[11] Rios, R., Duarte, S.: "Selection of ideal sites for the development of large-scale solar photovoltaic projects through analytical hierarchical process – geographic information systems (AHP-GIS) in Peru," *Renewable and Sustainable Energy Reviews*, 2021, 149, 111310.

[12] Zambrano-Asanza, S., Quiros-Tortos, J., and Franco, J.F.: "Optimal site selection for photovoltaic power plants using a GIS-based multi-criteria decision making and spatial overlay with electric load," *Renewable and Sustainable Energy Reviews*, 2021, 143, 110853.

[13] Segura-Salas, C.-S., Souza da Silveira, L.H.: "Eletrificação de regiões remotas: estudo de alternativas e aplicação no Pantanal Sul-mato-grossense [Electrification of remote areas: study of alternatives and application in the Pantanal Sul-mato-grossense]" (Margem da Palavra/Urutau, 2017, 1st edn.).

[14] Blanc, G.F.C., Ferronato, E.C.P., Santos, J.J.S., *et al.*: "Multicriteria environmental analysis for choosing alternative sources of electricity in isolated areas: the case of the Pantanal, Brazil," *Impact Assessment and Project Appraisal*, 2019, 37(6), pp. 471–479.

[15] Baird, C.M.: *Social Indicators: Statistics, Trends and Policy Development*, Nova Science Pub Inc., Hauppauge, NY, 2011, 1st edn.

[16] Krivoruchko, K.: *Spatial Statistical Data Analysis for GIS Users*, Esri, 2011, 1st edn.

[17] Segura-Salas, C.S.: "Report: Avaliação de tecnologias de armazenamento de energia e de soluções de gerenciamento da operação e manutenção para aplicação em sistemas isolados no Pantanal Sul-mato-grossense, no estado do Mato Grosso do Sul [Assessment of energy storage technologies and operation and maintenance management solutions for implementation in isolated systems in the South Pantanal of Mato Grosso do Sul, within the state of Mato Grosso do Sul - Final Report PD-00404-1609/2016]" (2021).

[18] Segura-Salas, C.S.: "Report: Elaboração de Metodologia de Suprimento de Energia Elétrica a Sistemas Isolados no Pantanal Sul-mato-grossense [Development of Methodology for Supplying Electricity to Isolated Systems in the South Pantanal of Mato Grosso do Sul—Final Report PD-0404-1502/2015]" (2016).

[19] Saaty, T.L.: *The Analytical Hierarchy Process, 1981*, pammc.areeo.ac.ir, 1981.

[20] Rahman, A., Farrok, O., and Haque, M.M.: "Environmental impact of renewable energy source based electrical power plants: solar, wind, hydroelectric, biomass, geothermal, tidal, ocean, and osmotic," *Renewable and Sustainable Energy Reviews*, 2022, 161, p. 112279.

[21] Hamed, T.A., Alshare, A.: "Environmental impact of solar and wind energy—a review," *Journal of Sustainable Development of Energy, Water and Environment Systems*, 2022, 10(2), p. 1090387.

[22] Lloyd, A.C., Cackette, T.A.: "Diesel engines: environmental impact and control," *Journal of the Air & Waste Management Association*, 2001, 51(6), pp. 809–847.

[23] Fan, W., Kang, N., Hebner, R., and Feng, X.: "Islanding detection in rural distribution systems," *Energies (Basel)*, 2020, 13(20), 5503.

[24] Franco, J.F., Rider, M.J., and Romero, R.: "A mixed-integer quadratically-constrained programming model for the distribution system expansion planning," *International Journal of Electrical Power & Energy Systems*, 2014, 62, pp. 265–272.

[25] Ilbahar, E., Cebi, S., and Kahraman, C.: "A state-of-the-art review on multi-attribute renewable energy decision making," *Energy Strategy Reviews*, 2019, 25, pp. 18–33.

[26] Shao, M., Han, Z., Sun, J., Xiao, C., Zhang, S., Zhao, Y.: "A review of multi-criteria decision making applications for renewable energy site selection," *Renewable Energy*, 2020, 157, pp. 377–403.

[27] Navarro, I.J., Yepes, V., and Marti, J.V: "A review of multicriteria assessment techniques applied to sustainable infrastructure design," *Advances in Civil Engineering*, 2019, 2019, p. 6134803.

228 *Clean energy for low-income communities*

[28] Miraftabzadeh, S.M., Colombo, C.G., Longo, M., and Foiadelli, F.: "K-means and alternative clustering methods in modern power systems," *IEEE Access*, 2023, 11, pp. 119596–119633.

[29] Rosenberg, M., Fletcher, J., Reynolds, M., French, T., and While, L.: "Identifying isolated microgrids in rural areas: an evolutionary algorithm approach for a graph clustering problem," in *2019 IEEE Congress on Evolutionary Computation (CEC)* (2019), pp. 2498–2505

[30] van Els, R.H., de Souza Vianna, J.N., and Brasil, A.C.P.: "The Brazilian experience of rural electrification in the Amazon with decentralized generation – the need to change the paradigm from electrification to development," *Renewable and Sustainable Energy Reviews*, 2012, 16(3), pp. 1450–1461.

[31] Saaty, R.W.: "The analytic hierarchy process—what it is and how it is used," *Mathematical Modelling*, 1987, 9(3–5), pp. 161–176.

[32] Weibull, W.: "A statistical distribution function of wide applicability," *Transactions of the American Society of Mechanical Engineers - Journal of Applied Mechanics*, 1955, 18, pp. 293–297.

[33] Agência Nacional de Energia Elétrica: "Resolução Normativa ANEEL No 1. 000, de 7 de dezembro de 2021—[Regulatory Resolution ANEEL No. 1,000, dated December 7, 2021]" (2021)

[34] Agência Nacional de Energia Elétrica: "Cadastro de atendimento aos sistemas intermitentes e Isolados—[National Agency of Electric Energy. Registration of intermittent and isolated systems]" (2021).

[35] Pereira, E.B.: *Solar and Wind Energy Resource Assessment*, Digital Press, 2000.

[36] CRESESB: "Potencial Solar—SunData," http://www.cresesb.cepel.br/index. php?section=sundata

[37] Coelho, S., Monteiro, M., and Karniol, M.: "Atlas de Bioenergia do Brasil. Projeto Fortalecimento Institucional do CENBIO" (2012).

[38] Agência Nacional de Energia Elétrica: "Resolução Normativa No 493, De 5 De Junho De 2012" (2012).

[39] Givler, T.: "Using HOMER software, NREL's micropower optimization model, to explore the role of gen-sets in small solar power systems: case study: SRI Lanka," Golden, Colo.: National Renewable Energy Laboratory, [2005], 2005.

[40] Agência Nacional de Energia Elétrica: "Módulo 8: Qualidade da Energia Elétrica."

[41] Segura-Salas, C.S., Silva, K.A., de Matos Gonçalves, A.M., do Nascimento, H.H.S.: "Off-grid photovoltaic systems implementation for electrification of remote areas: experiences and lessons learned in the Pantanal Sul-Mato-Grossense region of Brazil," *Brazilian Archives of Biology and Technology*, 2023, 66, pp. 1–16.

[42] Ahmad Pesaran, S.S., Kim, G.-H.: "Addressing the impact of temperature extremes on large format li-ion batteries for vehicle applications," in *Proceedings of the 30th International Battery Seminar* (2013).

[43] Sayfutdinov, T., Vorobev, P.: "Optimal utilization strategy of the LiFePO4 battery storage," *Applied Energy*, 2022, 316, p. 119080.

[44] Lu, Z., Yu, X.L., Wei, L.C., *et al.*: "A comprehensive experimental study on temperature-dependent performance of lithium-ion battery," *Applied Thermal Engineering*, 2019, 158, p. 113800.

[45] Chen, Y., Kang, Y., Zhao, Y., *et al.*: "A review of lithium-ion battery safety concerns: the issues, strategies, and testing standards," *Journal of Energy Chemistry*, 2021, 59, pp. 83–99.

[46] Feron, S., Cordero, R.R., andLabbe, F.: "Rural electrification efforts based on off-grid photovoltaic systems in the Andean region: Comparative assessment of their sustainability," *Sustainability*, 2017, 9(10), pp. 1–23.

[47] dos Santos Pereira, G.M., Weigert, G.R., Macedo, P.L., *et al.*: "Quasi-dynamic operation and maintenance plan for photovoltaic systems in remote areas: the framework of Pantanal-MS," *Renewable Energy*, 2022, 181, pp. 404–416.

[48] León, J., Martín-Campo, F.J., Ortuño, M.T., Vitoriano, B., Carrasco, L.M., Narvarte, L.: "A methodology for designing electrification programs for remote areas," *Central European Journal of Operations Research*, 2020, 28 (4), pp. 1265–1290.

[49] dos Santos Pereira, G.M., Weigert, G.R., Morais, P.S., *et al.*: "Transport route planning for operation and maintenance of off-grid photovoltaic energy systems in the Pantanal of Mato Grosso do Sul," in *2020 IEEE PES Transmission & Distribution Conference and Exhibition—Latin America (T&D LA)* (2020), pp. 1–6.

[50] Bosman, L.B., Leon-Salas, W.D., Hutzel, W., and Soto, E.A.: "PV system predictive maintenance: challenges, current approaches, and opportunities," *Energies*, 2020, 13, p. 1398.

[51] Hernández-Callejo, L., Gallardo-Saavedra, S., Alonso-Gómez, V.: "A review of photovoltaic systems: design, operation and maintenance," *Solar Energy*, 2019, 188, pp. 426–440.

[52] Osmani, K., Haddad, A., Lemenand, T., Castanier, B., and Ramadan, M.: "A review on maintenance strategies for PV systems," *Science of the Total Environment*, 2020, 746, p. 141753.

Chapter 8

Enhancing solar insolation in agricultural greenhouses by adjusting its orientation and shape

Gurpreet Khanuja[1], Rajeev Ruparathna[1] and David S-K. Ting[1]

A controlled environment greenhouse requires a large amount of heating during the winter months, which is conventionally supplied by environmentally damaging fossil fuels. To lessen the detrimental effect of fossil fuels on the environment, it is beneficial to use clean solar energy for heating these greenhouses. This paper aims to enhance solar insolation in a greenhouse located in Toronto, Ontario by manipulating greenhouse orientations, roof inclinations, and greenhouse shapes. Different greenhouse models were designed on SketchUp software and then simulated in TRNSYS software to determine the pattern of solar insolation available on different greenhouse models. Greenhouse orientation considered for this study included east-west orientation, north-south orientation, and distinct angles between these orientations. Different roof inclinations of 15°, 30°, 45°, and 60° were examined to observe the pattern of solar insolation availability on the greenhouse roofs. Further to this, typical shapes of a greenhouse (i.e., even, uneven, vinery, semi-circular, elliptical or arch, single span, and quonset) were also investigated to determine solar insolation on greenhouse surfaces.

The simulation results indicated that a 30° North of East greenhouse orientation maximizes solar radiation availability. Changing greenhouse orientation has a negligible impact on solar insolation availability, and the maximum percentage increase in solar insolation that can be achieved by changing the greenhouse orientation is only 0.6%. Solar insolation availability on the south roof and the south wall is the highest as compared to the other greenhouse roofs and walls. It can be further improved by changing the south-roof inclination of an even-span greenhouse from 15° to 60°. Furthermore, an east-west-oriented single-span greenhouse receives the maximum solar radiation, followed by an uneven-span greenhouse for a complete year. This investigation will benefit by capturing more

---

[1]Turbulence & Energy Lab, University of Windsor, Canada

232   *Clean energy for low-income communities*

solar insolation falling on a greenhouse surface, which can be harnessed to increase the inside air temperature of a greenhouse.

Keywords: Solar heating; Greenhouse orientations; Roof inclinations; Greenhouse shapes; Solar simulations

8.1   Introduction

Agriculture is a primary industry in many countries. It is an essential part of the food supply. Recently, United Nations (UN) projected that the human population would expand from roughly 8 billion today to 9.7 billion in 2050 and 10.4 billion by 2100 [1]. As the human population continues to grow rapidly, there is a need to increase the food supply to cater to the increasing demand for food. Food and Agriculture Organization (FAO) has estimated that to ensure food security in 2050, food production should increase by 35% [2]. This need for rapid expansion of agriculture can have a harmful impact on arable land since the quality of the land will degrade due to frequent use. United Nations Food and Agriculture Organization (UN-FAO) revealed that by 2050, arable land per person is projected to decrease to one-third of the amount available in 1970 [3]. Therefore, conventional farming will not be sufficient to support global food demand. Agriculture is ranked as the fifth-largest contributor to greenhouse gas (GHG) emissions. It accounts for 10% of the total national emissions, with an estimated 69 megatons of CO_2 equivalent being emitted. Throughout 1990–2021, these emissions have witnessed an increase from 49 megatons to 69 megatons of CO_2 equivalent [4]. Conventional agriculture is also sensitive to climate change [5,6]. It is estimated that for every 1 °C increase in atmospheric temperature, 10% of the arable land where we now grow food crops will be lost [7]. Numerous developed and underdeveloped nations are struggling with conventional agriculture due to unfavorable weather conditions stemming from their topographical limitations. Thus, the agricultural industry is actively pursuing scientific advancements to facilitate the production of fresh crops in close proximity to urban areas, which not only promotes the health and well-being of individuals but also supports the cultivation of local foods [8,9]. Recently, there has been a resurgence of interest in urban agriculture in many Organizations for Economic Co-operation and Development (OECD) countries, where new advancements, agro-architecture, environmental controls, phenomics, and automation have been employed to grow food in urban areas commercially [6]. Urban Agriculture has multiple advantages, such as providing crops throughout the year at a reasonable cost, quality, and freshness. It also aims to provide food security for the growing urban population [7], environmental sustainability, a lower carbon footprint of food production [10], and chemical-free food with no risks of pests and diseases [6]. Urban agriculture has been defined as an "industry that produces, processes, and markets food, on land and water dispersed throughout urban and peri-urban areas" [11]. Urban agriculture includes both conditioned agriculture as well as unconditioned agriculture in urban areas.

One of the major subsets of urban agriculture is controlled environment agriculture [12]. Controlled Environmental Agriculture (CEA) is a form of indoor farming that provides a regulated environment inside an energy-efficient greenhouse for growing crops. CEA provides conditioned growing spaces that provide control over environmental parameters such as inside air temperature, relative humidity, CO_2 concentrations, supplemental lighting, inside air velocity, etc. [12].

A controlled environment greenhouse requires heating during the night and winter months and cooling during the day and summer months. This is because the internal temperature of the greenhouse is not favorable for crop cultivation. Moreover, plants even require supplemental lighting when natural sunlight is not available. Consequently, maintaining a controlled environment inside the greenhouse constitutes very high operating costs for cold countries like Canada. In a recent study done for a greenhouse located in Quebec, the cost of installing a greenhouse approximates $450,000 to $500,000 depending upon the size and advancements considered in the facility, whereas the operating cost varies from $200,000 to $250,000 [13]. Another study conducted for a greenhouse in Southern California revealed that heating, ventilation, air conditioning, and dehumidification systems account for approximately 56% of the total operating cost [14]. This cost is further distributed between heating (33%), ventilation (5%), and cooling (18%) respectively [14,15]. Rorabaugh *et al.* [16] additionally demonstrated that the heating cost of northern greenhouses, such as in Canada, can be from 75% to 85% of the total operating cost. This total cost can vary based on the geographical location, facility type, automation, environmental control systems, and crop type [17]. The aforementioned operational cost associated with CEA is widely recognized as a significant hindrance in the agricultural industry and academia. Therefore, this study aims to explore energy-efficient methods for reducing the heating energy demand in these types of greenhouses by enhancing the amount of solar insolation available on different surfaces of the greenhouse. CEA can be successfully employed in low-income communities by using renewable energy for its operation. This will help to produce cheap crops locally and ensure food security. By minimizing the operational energy demand, the total CEA energy demand could be supplied by using renewable energy sources.

High heating costs can be reduced by utilizing clean and freely available solar energy [18]. This could be achieved by using a passive solar greenhouse, which operates entirely on the stored radiant energy from the sun. The function of a passive solar greenhouse is to store excessive solar energy during the daytime and use this stored energy during the nighttime to maintain a higher interior temperature than the ambient temperature. The interior air temperature of a passive solar greenhouse also depends upon ambient air temperature, amount, and duration of solar radiation intensity, transmitted solar radiation inside the greenhouse, overall heat transfer coefficient, covering material, wind speed and direction, and the type of crop grown [19]. In high northern latitudes, heating a greenhouse for about 8 months of the year is essential to ensure the growth and development of crops growing therein [20]. Therefore, to have enough heating energy available for almost a year, either supplemental heating is required or energy-efficient methods

234   *Clean energy for low-income communities*

can be implemented to capture more solar energy. Even though solar energy is abundantly available, it is important to understand the availability of solar radiation in a greenhouse throughout the year. A study shows that the thermal contribution of solar energy in Montreal varies from approximately 13% to 54% every month. During the coldest period of the year in this region, the direct contribution of solar energy for heating purposes fluctuated from 13% in the December and January months to roughly 20% in the November and February months [21]. Consequently, it is essential to optimize the availability of solar radiation for the year. This includes gaining more solar radiation during the winter months and less solar radiation during the summer months [22]. An increase in solar radiation availability inside the greenhouse will also increase the rate of photosynthesis for plants [18,21]. It is anticipated that solar radiation can be varied by changing greenhouse orientations and shapes [18]. Solar radiation availability also differs by changing the inclination of walls and roofs. Hence, the main objective of this paper is to analyze the solar radiation availability on different surfaces of a greenhouse. The study analyzes the impact of greenhouse orientation, roof inclinations, and greenhouse shapes on the accessibility of solar insolation. The findings of this study will inform the greenhouse orientation that maximizes the solar radiation for the given location and the solar radiation availability on different walls and roofs of a greenhouse. It will also inform the effect of different roof inclinations and the greenhouse shape, which help in increasing solar radiation.

8.2   Literature review

High operating costs will always be a barrier to the success of CEA facilities [17]. Hence, many studies have been conducted to identify energy-efficient measures for reducing the operating costs of a greenhouse. One of the primary constraints that greatly affects the design process of these CEA facilities revolves around the absence of specific and well-defined codes or standards that can be utilized to effectively design and construct these CEA facilities. However, different researchers tried to study greenhouses with different orientations and shapes. The available works of literature related to energy-efficient parameters were specific to locations close to the equator, especially in Asia and Europe. Mobtaker *et al.* [18] investigated how shapes can impact the energy consumption of a greenhouse in Tabriz, Iran. The result showed that the additional energy requirement to maintain the temperature desirable for the plant's growth was lowest in the east-west oriented single-span greenhouse with a north brick wall. Ali [19] developed a model for analyzing the effect of different orientations of greenhouses (most suitable for all year-round applications) on solar radiation availability. An east-west orientation received more solar radiation in January month and less solar radiation in July month, with small differences in received solar radiation during the 2 months. Gupta *et al.* [22] studied the energy-efficient greenhouse under cold climatic conditions in Northern India. Simulation results indicated that an arch-shaped greenhouse required 2.6% and 4.2% less heating as compared to gable and quonset shapes. An east-west oriented arch greenhouse required 2% less heating as compared to a north-south oriented one. Lawand *et al.* [21] designed and tested a greenhouse for colder regions. The greenhouse was oriented on an east-west axis, with the south-facing roof being transparent, and the inclined north-facing wall being insulated with a reflective cover on the interior face. The study was conducted at Laval University for one winter and showed a reduction in heating requirements by 30%–40% compared to a standard, double-layered plastic-covered greenhouse. Sethi [23] studied the effect of different greenhouse shapes on the hourly transmitted total solar radiation (i.e., beam, diffused, and ground reflected) for both east-west orientation and north-south orientation in Ludhiana, India. Results showed that an uneven-span shape greenhouse receives the maximum and a quonset shape receives the minimum solar radiation during each month of the year at all latitudes. East-west orientation is best suited for year-round use at all latitudes. Ahamed *et al.* [24] conducted a study on greenhouses located in the Canadian Prairies. The design parameters include the shape, orientation, angle of the roof, and width of the span. The simulation results proved that the uneven-span greenhouse receives the highest solar radiation, whereas the quonset shape receives the lowest solar radiation. Also, it shows that for northern latitudes, an east-west oriented greenhouse is more energy-efficient from a heating and cooling point of view. Table 8.1 summarizes the details of the literature studies conducted earlier.

Table 8.1 Summarizes the details of the published literature

Author's name	Year	Location	Greenhouse orientation	Roof inclination	Greenhouse shapes	Main focus
Lawand *et al.* [21]	1975	Canada	X	–	–	Calculating the heating requirement of the greenhouse
Gupta *et al.* [22]	2001	India	X	–	X	Calculating the heating requirement in the energy-efficient greenhouse
Ali [19]	2008	–	X	–	–	Analyzing the effect of different orientations on solar radiation availability
Sethi [23]	2008	India	X	–	X	Computing the transmitted total solar radiation (beam, diffused and ground reflected)
Mobtaker *et al.* [18]	2016	Iran	X	–	X	Calculating the additional energy requirement
Ahamed *et al.* [24]	2018	Canada	X	–	X	Calculating heating requirements for different greenhouses

236　*Clean energy for low-income communities*

8.3　Methodology

This paper focused on enhancing solar radiation availability in a greenhouse by changing the greenhouse orientation, roof inclination, and greenhouse shapes. Numerous cases were considered under each energy-efficient parameter, which are explained in Table 8.2. These cases are explained in more detail in their respective sections.

Table 8.2　Indicates the three main energy-efficient parameters of this study with different cases under each parameter

Parameter 1 Greenhouse orientation	Parameter 2 Roof inclination	Parameter 3 Greenhouse shape
1. East–West Orientation	1. 15° roof inclination	1. Even shape
2. 60° North of East	2. 30° roof inclination	2. Uneven shape
3. 45° North of East	3. 45° roof inclination	3. Vinery shape
4. 30° North of East	4. 60° roof inclination	4. Semicircular shape
5. North-South Orientation		5. Elliptical or arch shape
6. 30° North of West		6. Single span
7. 45° North of West		7. Quonset shape
8. 60° North of West		

8.3.1　Model development

Different greenhouse models were designed in SketchUp software and then imported to TRNSYS-18 (version 18.00.0008) for further simulation. TRNSYS is an energy simulation software used to simulate the behavior of a transient system. SketchUp software is a plug-in feature to TRNSYS software to import 3-D drawings of buildings in TRNBuild. TRNSYS-18 automatically reads the data of 3-D drawings from SketchUp software. SketchUp is a popular software for drafting building models and enabling the creation of detailed greenhouse models. The greenhouse models designed for this study have a ground surface area of 2,000 m^2 (L = 50 m and B = 40 m), a side wall height of 5 m, and a maximum vertical height of 16.5 m. To obtain comparable results for all the greenhouse models being evaluated, the main dimensions such as length, breadth, side wall height, and maximum height of the greenhouses were kept constant across all models. The basic design of a greenhouse is presented in Figure 8.1.

8.3.1.1　Greenhouse orientations

Previous studies presented two main orientations, i.e., the east-west orientation and the north-south orientation [18,23,24]. However, no study was executed for other orientations other than these. Therefore, the axis of the even-span greenhouse (as shown in Figure 8.1) was oriented at distinct angles of 15° increments from east to west to understand the pattern and effect of solar radiation availability on these greenhouses. Figure 8.2 indicates the different greenhouse orientations considered for this study.

Enhancing solar insolation in agricultural greenhouses 237

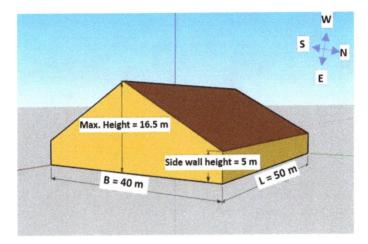

Figure 8.1 An east-west oriented greenhouse model developed in SketchUp software

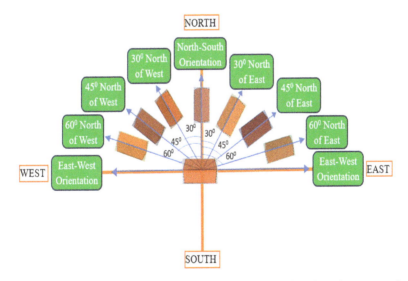

Figure 8.2 Different greenhouse orientations considered in this research

8.3.1.2 Roof inclinations

The impact of roof inclination on an even span greenhouse from 15° to 60° with an augmentation of 15° was studied. This was the minimum difference in inclination to see the observable changes in solar radiation availability by changing the roof inclinations. Roof inclination beyond 60° was not considered because the maximum height of the greenhouse increased drastically. Figure 8.1 shows the main dimensions, such as length (L), breadth (B), side wall height, and maximum height,

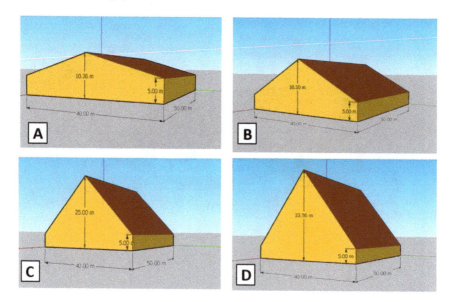

Figure 8.3 Different roof inclinations considered for this study. (A) 15° Roof inclination. (B) 30° Roof inclination. (C) 45° Roof inclination. (D) 60° Roof inclination.

of the greenhouse model. For these models, the maximum height changes as the roof inclination changes. The dimensions of the greenhouse models considered for roof inclinations are defined above in Figure 8.3.

8.3.1.3 Greenhouse shapes

The commonly identified greenhouse shapes based on previous literature reviews include even span, uneven span, vinery, semi-circular, elliptical or arch shape, single span, and quonset shape [18,22–24]. The ground surface area and the maximum height were fixed for all the greenhouse shapes to compare the solar radiation availability. The inclination of different walls and roofs was decided based on the results obtained for the different roof inclinations and to meet the maximum height of 16.5 m for all the greenhouses. The greenhouse model design for different greenhouse shapes is shown in Figure 8.4.

Greenhouse models for different greenhouse shapes are explained in Table 8.3. It includes the dimensions of a greenhouse, maximum height, and inclination of different roofs and walls, as applicable.

8.3.2 *The TRNSYS-18 model*

The basic simulation model of TRNSYS-18 is shown in Figure 8.5. Once the greenhouse models were developed in the SketchUp software, these files were saved as a .idf file. With the help of the simulation studio of TRNSYS, the above .idf file was then loaded into the simulation studio for simulations. The description

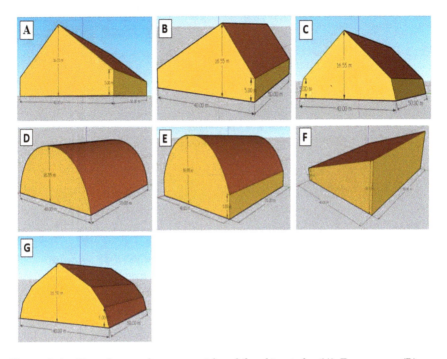

Figure 8.4 Greenhouse shapes considered for this study. (A) Even span. (B) Uneven span. (C) Vinery shape. (D) Semi-circular. (E) Elliptical or arch. (F) Single span. (G) Quonset shape.

Table 8.3 Represents different greenhouse shapes and model descriptions including length (L), breadth (B), maximum height, and roof and wall inclination considered for parameter 3

Cases	Greenhouse shapes	Model description
Case A	Even span	$L = 50$ m and $B = 40$ m; maximum height $= 16.5$ m Roof inclination $= 30°$
Case B	Uneven span	$L = 50$ m and $B = 40$ m; maximum height $= 16.5$ m South roof inclination $= 22°$; North roof inclination $= 45°$
Case C	Vinery shape	$L = 50$ m and $B = 40$ m; maximum height $= 16.5$ m Wall inclination $= 70°$; roof inclination $= 32°$
Case D	Semi-circular shape	$L = 50$ m and $B = 40$ m; maximum height $= 16.5$ m
Case E	Elliptical or arch-shape	$L = 50$ m and $B = 40$ m; maximum height $= 16.5$ m
Case F	Single span	$L = 50$ m and $B = 40$ m; maximum height $= 16.5$ m Roof inclination $= 16°$
Case G	Quonset shape	$L = 50$ m and $B = 40$ m; maximum height $= 16.5$ m Roof inclination 1 $= 70°$; roof inclination 2 $= 45°$; roof inclination 3 $= 26°$

240 *Clean energy for low-income communities*

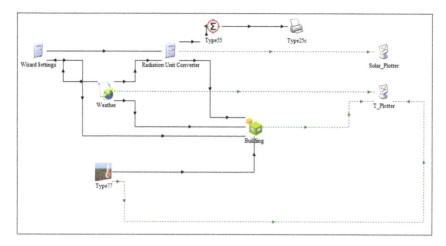

Figure 8.5 A complete TRNSYS model developed in TRNSYS-18 software indicating the input–outputs of each component

of the greenhouse model was stored in the building component. The building component is a type-56 for multi-zone building modeling, which models the thermal behavior of a building. To use this component, a pre-processing program known as TRNBuild was to be executed, which reads in and processes a file containing the building description. TRNBuild generates an information file describing the outputs and required inputs of Type-56 [25]. This building gets weather input such as total solar radiation, direct solar radiation, dry bulb temperature, slope, and the azimuth of surfaces, etc. from the weather component (Type-15). The weather file selected for this analysis was the typical meteorological year (.tmy) file for the city of Toronto, Ontario [26]. TMY files contain the generated values from a data bank for a specific location for at least 12 years [27]. The output from the weather component gives the solar radiation flux available on the different surfaces of the greenhouse models in KJ $h^{-1}m^{-2}$. A radiation unit converter was added between a weather component and a building component. The function of this converter was to interpolate radiation data, calculate the angle of incidence, total, and beam solar radiation input related to the position of the sun, and estimate insolation on several surfaces with either fixed or variable orientation. Type-77 component models the vertical temperature distribution of the ground with details like the mean surface temperature for the year and the thermal properties of the soil, which also affect the inside temperature of the building. Type-55 is an integrator that was used to integrate the total solar radiation available for a year. Different output components were used for this analysis. Type-25c is a printer to print the numerical results on an Excel file for further analysis, whereas the solar plotters and T_plotters are online plotters to plot the graphs obtained from the results.

Enhancing solar insolation in agricultural greenhouses 241

The basic calculation that was used to calculate the total solar radiation on different surfaces of a greenhouse is as follows: -

Total Solar Radiation in MWh for a year =
$[(IT_1 * A_1)+(IT_2 * A_2)+(IT_3 * A_3)+-----+(IT_n * A_n)] * 10^{-6}$

Where, IT_1 = Solar radiation on the surface – 1 of the greenhouses (W/m^2)
A_1 = Area of surface – 1 of the greenhouses (m^2)
IT_2 = Solar radiation on the surface – 2 of the greenhouses (W/m^2)
A_2 = Area of surface – 2 of the greenhouses (m^2)
IT_n = Solar radiation on the nth surface of the greenhouse (W/m^2)
A_n = Area of the nth surface of the greenhouse (m^2)

8.4 Results

The results obtained from the TRNSYS file were arranged based on the numerous cases considered. This study was important to apprehend the pattern of solar radiation on different walls and roofs of a greenhouse during different months of the year. The result for each parameter is explained in the following subsections.

8.4.1 Impact of greenhouse orientations on solar radiation availability

The results shown in Figure 8.6 indicate that solar radiation can be increased by orienting the greenhouse at 30° North of East for the location of Toronto, Ontario. For this location, the least solar radiation is received by the east-west orientation. However, the variation in solar radiation availability for all the greenhouse orientations is insignificant. The percentage difference between the maximum solar radiation availability and the minimum solar radiation availability for different

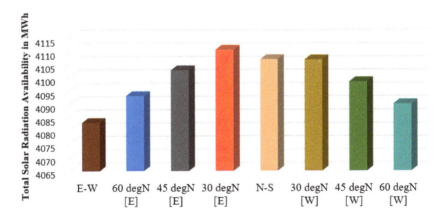

Figure 8.6 Graph for comparison of solar radiation availability at different orientations of the greenhouse

greenhouse orientations is only 0.6%. This implies that the greenhouse orientations barely alter the solar radiation availability. However, based on the literature review, an east-west-oriented greenhouse was the preferred orientation for the location in India [22,23]. This may be due to changes in the location and solar radiation patterns for that location. Therefore, other factors such as space availability, radiation on different walls, etc. should also be studied to finalize the orientation of a greenhouse. Hence, analysis was done to check the solar radiation availability on different walls and roofs of a greenhouse.

The graphs shown in Figures 8.7 and 8.8 explain the solar radiation availability on the different roofs and walls of a greenhouse. The result shows that the south roof receives the maximum solar radiation throughout the year. This is almost 50% more solar radiation as compared to the east roof and the west roof during the winter months. The results also show that the north roof receives the least solar radiation throughout the year. This underlines the objective of optimizing the availability of solar radiation during the winter months when a large amount of heating is required to maintain the inside air temperature. Also, the south wall receives 70% more solar radiation as compared to other walls during the winter months.

8.4.2 Impact of roof inclinations on solar radiation availability

Figures 8.9 and 8.10 show the solar radiation pattern for different inclinations of the south roof and north roof. Solar radiation availability during the winter months is at its maximum when the south roof inclination is 60° and the north roof inclination is 15°.

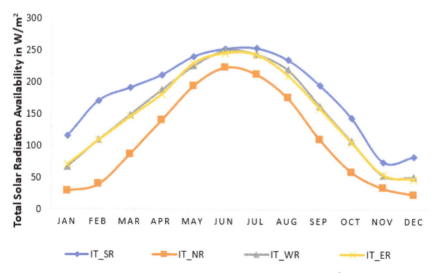

Figure 8.7 Comparison of solar radiation availability in W/m² on different roofs of a greenhouse where SR = South roof, NR = North roof, WR = West roof, ER = East Roof

Enhancing solar insolation in agricultural greenhouses 243

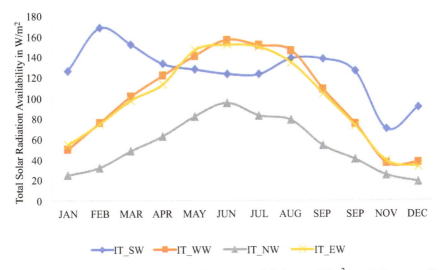

Figure 8.8 Comparison of solar radiation availability in W/m² on different walls of a greenhouse where SW = South Wall, WW = West Wall, NW = North Wall and EW = East Wall

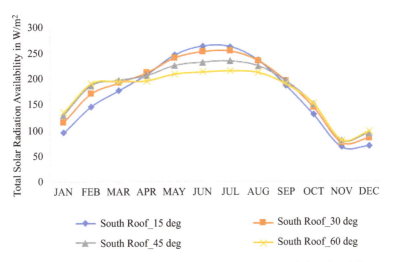

Figure 8.9 Graph for comparison of solar radiation availability for different roof inclinations of the south roof

For the south roof, the solar radiation availability increases during the winter months and decreases during the summer months as the roof inclination changes from 15° to 60°. However, for the north roof, the maximum solar radiation availability is seen at a 15° roof inclination throughout the year. Since the prime objective of this study is to optimize solar radiation availability during the winter

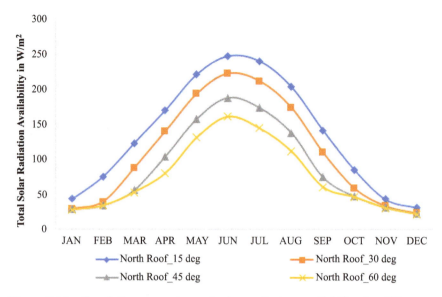

Figure 8.10 Graph for comparison of solar radiation availability for different roof inclinations of the north roof

months, the focus will be on south roof inclination rather than north roof inclination. This is because the south roof receives more solar radiation flux, which is almost 2 to 3 times more as compared to the north roof during the winter months. Hence, it can be concluded that the solar radiation flux on the south roof can be increased by approximately 40% by increasing the south roof inclination from 15° to 60°. This indicates that roof inclination is an important parameter for increasing solar radiation availability.

8.4.3 Impact of greenhouse shapes on solar radiation availability

Figure 8.11 shows the variation in total solar radiation availability across the different shapes of a greenhouse. For the selected location, a single-span greenhouse receives the maximum solar radiation of about 4,574 MWh/year and a semi-circular greenhouse receives the least solar radiation of about 3,916 MWh/year. This is because a single-span greenhouse has the highest available south roof area, because of which it can capture more solar radiation. The solar radiation percentage difference between the maximum and minimum solar radiation availability for different greenhouse shapes is approximately 17%. Mobtaker *et al.* [18] also investigated the different shapes of greenhouses for the climatic conditions of Tabriz, Iran. The result showed that the additional energy requirement to maintain a desirable temperature for the plant's growth was lowest in an east-west oriented single-span greenhouse. This is because an east-west oriented single-span greenhouse receives the highest solar radiation. As indicated in Figure 8.11, this shape

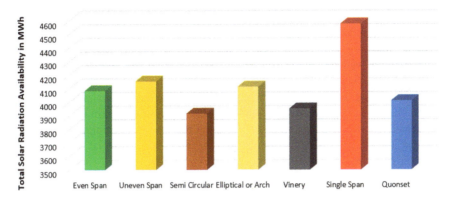

Figure 8.11 Graph for comparison of total solar radiation availability (in MWh) on different shapes of a greenhouse

reduces the additional energy required to maintain the optimum temperature for plant growth. Also, as per Ahamed *et al.*'s [24] study, an uneven-span greenhouse receives maximum solar radiation, this is because they did not consider a single-span greenhouse in their study. According to the current results, an uneven-span greenhouse receives the second highest solar radiation after a single-span greenhouse.

8.5 Conclusions

Based on the results, it is apparent that the solar radiation availability in the greenhouse can be varied by changing the greenhouse orientations, roof inclinations, and greenhouse shapes. Some parameters have a major impact on solar insolation availability, such as greenhouse shapes and roof inclinations. However, some parameters do not have much effect on the solar insolation falling on the greenhouse, such as greenhouse orientation. Therefore, these parameters can be executed in pairs or together to meet the requirements for a greenhouse. However, the following conclusions can be drawn from the above set of results, for the greenhouse located in the city of Toronto, Ontario:

1. 30° North of East is the best orientation to increase solar radiation availability. But it is also important to consider other parameters like solar radiation availability on different walls, space availability, etc.
2. The orientation of the greenhouse does not have much impact on the variation in solar radiation availability. The solar radiation percentage difference between the maximum and minimum solar radiation availability for different greenhouse orientations is only 0.6%.
3. The south roof receives the maximum solar radiation flux throughout the year. This percentage is almost 50% more solar radiation as compared to the east and west roofs.

246 *Clean energy for low-income communities*

4. Similarly, the south wall receives the maximum solar radiation during the winter months. This percentage is approximately 70% higher than the other walls.
5. The north wall and the north roof always receive the least solar radiation throughout the year.
6. Roof inclination plays a major role in optimizing solar radiation availability. The solar radiation availability on the south roof is at its maximum during the winter months when the roof inclination is 60° and at its minimum when the roof inclination is 15°. This increment is approximately equal to 40% more solar radiation as compared to a 15° roof inclination.
7. The solar radiation percentage difference between the maximum and minimum solar radiation availability for different greenhouse shapes is approximately 17%.

The study of greenhouse orientations, roof inclinations, and greenhouse shape was vital to understanding how these parameters affect solar radiation availability. The studies mentioned above are limited to countries close to the equator, especially those located in Asia and Europe. Therefore, a modification was required to analyze locations with cold climatic conditions, such as Canada. Most of the existing literature discusses the inside air conditions, heating, and cooling requirements. To make these greenhouses in cold climates more sustainable, it is important to optimize the solar insolation falling on the roofs and walls of the greenhouse. Future work will be required to extend the findings of this research and will include the investigation of the interior environment of a greenhouse by using combinations of these parameters. A study of this nature should consider analyzing the greenhouse's interactions with the external environmental conditions, soil conditions inside a greenhouse, inputs from the weather file, location parameters, and evapotranspiration from plants and soil. Furthermore, a detailed analysis is required to calculate the amount of solar radiation transmitted inside a greenhouse surface that will regulate the inside air temperature of the greenhouse. This will require the identification of the thermal properties of the construction material. Therefore, a more complex study is required to understand the dynamic model of a greenhouse to ensure that this will lead to a reduction in the heating costs of a greenhouse.

References

[1] United Nations, Department of Economic and Social Affairs, Population Division, (2013) "World Population Prospects: The 2012 Revision," Volume I: Comprehensive Tables ST/ESA/SER.A/336. Available from: https://population.un.org/wpp/Publications/Files/WPP2012_Volume-I_Comprehensive-Tables. pdf.
[2] FAO (2009) *How to Feed the World in 2050*, UN Food and Agriculture Organization, Rome.
[3] United Nations Food and Agriculture Organization (FAO), "Database on Arable Land," 13 September 2016. http://data.worldbank.org/indicator/AG.LND.

ARBL.HA.PC?end%20&hx003D;2013&hx0026;start%20&hx003D;1961&hx0026;view&hx003D;chart.

[4] Greenhouse gas emissions and agriculture—Government of Canada Greenhouse gas emissions and agriculture—agriculture.canada.ca

[5] K. Benke and B. Tomkins, (2017) "Future food-production systems: Vertical farming and controlled environment agriculture," *Sustainability: Science, Practice and Policy*, vol. 13(1), pp. 13–26.

[6] C. A. O'Sullivan, G. D. Bonnett, C. L. McIntyre, Z. Hochman, and A. P. Wasson, (2019) "Strategies to improve the productivity, product diversity, and profitability of urban agriculture," *Agricultural Systems* vol. 174, pp. 133–144.

[7] D. Despommier, (2011) "The vertical farm: Controlled environment agriculture carried out in tall buildings would create greater food safety and security for large urban populations," *Journal of Consumer Protection and Food Safety* vol. 6(2), pp. 233–236.

[8] K.H. Brown and A. L. Jameton, (2000) "Public health implications of urban agriculture," *Journal of Public Health Policy* vol. 21(1), pp. 20–39.

[9] J. Dixon, A. M. Omwega, S. Friel, C. Burns, K. Donati, and R. Carlisle, (2007) "The health equity dimensions of urban food systems," *Journal of Urban Health: Bulletin of the New York Academy of Medicine* vol. 84, pp. 118–129.

[10] F. C. Coelho, E. M. Coelho, and M. Egerer, (2018) "Local food: Benefits and failings due to modern agriculture," *Science Agriculture* vol. 75(1), pp. 84–94.

[11] J. Smit, A. Ratta, and J. Nasr, (2001) *Urban Agriculture: Food, Jobs, and Sustainable Cities*. The Urban Agriculture Network, Inc, Washington DC, USA.

[12] N. Engler and M. Krarti (2021) "Review of energy efficiency in controlled environment agriculture," *Journal Renewable and Sustainability Energy Reviews* vol. 141, 110786.

[13] J. Eaves and S. Eaves (2018), "Comparing the profitability of a greenhouse to a vertical farm in Quebec," *Canadian Journal of Agricultural Economics* vol. 66, pp. 43–54.

[14] G. Schimelpfenig and D. Smith (2021) *Controlled Environment Agriculture Market Characterization Report, Supply Chains, Energy Sources and Uses, and Barriers to Efficiency*, Resource Innovation Institute and United States Department of Agriculture.

[15] F. Brosseau and G. Hemery (1970) "Certains aspects économiques de la production de tomates en serres chaudes." *Semaine du Cultivateur. St. Hyacinthe, Gouvernement du Québec, Ministère de l'Agriculture et de la Colonisation*, pp. 146–159.

[16] P. A. Rorabaugh (2015) *Introduction to Hydroponics and Controlled Environment Agriculture*. University of Arizona, Controlled Environment Agriculture Center, Tucson, AZ. https://ceac.arizona.edu/resources/intro-hydroponics-cea.

[17] American Council for an Energy-Efficient Economy – Emerging Opportunities series

248 *Clean energy for low-income communities*

[18] H. G. Mobtaker, Y. Ajabshirchi, S. F. Ranjbar, and M. Matloobi (2016) "Solar energy conservation in a greenhouse: Thermal analysis and experimental validation," *Renewable Energy* vol. 96, pp. 509–519.

[19] S. A. Ali, (2008) *Modeling of Solar Radiation Available at Different Orientations of Greenhouse*, Agriculture Engineering Department.

[20] M. S. Ahamed, H. Guo, and K. Tanino (2017) "A quasi-steady state model for predicting the heating requirements of conventional greenhouses in cold regions," *Information Processing in Agriculture*, vol. 5, pp. 33–46.

[21] T. A. Lawand, R. Alward, B. Saulnier, and E. Brunet (1975) "The development and testing of environmentally designed greenhouse for colder regions," *Journal of Solar Energy* vol. 17, pp. 307–312.

[22] M. J. Gupta and P. Chandra (2001) "Effect of greenhouse design parameters on conservation of energy for greenhouse environmental control," *Energy* vol. 27, pp. 777–794.

[23] V. P. Sethi (2008) "On the selection of shape and orientation of greenhouse: Thermal modelling and experimental validation," *Solar Energy* vol. 83, pp. 21–38.

[24] M. S. Ahamed, H. Guo, and K. Tanino (2018) "Energy-efficient design of greenhouse for Canadian Prairies using a heating simulation model," *International Journal of Energy Research* vol. 42, pp. 2263–2272.

[25] Multizone Building Modeling with Type56 and TRNBuild – TRNSYS-18 "A Transient System Simulation program" Volume 5.

[26] Weather Data—TRNSYS-18 "A Transient System Simulation Program" Volume 8.

[27] Wikipedia The free encyclopedia for a typical meteorological year (Typical meteorological year—Wikipedia).

Chapter 9

Recent advances in biofuels production: industrial applications

Kang Kang[1], Sophia Quan He[2] and Yulin Hu[3]

Biofuels, bioenergy, and biomaterials play an important role in the global transition to green energy and sustainable materials. This chapter provides a brief overview and update on recent advancements in biofuel production through various biomass conversion technologies, emphasizing their industrial applications. It explores both thermochemical and biological pathways, addressing the high potential of biomass as a renewable energy source that can enhance energy security, reduce reliance on fossil fuels, and mitigate environmental impacts. The discussion includes detailed examinations of combustion, torrefaction, pyrolysis, and gasification processes, along with their advantages and challenges, particularly in low-income communities. The chapter also reviews studies on anaerobic digestion and its role in biofuel production. Through this comprehensive analysis, the chapter aims to contribute to a sustainable and decarbonized economy by providing insights into the production and application of biofuels across different industries.

Keywords: Biomass; Biofuels; Thermochemical conversion; Industrial applications

9.1 Introduction

Biomass resources that are made up of a wide array of organic materials stand as a renewable and abundant resource across various regions, including those nations inhabited by low-income communities. As suggested by the International Energy Agency (IEA), the use of biofuels accounts for over 3% of the transport energy demand worldwide, primarily for on-road vehicle applications. For the past 5 years, the expansion of biofuels has steadily increased by 5% per year. Specifically, Asia surpasses Europe and becomes the leading biofuel producer in the forecast period of 2021–2026 due to supportive domestic policy, growing

---

[1]Biorefining Research Institute and Department of Chemical Engineering, Lakehead University, Canada
[2]Department of Engineering, Dalhousie University, Canada
[3]Faculty of Sustainable Design Engineering, University of Prince Edward Island, Canada

liquid fuel demand (i.e., ethanol, biodiesel, renewable diesel, and bio-jet fuel), and export-driven production [1].

To fill the growing demand for biofuel, different biomass conversion approaches have been developed, like thermochemical and biological conversion. A summary of the currently available biomass conversion methods is depicted in Figure 9.1. These technologies provide a viable opportunity to harness biomass resources, contributing to energy security, reducing dependency on fossil fuels, and mitigating the environmental impacts associated with traditional energy production methods. The conversion of biomass into biofuels represents a pivotal advancement in the realm of sustainable energy production technologies, offering promising pathways for generating clean energy that is accessible to low-income communities. Especially in the case where no other renewable resources are available, but biomass is abundant. A variety of sources of biomass have been considered promising raw materials to be converted and further refined into valuable biofuels and other bioproducts. The potential sources of biomass range from agricultural and forestry residue, animal manure, and sewage sludge, to industrial processing and municipal solid waste [3]. Aside from the abundance of biomass sources, it also has several benefits, including being sustainable, renewable, and most importantly, carbon neutral [4].

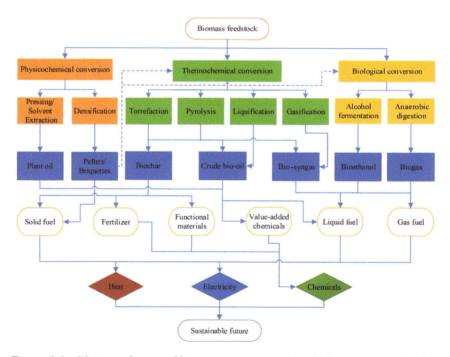

Figure 9.1 Major pathways of biomass conversion into fuels, energy, and value-added products [2]

Recent advances in biofuels production: industrial applications 251

In comparison, thermochemical conversion technologies normally feature fast reaction kinetics and versatility in producing a variety of useful products (i.e., biochar, bio-syngas, and bio-oil). On the contrary, biological conversion technologies require lower energy input, and well-established large-scale applications (e.g., biogas and bioethanol) [2]. However, the reaction is considerably longer than the thermochemical conversion. In the following sections, this chapter will dive into the recent progress of different conversion technologies and then address several critical challenges in product upgrading and applications.

9.2 Thermochemical conversion of biomass

This section dives into the essence of thermochemical processes, including combustion, torrefaction, pyrolysis, and gasification, each of which plays a crucial role in converting biomass into valuable biofuels and energy carriers. The importance of these technologies cannot be overstated, especially in the context of providing clean energy solutions to low-income communities. These communities often face significant barriers to accessing reliable and affordable energy sources, which are crucial for heating, cooking, and powering local industries. Thermochemical conversion processes can be adapted to small-scale operations, making them suitable for decentralized energy production, which is particularly beneficial for remote or rural areas with limited access to the central power grid [5].

Combustion is the most conventional form of biomass conversion and offers a straightforward approach to energy production through the burning of organic materials. However, more advanced techniques like torrefaction, pyrolysis, and gasification allow the production of more refined biofuels, such as bio-oil and bio-syngas, which can be utilized in a variety of applications, from domestic heating to powering vehicles and generating electricity [6]. These thermochemical technologies not only provide pathways to clean energy but also contribute to the circular economy by utilizing waste materials as feedstocks, thereby reducing environmental pollution, and enhancing sustainability. By focusing on the development of these conversion processes, it is possible to make significant contributions to energy equity, improving the quality of life in low-income communities, and moving toward a more sustainable and resilient energy future. A separate review section for each technology is provided in the following content.

9.2.1 Combustion

9.2.1.1 General description
Biomass combustion refers to the process of burning biomass in different forms, such as wood chips, briquettes [7], pellets [8], etc., to generate heat and power. This process is one of the oldest forms of energy conversion, utilizing the chemical energy stored in biomass through photosynthesis and converting it into thermal energy. The heat generated can be used directly for heating purposes or indirectly to produce steam for electricity generation in power plants. The field of biomass combustion has seen significant advancements in recent years, driven by the need

252 *Clean energy for low-income communities*

for more efficient, sustainable, and cleaner energy sources. Biomass combustion can significantly contribute to clean energy access in low-income communities by providing a renewable and locally available energy source. By utilizing locally sourced biomass, communities can potentially reduce their reliance on expensive and often environmentally harmful fossil fuels. Small-scale biomass combustion systems can be tailored to the specific needs and resources of a community, providing a decentralized energy solution that enhances energy security and self-sufficiency. Moreover, advancements in combustion technology have led to more efficient and cleaner burning systems that can minimize health and environmental impacts, making biomass combustion a more viable and sustainable option for these communities.

9.2.1.2 Recent advancements

In particular, research and development efforts have focused on improving combustion technologies to increase efficiency, reduce emissions, and enable the use of a broader range of biomass feedstocks [9]. Zhu *et al.* [10] integrated CO_2 capture using an amine-based wet scrubber with biomass combustion for power generation using a conventional Rankine cycle, Uchino *et al.* [11] designed a biomass power plant combined with a $Ca(OH)_2/CaO$-based thermochemical energy storage cycle, and Chen *et al.* [12] integrated concentrated solar power (CSP) with a biomass-based combined heat and power generation plant. Innovations include the computational fluid dynamics (CFD) modeling of advanced combustion systems such as fluidized beds, and burners that can handle varying qualities of biomass with higher efficiencies [13,14]. Additionally, there have been significant improvements in emission control technologies, for example, using electrostatic precipitators to achieve better particulate matter emission control from small residential boilers [15]. For the particulate control, Cheng *et al.* [16] applied the additive of phosphoric acid-modified kaolin, and Guerrero *et al.* [17] used inert porous materials (i.e., silicon carbide), while the formation mechanism and reduction strategies of NO_x emissions have been previously discussed by Ma *et al.* [18], Osman [19], Archan *et al.* [20], and Krzywanski *et al.* [21].

9.2.1.3 Technical challenges

While the technology of biomass combustion should be revisited to explore its potential, several challenges need to be addressed to fully realize the benefits. One of the primary challenges is the variability in biomass feedstock, which can affect combustion efficiency and emissions. In the long term, developing combustion systems that can adapt to different types of biomass with minimal efficiency loss would be ideal. Emissions control remains another significant challenge, as biomass combustion can produce pollutants that impact air quality and human health. Continued research and development into more efficient combustion technologies and advanced emissions control systems are essential to mitigating these impacts. In addition, establishing clear guidelines and sustainability criteria for biomass production and harvesting is necessary to ensure that biomass combustion contributes positively to energy transition and environmental conservation [22].

Compared to other processes, the main advantage of biomass combustion lies in its simplicity and directness, allowing for the immediate generation of heat and power without the need for intermediate processing or conversion steps. This can result in lower capital costs and simpler operational requirements, making it an attractive option for small-scale and community-level applications for low-income communities. However, biomass combustion also has its drawbacks. Compared to other technologies, it can be less efficient and produce higher levels of emissions, particularly particulate matter, and other pollutants, if not properly managed [9]. Technologies like pyrolysis, gasification, and torrefaction, on the other hand, operate at lower temperatures and in oxygen-limited environments, which can lead to lower emissions and the production of more valuable by-products such as bio-char from pyrolysis and syngas from gasification. Thus, in the following sections, the general description, recent developments, and technical challenges of torrefaction, pyrolysis, and gasification are discussed.

9.2.2 Torrefaction
9.2.2.1 General description
Torrefaction is a thermochemical process that serves as a critical step in the transformation of biomass into a more energy-dense and transportable form, facilitating its use as a sustainable energy source [23]. The process involves heating biomass in an inert or reduced oxygen environment to temperatures ranging from 200 °C to 300 °C. This treatment leads to the decomposition of hemicellulose and the partial breakdown of cellulose and lignin, resulting in a product with reduced moisture content and increased calorific value known as torrefied biomass. Simonic *et al.* [24] charged mixed waste wood, oak waste wood, and sewage sludge to the torrefaction system, and the results showed that the torrefied feedstock was more hydrophobic, had enhanced grindability, and had a higher energy density after torrefaction treatment. Wang *et al.* [25] also reported a positive impact of torrefaction at 225 °C–275 °C for 30–60 min on the biomass properties in terms of grindability, mechanical properties, as well as hydrophobicity. Comparatively, torrefaction has several benefits over other biomass conversion technologies. Its primary advantage lies in the lower reaction temperature required and the increased homogeneity and energy density of the output, which makes torrefied biomass easier to grind, transport, and store. Unlike combustion, torrefaction does not involve the direct burning of biomass, thereby reducing emissions of particulates and greenhouse gases. When compared to pyrolysis and gasification, torrefaction is a less severe thermal process, which means it requires lower energy input and is technically less complex, making it more economically feasible for large-scale applications. In terms of contributing to clean energy for low-income communities, biomass torrefaction presents a promising avenue. Torrefied biomass pellets can be used as a cleaner and more efficient fuel for cooking and heating, replacing traditional biomass and fossil fuels. This not only helps in reducing indoor air pollution, which is a significant health hazard in many low-income areas but also contributes to reducing the carbon footprint associated with household energy use [26].

9.2.2.2 Recent advancements

Recent advances in biomass torrefaction research and applications have focused on optimizing process parameters, enhancing the energy density of the end product, and integrating torrefaction into existing biomass processing chains [27]. Majamo and Amibo [28] adopted a central composite design (CCD) to optimize the torrefaction process, and the studied variables were temperature, residence time, acid concentration, and particle size. The optimized reaction conditions were found to be 300 °C, 31.89 min, 0.75 g/L acid concentration, and 0.2 mm particle size, which resulted in the highest heating value (HHV) of 25.05 MJ/kg. Several comprehensive reviews detailing the major reaction parameters in torrefaction have been published, such as Thengane *et al.* [29], Negi *et al.* [30], and Adeleke *et al.* [31]. Efforts have also been made in the development of continuous torrefaction reactors that improve the efficiency and scalability of the process [32]. Aside from these, alternative energy sources like solar energy [33,34] and microwave irradiation have been employed in biomass torrefaction [35,36]. Researchers have also been exploring the use of catalysts and varying process atmospheres to enhance the yield and quality of the torrefied product [37]. On the application front, torrefied biomass is increasingly being considered as a viable feedstock for co-firing in coal-fired power plants, offering a cleaner, renewable alternative to fossil fuels [38]. Furthermore, torrefaction can act as a pretreatment step for other biomass conversion technologies. For instance, torrefied biomass has been shown to be more reactive in gasification processes and yields higher quality bio-oil during pyrolysis due to its reduced moisture content and increased porosity [39]. He *et al.* [40] pretreated biomass at 200 °C–320 °C for 0.5–1.5 h using torrefaction, followed by pyrolysis at 550 °C, and the results found a strong correlation between the torrefaction severity (temperature × residence time) and bio-oil yield, and this correlation fitted well with a quadratic function. In a similar study, Valizadeh *et al.* [41] observed that the higher torrefaction temperature led to an increase in the phenolic content in bio-oil obtained from torrefied biomass but a reduction in the carboxylic acid contents in bio-oil. This synergy can enhance the overall efficiency and output of biomass conversion into biofuels and bioenergy.

9.2.2.3 Technical challenges

The torrefaction technology also faces several challenges. For example, the initial capital investment for torrefaction plants is significant, which can be a barrier to deployment in low-income regions without financial incentives or subsidies [42]. The variability in biomass feedstock types and qualities can also affect the consistency of the torrefied product, posing a challenge for its use in more stringent applications such as co-firing in power plants. Furthermore, research and development efforts must continue to focus on improving efficiency and reducing the costs of torrefaction technology. Moreover, community engagement and education are crucial to ensuring the sustainable and effective use of torrefied biomass in low-income areas, ensuring that the benefits of this technology are realized at the grassroots level.

Recent advances in biofuels production: industrial applications 255

In conclusion, the torrefaction of biomass should be considered an effective way of improving the grindability and energy density of raw biomass rather than an ultimate conversion technology. With ongoing advancements in research and strategic efforts to overcome existing challenges, torrefaction has the potential to play a significant role in the global transition toward cleaner, renewable energy sources, particularly in enhancing the energy security and environmental conditions of low-income communities. In the following sections, the biomass conversion technologies, i.e., pyrolysis and gasification, are discussed, focusing on the technical advancements and challenges.

9.2.3 Pyrolysis

9.2.3.1 General description

Biomass pyrolysis is a thermochemical decomposition process that occurs in the absence of oxygen, converting biomass into bio-oil, biochar, and syngas at a wide temperature range from 300 °C to 950 °C [43]. This process offers a promising pathway toward sustainable bioenergy production. Dependent on the heating rate, residence time, and other reaction parameters, the pyrolysis process can be tuned to enhance the yield and quality of specific products, such as bio-oil [44] or biochar [45]. Compared to other processes, pyrolysis offers distinct advantages. The production of liquid bio-oil presents a versatile energy carrier that can be upgraded to fuels or chemicals, offering a higher value product than the heat or power typically generated through combustion [46]. Biomass pyrolysis can produce biochar with applications in soil amendment, catalysis, and carbon sequestration. Compared to gasification, pyrolysis operates at lower temperatures and does not require a gasifying agent, making it a less energy-intensive process [47]. For low-income communities, the versatility of pyrolysis, particularly in producing high-quality biochar, holds considerable promise for clean energy and sustainable agricultural practices. Biochar can be used as a soil enhancer to increase agricultural productivity and water retention, reduce the need for chemical fertilizers, and improve food security. Additionally, the energy products from pyrolysis, such as bio-oil and syngas, can be utilized for local power generation, offering a renewable alternative to fossil fuels, and enhancing energy access in remote areas.

9.2.3.2 Recent advancements

Recent progress in biomass pyrolysis research and applications has been marked by significant advancements in co-pyrolysis technologies. Co-pyrolysis involves the simultaneous thermal decomposition of biomass with other materials, such as plastics or other waste materials, enhancing the quality and yield of pyrolysis products. This approach not only improves the economic viability of the pyrolysis process by taking advantage of the synergistic effect of different raw materials during the reaction but also contributes to waste management solutions. However, more studies are needed to further understand the co-pyrolysis mechanism. A summary of recent studies on co-pyrolysis of biomass and other materials is

256 *Clean energy for low-income communities*

Table 9.1   Summary of recent studies on biomass co-pyrolysis with other materials

Biomass	Co-pyrolyzed material	Reactor technology/ temperature	Research high lights	References
Groundnut shell, bagasse, rice husk, woody residues, sawdust	Low-density polyethylene (LDPE)	Microwave reactor	The energy efficiency of co-pyrolysis is higher than pyrolysis of biomass (63%–68% vs. 51%–57%), and the co-pyrolysis produced high-quality bio-oil (38–42 MJ/kg)/–	[48]
Corn stalk (CS)	Sewage sludge (SS)	Fluidized bed/ 450 °C–850 °C	The heating value of bio-oil from co-pyrolysis improved greatly. Synergistic effects originated from the interaction between radicals released by CS and metallic elements in SS.	[49]
Lignin	SS	Fixed bed/ 400 °C–800 °C	Co-pyrolysis of SS and lignin enhanced the decomposition at a reduced temperature. Biochar produced at 600 °C showed the best adsorption performance for methylene blue (MB).	[50]
Pine wood sawdust	SS	Thermogravimetric analyzer/ 30 °C–900 °C	Co-pyrolysis of SS with sawdust increased the bio-oil yield while reducing gas production but did not impact char yield significantly.	[51]
Rice straw (RS) and Sawdust (SD)	bituminous coal	Thermal-mass spectrometry (TG-MS)/ 100 °C–1,000 °C	H_2 was mainly produced from biomass pyrolysis at low temperatures. However, the production of H_2 did not improve the pyrolysis of coal.	[52]

provided in Table 9.1. Additionally, there is a growing focus on refining bio-oil into transportation fuels and enhancing the properties of biochar for use in agriculture and carbon sequestration [53].

9.2.3.3 Technical challenges

However, the widespread adoption of biomass pyrolysis technologies faces several challenges. The high initial capital and operating costs of pyrolysis plants can be prohibitive, particularly in low-income regions. The quality and consistency of bio-oil produced through pyrolysis can vary significantly based on the feedstock and process conditions, requiring further refinement for use as a transportation fuel. Moreover, the logistics of biomass collection, storage, and transportation also pose challenges, particularly in regions with dispersed biomass resources.

To address these challenges, targeted research is needed to improve efficiency and reduce the costs of pyrolysis technology. Developing decentralized pyrolysis systems that can operate at smaller scales could enhance the viability of this technology in low-income and rural areas [2]. Educational initiatives and capacity-building programs are essential to increasing awareness of the benefits of pyrolysis products, particularly biochar, in enhancing soil fertility and sequestering carbon. In conclusion, with continued advancements in technology and strategic efforts to overcome existing barriers, biomass pyrolysis has the potential to play a significant role in the transition toward a more sustainable and resilient energy future.

Aside from solid fuel and liquid fuel that can be produced from biomass conversion via torrefaction and pyrolysis, gaseous fuel, including bio-syngas and H_2, can be produced by the gasification of various types of biomass and organic waste. The general description, recent advancements, and technical challenges faced by biomass gasification are covered in the following sections.

9.2.4 Gasification

9.2.4.1 General description

Gasification is a thermochemical process that converts organic materials into a combustible gas mixture, commonly known as bio-syngas, through partial oxidation at high temperatures (above 700 °C) [54]. Bio-syngas, primarily composed of hydrogen, carbon monoxide, and smaller amounts of carbon dioxide and methane, can be used as a fuel for generating electricity, as a feedstock for producing liquid biofuels, or for direct heating applications. The structures of several commonly seen biomass gasification reactors are shown in Figure 9.2 [55]. As indicated in Figure 9.2, the most commonly used gasifiers include fixed beds (updraft and downdraft gasifiers), entrained flow bed gasifiers, and fluidized bed gasifiers. Among them, updraft and downdraft gasifiers are the most conventional reactors used in gasification. One of the main technical challenges of updraft gasifiers is that the tar concentration in the product gas is much higher than that of other gasifiers, which thus causes problems in downstream applications like gas turbines. Entrained flow and fluidized bed gasifiers are used for large-scale applications. For example, several companies have built entrained flow gasifiers such as GE Energy Gasifier (formerly Chevron Texaco), Shell Gasifiers, and Siemens Gasifiers. The industrially available fluidized bed gasifiers have been developed and built by KBR Transport Gasifier,

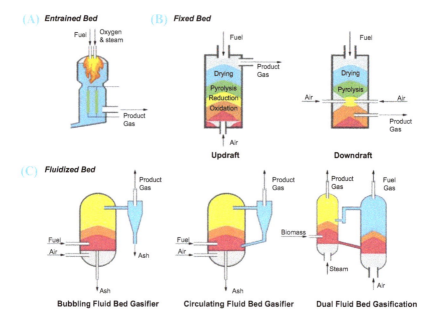

Figure 9.2 Schematics of different gasification reactors: (A) entrained flow, (B) updraft and downdraft fixed beds, and (C) left to right: bubbling fluidized bed, circulating fluidized bed, and dual fluidized beds [31]

Great Point Energy, and High Temperature Winkler (HTW) Gasifier. The application of biomass gasification in low-income communities is challenged by the technology's complexity and high operational temperatures, which demand robust and often expensive engineering solutions [56]. Despite these challenges, there are examples of biomass gasification plants contributing to clean energy in various countries. For instance, in India, gasification units provide electricity to remote rural communities, enhancing energy access and reducing dependence on traditional biomass and fossil fuels [57]. In Africa, gasification projects have been initiated to power agricultural processing, providing a sustainable energy source for local industries [58].

9.2.4.2 Recent advancements

Recent advances in biomass gasification research and applications have focused on improving the efficiency and environmental performance of the gasification process. A significant area of progress is in the cleaning and upgrading of bio-syngas to remove contaminants such as tar, particulates, and acid gases, which can hinder the end-use applications of syngas [59]. Specifically, the developments in gas cleaning technologies, such as advanced scrubbing, thermal cracking, and catalytic reforming, have enhanced the quality of syngas, making them more compatible with downstream processes for biofuel synthesis and power generation. Additionally, research into integrated gasification combined cycle (IGCC) systems has opened new pathways for the efficient utilization of biomass gasification products [60]. Compared to

Recent advances in biofuels production: industrial applications 259

other thermochemical conversion technologies, gasification provides a more versatile and cleaner product in the form of syngas, which can be further processed into a wide range of fuels and chemicals. Compared to combustion, gasification converts biomass into a more valuable and flexible energy carrier. While pyrolysis also produces a gas product (bio-syngas), the gas yield and quality from gasification are generally higher due to the complete conversion of biomass.

9.2.4.3 Technical challenges

The primary challenges in biomass gasification technology include the high capital and maintenance costs associated with advanced gasification systems and the need for skilled operation and maintenance. The tar and particulate matter produced during gasification can pose significant technical hurdles for gas cleaning and conditioning systems, impacting the overall efficiency and cost-effectiveness of the process [59]. To make biomass gasification more accessible and beneficial for low-income communities, there is a need to develop simplified, low-cost gasification systems tailored for small-scale, local applications that could improve technology accessibility. Also, leveraging local feedstocks and integrating gasification systems with community-based waste management and agricultural practices can enhance sustainability and economic viability. In summary, for biomass gasification, challenges related to system complexity, cost, and feedstock variability need to be addressed to enhance its applicability in low-income communities.

In addition to the thermochemical conversion of biomass, biological conversion, such as anaerobic digestion of biomass, is another well-established and promising technology for producing heat and power, particularly for low-income countries. In the following sections, the recent development and application of anaerobic digestion are discussed.

9.3 Anaerobic digestion of biomass

Anaerobic digestion is one of the biochemical biomass conversion technologies, during which microorganisms are capable of decomposing complex organic materials in the absence of oxygen to produce biogas as the main product and digestate as the by-product. Various sources of organics can be utilized as feedstock in anaerobic digestion, such as municipal waste, animal manure, and industrial waste [61]. As indicated in Figure 9.3, there are four main steps involved in the anaerobic digestion of organics to produce biogas, including hydrolysis, acidogenesis, acetogenesis, and methanogenesis [62].

To date, although anaerobic digestion has been determined to be a highly promising approach to valorize organic waste into biomethane, the production rate and efficiency of biomethane production and recovery are low, which consequently limits the engineering application of anaerobic digestion [63]. The primary causes of such a low biomethane production rate are (i) the low inter-species electron transfer efficiency of H_2 and HCO_2^-; (ii) the accumulation of volatile fatty acids

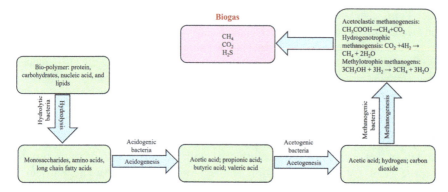

Figure 9.3 A schematic diagram describing the major steps involved in anaerobic digestion

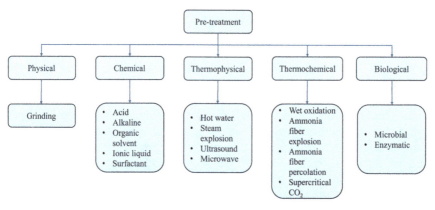

Figure 9.4 A summary of available pre-treatments used to enhance anaerobic digestion

and ammonia during the reaction, which leads to reduced activity of methanogenic bacteria [64]; (iii) the presence of inhibitory chemicals like phenols, polymeric aromatic hydrochar [65], sulfide, and heavy metals [66] in the growing medium for bacteria and potentially results in toxicity toward microbes; and (iv) the prolonged start-up time and instability during operation, which are another critical issue limiting the wide commercial applications of anaerobic digestion. To address the challenges listed above, different strategies have been adopted, like (i) conducting a biomass pre-treatment [67]; (ii) using co-substrate; and (iii) adding conductive additives (e.g., granular activated carbon, magnetite, and biochar) [68]. All these approaches are helpful for not only enhancing the degradability of the substrate but also increasing the hydrolysis rate and methanogen metabolism.

Till now, there are a wide range of pre-treatment approaches (Figure 9.4) that have been developed and, in particular, have been widely used for lignocellulose when

Recent advances in biofuels production: industrial applications   261

using it as the feedstock in anaerobic digestion. After pre-treatment, the digestibility of lignocellulosic biomass is increased by altering the lignin content, the interlinkage of cellulose and hemicellulose, the crystallinity of cellulose, and enhancing the accessible surface area for microbes [68]. The most technical challenges relating to the use of pre-treatment are the lack of identification of the most economic and favorable method and the optimal operating conditions for pre-treatment. Previously, Yang and Wayman [69] reported that the utilization of a pre-treatment step prior to anaerobic digestion accounts for more than 20% of the entire production cost.

Alternatively, anaerobic co-digestion of a mixture of organic waste has been explored. Shahbaz *et al.* [70] used a mixture of municipal solid waste and food waste as the co-substrate in co-digestion, and the effect of the C/N ratio on biogas production was investigated. The results showed that the increase in C/N ratio led to a decrease in biogas production, which could be due to the lack of sufficient availability of organic N for microbial growth. Retfalvi *et al.* [71] conducted anaerobic co-digestion of used cooking oil, maize silage, mill residue, and *Chlorella vulgaris*, Ponsa *et al.* [72] used the co-substrate of the organic fraction of municipal solid waste, vegetable oil, animal fats, cellulose, and protein in the anaerobic co-digestion, and Masih-Das and Tao [73] observed that the use of the co-substrate of food waste and liquid dairy manure or manure digestate was capable of stabilizing the stability during the biomethane formation in the anaerobic co-digestion process. A summary of recent studies and associated main findings is provided in Table 9.2.

Table 9.2   A summary of recent studies regarding anaerobic co-digestion

Co-substrate	Operating conditions in anaerobic co-digestion	Key findings	References
Palm oil mill effluent and rumen fluid	37 °C± 3 °C; 100 rpm; HRT of 6–20 days; OLR: 1.65–4.06 g COD/ (L·d)	• Co-digestion was helpful for maintaining the pH of the medium and COD removal. • The highest CH_4 content was 61.8% at 6 days HRT with 4.06 g COD/(L·d) using 10% rumen fluid and 90% palm oil mill effluent.	[74]
Cattle manure and glycerine phase	39 °C; 60 rpm; HRT of 30 day; OLR: 2.3 volatile solid/(L·d)	• The addition of 10% of glycerine phase to the substrate produced 3.1 times more biogas and 10% more CH_4 content than pure cattle manure as the substrate.	[75]

(Continues)

262 *Clean energy for low-income communities*

Table 9.2   (Continued)

Co-substrate	Operating conditions in anaerobic co-digestion	Key findings	References
Organic waste (fruits, food, and vegetable waste), methanogens-activated sludge, and tea powder waste	37 °C; HRT of 60 days	• Biogas formation was mainly related to the concentration of NH_3-N and volatile fatty acids. • The optimal feed ratio was 1:2:1 organic waste: tea powder waste: methanogens activated sludge for producing CH_4. • Using co-substrate led to ~65% biogas yield than ~45% obtained from mono-substrate.	[76]
Catering food waste and *Parthenium hysterophorus*	30 °C; HRT of 60 days	• The mixture of 60% of catering food waste and 40% of *Parthenium hysterophorus* led to the highest biogas production rate of 559 mL/(L·d) and a yield of 5,532 mL/L. • The maximum removal efficiency of organic matter was 38.9%.	[77]

*HRT: hydraulic retention time; OLR: Organics loading rate; COD: Chemical oxygen demand.

In addition, adding conductive additives (e.g., granular activated carbon, magnetite, and biochar) is another method that has been utilized by the researchers. Among them, biochar has been extensively applied to enhance the anaerobic digestion process by limiting the formation of inhibitory chemicals and promoting CH_4 formation. The relevant review articles have been published, such as Fagbohungbe *et al.* [78], Pan *et al.* [79], Masebinu *et al.* [80], Chiappero *et al.* [81], Luz *et al.* [82], Zhao *et al.* [83], Qiu *et al.* [84]. The influence of biochar addition to the anaerobic digestion process can be reflected in (i) biogas production;(ii) the mechanism of intermediate products in aerobic digestion; and (iii) enhancing microbial activity. The positive impacts of adding biochar to anaerobic digestion relate to the higher specific surface area, high porosity, low bulk density, high thermal and chemical stability, and great adsorption performance. In a previous study, Luo *et al.* [85] added bio-stable biochar at a concentration of 10 g/L to the mesophilic anaerobic digestion at

Recent advances in biofuels production: industrial applications 263

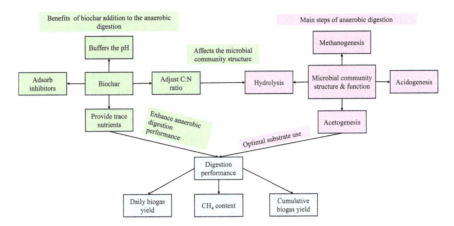

Figure 9.5   A summary of the positive effects of biochar additives in anaerobic digestion [87]

35 °C, and it was found that a dual role of biochar included: (i) shortening the lag phase of methanogenesis and leading to an increase in CH_4 production by 70.6%; (ii) promoting the production and degradation of volatile fatty acids. As earlier mentioned, the formation of inhibitory chemicals is one of the critical problems in anaerobic digestion. While biochar could act as a buffering agent by reducing the inhibitory influences of these chemicals. For instance, Cai *et al.* [86] have comprehensively reviewed the buffering effect of biochar on ammonia toxicity in anaerobic digestion, and the four associated mechanisms are: (i) facilitating the direct interspecies electron transfer (DIET) pathway; (ii) helping the growth and attachment of microorganism by avoiding the washing out effect of microorganisms; (iii) providing the required nutrients for microorganisms; and (iv) adsorbing ammonia and adjusting the pH of the growth medium. Aside from ammonia, biochar has been reported to have the ability to control the inhibitory influences caused by volatile fatty acids, acetic acids, other organic acids, sulfides, cations (e.g., Ca, Na, K, and Mg), heavy metals, and phenolics [87]. Other beneficial impacts of biochar additives in anaerobic digestion were previously reviewed by Fagbohungbe *et al.* [78], Pan *et al.* [79], Masebinu *et al.* [80], Kumar *et al.* [88], and Khalid *et al.* [89]. The overall advantageous influences of biochar addition are displayed in Figure 9.5.

9.4   Industrial applications of biofuels

Various types of biofuels, including solid, liquid, and gaseous biofuels, have been produced by the thermochemical conversion approaches discussed earlier and utilized in different industries. In this section, the industrial applications of biofuels for domestic heating, power plants, fuel oil, etc. are covered.

9.4.1 Biomass-based energy for heating
9.4.1.1 Domestic heating

One area of renewable energy and biofuels that shows early promise is biomass pellets for domestic heating. Heating using biomass like wood is not a new concept for many households, particularly in low-income countries [90]. The most common biomass heating method is the burning of wood in log form on an open fire, which is normally done in conjunction with a back boiler to provide room heat and hot water or central heating [91]. Due to the possibility of overheating or boiling the water, the back boiler is restricted in Ireland, UK, for safety concerns. In modern times, a closed system instead of an open fire system that utilizes wood pellets has been developed and then widely used for domestic heating. Wood pellets are made from the compression or densification of sawdust, shaving, bark, and chips into a cylindrical shape, and the associated process flow diagram to prepare wood pellets is shown in Figure 9.6. The process to manufacture wood pellets involves loading the dried biomass particles in a screw auger under certain pressure and temperature, followed by compressing them to pass through a die extruder to form the desired shape. In some cases, the binder is added together with biomass particles in order to enhance the mechanical stability of the resulting fuel pellets [92]. Several benefits that can be offered by biomass heating, especially for low-income off-grid households, (i) low-carbon fuel; (ii) cheaper than fuel oil; (iii) can be purchased in a low quantity (i.e., like a few bags of wood pellets) [90].

The environmental benefit of using wood pellets is the net CO_2 emission since the CO_2 emitted upon biomass burning can be reused by the wood through photosynthesis during its growth stage [93]. In addition, the economic advantage of wood pellets is related to their relatively low price compared to fuel oil. More importantly, the heating value of wood pellets is found to be higher than that of fuel oil. When compared with the use of liquefied petroleum gas (LPG) and electric-powered systems, the cost of using wood pellets for domestic heating is still cheaper, as previously reported in the UK. Compared with other alternative

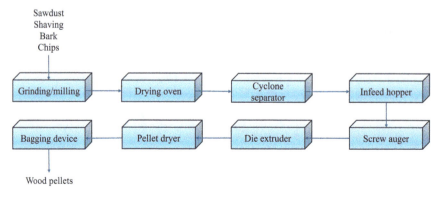

Figure 9.6 A flow diagram to make wood pellets [92]

Recent advances in biofuels production: industrial applications   265

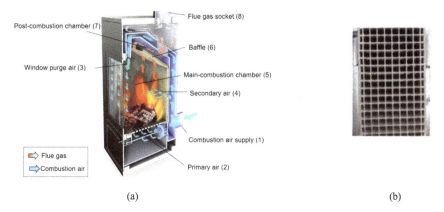

Figure 9.7   (a) A schematic diagram of a staged-air wood stove and (b) a honeycomb catalyst used in the catalytic wood stove for emission control (license: CC BY 4.0 DEED)

renewable energy sources (i.e., solar thermal systems), wood pellets can provide a higher level of comfort and a more uniform thermal environment since the wood pellet furnaces or boilers can operate continuously, but the operation of the solar system depends on the surrounding environment and climate. Other advantages of wood pellets are their widespread accessibility and the versatility of the technology. However, the suitability and utilization of wood pellets have raised a few concerns when compared with other alternative renewable energy sources for domestic heating. For example, one of the environmental concerns is the emission (e.g., nitrogen oxides, NO_x; particulate matter, PM; and benzene, toluene, ethylbenzene, xylenes, and trimethylbenzene, BTEXT) released upon the wood pellet combustion in the stove [9]. The NO_x and BTEXT emissions obtained from the combustion of different types of wood pellets have previously been discussed by Ozgen et al. [9] and Schmidt et al. [94]. The formation of NO_x during biomass combustion can be explained by three governing mechanisms, (i) thermal NO_x: N_2 comes from the air and is oxidized at temperatures >1,300 °C; (ii) fuel NO_x: the N present in the biomass fuel is oxidized; and (iii) prompt NOx: the reactions with N_2 from the air in the flame front. To reduce such emissions in the wood pellet stove for domestic heating [18], a staged-air stove has been designed, as illustrated in Figure 9.7, and the underlying mechanism is to apply to stage air to control NOx emissions. Besides, a catalytic stove is another emission control method, during which the oxidation catalysts are used to remove unburnt combustion products (e.g., CO, hydrocarbons, and soot). The catalysts (see Figure 9.7) often constructed using an iron-alloy ceramic (e.g., Al_2O_3 and ZrO_2) with noble (Pd and Pt) or transition metals (e.g., Cu and Ni) to prepare heterogeneous catalysts, and are exhibited in either a packed bed, honeycomb, or network/wire mesh shape [95].

266 *Clean energy for low-income communities*

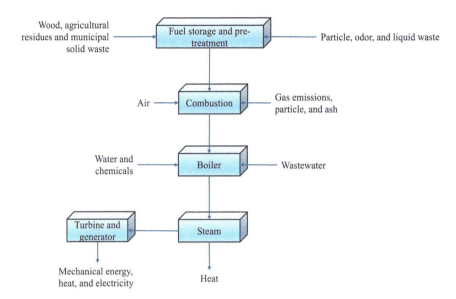

Figure 9.8 A schematic diagram of a biomass-fired power plant

9.4.1.2 Biomass-fired power plant

Except for small-scale domestic heating, biomass fuel pellets can also be used for large-scale applications like power plants. Biomass-fired power plants work on a similar principle that is similar to a coal or natural gas-powered plant, where biomass is converted to a useful stream, heat, and combustible gases. The schematic diagram for using biomass for power generation is displayed in Figure 9.8. Typically, biomass particles are delivered to the plant and stored either in silos or storage piles if they are not immediately loaded into the combustor. Afterward, biomass undergoes a series of pre-treatments, including grinding/milling, separation, drying, and charging into a combustor, in which biomass oxidation occurs under an excessive amount of air in order to fully combust biomass to generate more heat and reduce the emission of CO and particles. Owing to the exothermic nature of the combustion reaction, the cooled water is turned into high-temperature and high-pressure steam by heat exchangers to power a Rankine power cycle and finally generate electricity. However, due to their relatively low energy content, it should be mentioned that biomass-fired power plants typically have a lower electricity generation capacity than coal-fired power plants, i.e., coal-fired power plants are typically an order of magnitude larger than biomass-fired power plants. Table 9.3 summarizes the current commercial biomass-fired system and the associated electricity generation capacity.

9.4.2 Use of bio-oil as fuel oil

Bio-oil is a liquid fuel that can be a replacement for petroleum crude oil and is typically produced from fast pyrolysis or hydrothermal liquefaction (HTL). The

Recent advances in biofuels production: industrial applications 267

Table 9.3 A list of commercial biomass-fired power plants

Commercial plants	Location	Fuel source	Electricity generation capacity
Draw Power Station	North Yorkshire, England	Wood pellets	3,906 MW
Amager Bakke	Amager, Denmark	Municipal solid waste	0–63 MW electricity and 157–247 MW district heating
Atikokan Generating Station	Atikokan, Canada	Wood pellets	900 million KW/year
Avedore Power Station	Avedore, Denmark	Straw and wood pellets	793 MW of electricity and 918 MW of heat

Table 9.4 A summary of several major properties of pyrolysis bio-oil and heavy fuel oil [96,97]

	Grade G – Pyrolysis bio-oil	Grade D – Pyrolysis bio-oil	Heavy fuel oil
Gross heat of combustion, Mg/kg	15	15	43
Water content, wt.%	30	30	0.32
Solid content, wt.%	2.5	0.25	
Kinetic viscosity at 40 °C, mm^2/s	125	125	130
Density at 20 °C, kg/dm^3	1.1–1.3	11–1.3	/
S content, wt.%	0.05	0.05	/
Ash content, wt.%	0.25	0.15	/
pH	2-3	2-3	/
Flash point, °C	45	45	110
Pour point, °C	−9	−9	/

typical properties of bio-oil obtained from fast pyrolysis are compared with those of heavy fuel oil, and the results are summarized in Table 9.4.

With the proper upgrading process, bio-oil obtained from either fast pyrolysis or HTL can be upgraded to hydrocarbon fuels, and the available upgrading approaches include hydrodenitrogenation (HDN), hydrodeoxygenation (HDO), catalytic cracking, thermal cracking, and solvent addition [98]. A number of review articles have been published on the upgrading of bio-oil obtained from fast pyrolysis, such as Zhang *et al.* [99], Mostafazadeh *et al.* [100], Sharifzadeh *et al.* [101], and Chen *et al.* [102], and from HTL, such as Galadima and Muraza [103], Djandja *et al.* [104], Yang *et al.* [105], and Shakya *et al.* [106]. However, bio-oil upgrading suffers from several technical challenges that must be addressed:

- Even though HDO has been identified as the most effective bio-oil upgrading approach to remove heteroatoms, lower the acid value, and increase heating

268   *Clean energy for low-income communities*

value, the requirements of high operating temperature, particularly the high H_2 pressure and catalyst deactivation, considerably increase the operating cost and thus reduce the economic competitiveness of bio-oil. Thus, in future studies, the exploration of cheap and highly stable HDO catalysts must be performed, together with research on conducting HDO at a relatively low H_2 pressure without affecting the yield and quality of upgraded bio-oil.

- The oxygenates derived from lignin are significantly more difficult to upgrade when compared to those derived from cellulose and hemicellulose, which is mainly due to the variations in the chemical structure. More importantly, phenolics, as one of the dominant intermediates derived from lignin degradation, could cause coke formation and subsequently lead to catalyst deactivation in HDO treatment. One of the solutions to tackle this challenge could be the lignin isolation pre-treatment.
- Current studies are only limited to the batch reactor used in HDO or other bio-oil upgrading treatments, and a few studies have employed continuous flow systems, especially at a pilot scale.
- Furthermore, there is a lack of life cycle assessment results for biomass conversion, bio-oil production, and subsequent upgrading stages [98,107].

In addition to bio-oil upgrading to hydrocarbon fuels, bio-oil could be directly used as fuel oil, which is another industrial application of biomass-derived fuel. The commercial examples include: (i) Energy Gardens in the United Kingdom, which utilize woody biomass-derived bio-oil to provide heating for public spaces like parks, community centers, and schools, (ii) Varnamo Biomass Gasification Plant in Sweden, which generates heat for district heating and commercial buildings using bio-oil as a fuel source in the gasification process; and (iii) Bioenergy Villages in Germany, where bio-oil (originally derived from wood chips, agricultural waste, and other organic waste) is incorporated into a decentralized heating system for rural communities.

The results released by several studies in Europe found that the total cost of using bio-oil to generate the same amount of heat was lower by 76%, 65%, and 50% than that obtained from burning fossil fuel-based boilers in Austria, Finland, and Belgium, respectively. Moreover, past literature has suggested that it is undoubtedly true that bio-oil can be utilized as a fuel source in oil boilers and heavy engines; however, it must be mentioned that certain modifications to the boilers or engines are required because of the differences in the physicochemical properties of bio-oil and fuel oil. For example, the air-fuel ratio needs to be recalculated, the construction materials are required to be changed since the acid value of bio-oil is higher than that of fuel oil, and a filtration system might be necessary as well to remove the solid particles from bio-oil [97]. Such limitations lead to the use of bio-oil as a co-feed with coal, diesel, and natural gas to limit modifications to engines or boilers and enhance engine efficiency. One successful example is the co-firing of bio-oil with natural gas in a 350 MW natural gas-fired power plant in the Netherlands, and a bio-oil loading of $>1\%$ of the co-feed can be converted to 25 MWh of electricity. Other industrial attempts to co-fire bio-oil with other fossil fuels have been conducted by Red Arrow Products Company, Ensyn Technologies Ltd, Stork Technical Service (Netherlands), and Dreizler GmbH (Germany) [96].

9.4.3 Use of biogas

Biogas that is obtained from the anaerobic digestion of organic waste has also been industrialized. Overall, biogas can be applied for: (i) heat generation via direct combustion, (ii) electricity production by fuel cells or microturbines, (iii) combined heat and power generation; and (iv) vehicle fuels. The critical technical bottleneck for the widespread use of biogas is associated with its low energy content [108]. China has a long history in the commercial applications of anaerobic digestion and biogas, and so far, household-scale digesters, biogas septic tanks, and biogas plants have been designed for handling municipal, industrial, and agricultural waste. For the household-scale digester, it can treat 8–20 pigs, 1–2 cows, or 150–200 chickens, leading to a production of 0.8–20 m³ of biogas/day. Besides, biogas septic tanks have been widely used for wastewater treatment at an atmospheric temperature for handling the sewage, thereby lowering environmental pollution and enhancing sanitation [109]. Furthermore, for the biogas plants, except for China, other countries like Canada and the US have also utilized this technology to treat organic waste derived from industrial and municipal solid water to form biogas and digestate (a biofertilizer). For example, Cavendish Farm, a leading potato product, has built an anaerobic digestion plant in Charlottetown, Canada, to handle their potato processing waste. In addition, there are more than 2,400 sites of biogas plants available in the US across 50 states (see Figure 9.9). Specifically, there are 473 anaerobic digestion plants built on the farms, 1,269 anaerobic digestion plants

Figure 9.9 Operating anaerobic digestion and biogas plants in the US (source: American Biogas Council, https://americanbiogascouncil.org/biogas-market-snapshot/)

270　*Clean energy for low-income communities*

built on water resource recovery facilities, 102 anaerobic digestion and biogas plants built to digest food waste, as well as 566 landfill gas projects. In Europe, approximately 10,000 operating anaerobic digestion and biogas plants are in Germany alone.

9.5　Conclusions

In this chapter, the recent advances regarding the application of biofuels in different industrial areas are discussed. In general, biofuels can be categorized as solid, liquid, and gaseous biofuels, and they can be produced from various types of biomass by torrefaction, pyrolysis, gasification, and anaerobic digestion. The description of these biomass conversion technologies is covered, followed by a detailed discussion about a few industrial applications of biofuels, including biomass heating for both small-scale (domestic heating) and large-scale (biomass-fired power plants) applications and using bio-oil as fuel oil, anaerobic digestion, and biogas plants. In short, this chapter offers an overall understanding of the production and applications of biofuels, providing valuable knowledge and information for fostering a sustainable and decarbonized economy.

Acknowledgments

The authors are grateful for the financial support from the Discovery Grant, National Science and Engineering Research Council (NSERC), Canada.

References

[1] Biofuels – Renewables 2021, Int. Energy Agency. (2021) 1–6. https://www. iea.org/reports/renewables-2021/biofuels?mode=transport&region=World& publication=2021&flow=Consumption&product=Ethanol.

[2] K. Kang, N.B. Klinghoffer, I. ElGhamrawy, and F. Berruti, Thermo-chemical conversion of agroforestry biomass and solid waste using decentralized and mobile systems for renewable energy and products, *Renew. Sustain. Energy Rev.* 149 (2021) 111372. https://doi.org/https://doi.org/10. 1016/j.rser.2021.111372.

[3] R.K. Srivastava, N.P. Shetti, K.R. Reddy, E.E. Kwon, M.N. Nadagouda, and T.M. Aminabhavi, Biomass utilization and production of biofuels from carbon neutral materials, *Environ. Pollut.* 276 (2021) 116731. https://doi. org/https://doi.org/10.1016/j.envpol.2021.116731.

[4] N. Khan, K. Sudhakar, and R. Mamat, Role of biofuels in energy transition, green economy and carbon neutrality, *Sustainability.* 13 (2021) 12374. https://doi.org/10.3390/su132212374.

[5] T. Kirch, P.R. Medwell, C.H. Birzer, and P.J. van Eyk, Small-scale autothermal thermochemical conversion of multiple solid biomass feedstock,

Renew. Energy. 149 (2020) 1261–1270. https://doi.org/https://doi.org/10.1016/j.renene.2019.10.120.

[6] P. Manara, A. Zabaniotou, Towards sewage sludge based biofuels via thermochemical conversion – A review, *Renew. Sustain. Energy Rev.* 16 (2012) 2566–2582. https://doi.org/https://doi.org/10.1016/j.rser.2012.01.074.

[7] S. Velusamy, A. Subbaiyan, S. Kandasamy, M. Shanmugamoorthi, and P. Thirumoorthy, Combustion characteristics of biomass fuel briquettes from onion peels and tamarind shells, *Arch. Environ. Occup. Health.* 77 (2022) 251–262. https://doi.org/10.1080/19338244.2021.1936437.

[8] I. Mian, X. Li, O.D. Dacres, *et al.*, Combustion kinetics and mechanism of biomass pellet, *Energy.* 205 (2020) 117909. https://doi.org/https://doi.org/10.1016/j.energy.2020.117909.

[9] S. Ozgen, S. Cernuschi, and S. Caserini, An overview of nitrogen oxides emissions from biomass combustion for domestic heat production, *Renew. Sustain. Energy Rev.* 135 (2021) 110113. https://doi.org/https://doi.org/10.1016/j.rser.2020.110113.

[10] Y. Zhu, W. Li, J. Li, H. Li, Y. Wang, and S. Li, Thermodynamic analysis and economic assessment of biomass-fired organic Rankine cycle combined heat and power system integrated with CO2 capture, *Energy Convers. Manag.* 204 (2020) 112310. https://doi.org/https://doi.org/10.1016/j.enconman.2019.112310.

[11] T. Uchino, T. Yasui, and C. Fushimi, Design of biomass power plant integrated with thermochemical heat storage using $Ca(OH)_2/CaO$ and evaluation of the flexibility of power generation: Dynamic simulation and energy analysis, *Energy Convers. Manag.* 243 (2021) 114366. https://doi.org/https://doi.org/10.1016/j.enconman.2021.114366.

[12] H. Chen, K. Xue, Y. Wu, G. Xu, X. Jin, and W. Liu, Thermodynamic and economic analyses of a solar-aided biomass-fired combined heat and power system, *Energy.* 214 (2021) 119023. https://doi.org/https://doi.org/10.1016/j.energy.2020.119023.

[13] J.D. Smith, V. Sreedharan, M. Landon, and Z.P. Smith, Advanced design optimization of combustion equipment for biomass combustion, *Renew. Energy.* 145 (2020) 1597–1607. https://doi.org/https://doi.org/10.1016/j.renene.2019.07.074.

[14] M. Yang, J. Zhang, S. Zhong, *et al.*, CFD modeling of biomass combustion and gasification in fluidized bed reactors using a distribution kernel method, *Combust. Flame.* 236 (2022) 111744. https://doi.org/https://doi.org/10.1016/j.combustflame.2021.111744.

[15] A. Jaworek, A.T. Sobczyk, A. Marchewicz, A. Krupa, and T. Czech, Particulate matter emission control from small residential boilers after biomass combustion. A review, *Renew. Sustain. Energy Rev.* 137 (2021) 110446. https://doi.org/https://doi.org/10.1016/j.rser.2020.110446.

[16] W. Cheng, Y. Zhu, J. Shao, *et al.*, Mitigation of ultrafine particulate matter emission from agricultural biomass pellet combustion by the additive of

phosphoric acid modified kaolin, *Renew. Energy.* 172 (2021) 177–187. https://doi.org/https://doi.org/10.1016/j.renene.2021.03.041.

[17] F. Guerrero, A. Arriagada, F. Muñoz, P. Silva, N. Ripoll, and M. Toledo, Particulate matter emissions reduction from residential wood stove using inert porous material inside its combustion chamber, *Fuel.* 289 (2021) 119756. https://doi.org/https://doi.org/10.1016/j.fuel.2020.119756.

[18] W. Ma, C. Ma, X. Liu, *et al.*, Nox formation in fixed-bed biomass combustion: Chemistry and modeling, *Fuel.* 290 (2021) 119694. https://doi.org/https://doi.org/10.1016/j.fuel.2020.119694.

[19] A.I. Osman, Mass spectrometry study of lignocellulosic biomass combustion and pyrolysis with NOx removal, *Renew. Energy.* 146 (2020) 484–496. https://doi.org/https://doi.org/10.1016/j.renene.2019.06.155.

[20] G. Archan, A. Anca-Couce, M. Buchmayr, C. Hochenauer, J. Gruber, and R. Scharler, Experimental evaluation of primary measures for NOX and dust emission reduction in a novel 200 kW multi-fuel biomass boiler, *Renew. Energy.* 170 (2021) 1186–1196. https://doi.org/https://doi.org/10.1016/j.renene.2021.02.055.

[21] J. Krzywanski, T. Czakiert, A. Zylka, *et al.*, Modelling of SO_2 and NO_x emissions from coal and biomass combustion in air-firing, oxyfuel, iG-CLC, and CLOU conditions by fuzzy logic approach, *Energies.* 15 (2022) 8095. https://doi.org/10.3390/en15218095.

[22] A.B.M. Abdul Malek, M. Hasanuzzaman, and N.A. Rahim, Prospects, progress, challenges and policies for clean power generation from biomass resources, *Clean Technol. Environ. Policy.* 22 (2020) 1229–1253. https://doi.org/10.1007/s10098-020-01873-4.

[23] T.A. Mamvura, G. Danha, Biomass torrefaction as an emerging technology to aid in energy production, *Heliyon.* 6 (2020) e03531. https://doi.org/https://doi.org/10.1016/j.heliyon.2020.e03531.

[24] M. Simonic, D. Goricanec, and D. Urbancl, Impact of torrefaction on biomass properties depending on temperature and operation time, *Sci. Total Environ.* 740 (2020) 140086. https://doi.org/https://doi.org/10.1016/j.scitotenv.2020.140086.

[25] L. Wang, L. Riva, Ø. Skreiberg, *et al.*, Effect of torrefaction on properties of pellets produced from woody biomass, *Energy & Fuels.* 34 (2020) 15343–15354. https://doi.org/10.1021/acs.energyfuels.0c02671.

[26] D. Maxwell, B.A. Gudka, J.M. Jones, and A. Williams, Emissions from the combustion of torrefied and raw biomass fuels in a domestic heating stove, *Fuel Process. Technol.* 199 (2020) 106266. https://doi.org/https://doi.org/10.1016/j.fuproc.2019.106266.

[27] W.-H. Chen, B.-J. Lin, Y.-Y. Lin, *et al.*, Progress in biomass torrefaction: Principles, applications and challenges, *Prog. Energy Combust. Sci.* 82 (2021) 100887. https://doi.org/https://doi.org/10.1016/j.pecs.2020.100887.

[28] S.L. Majamo, T.A. Amibo, Modeling and optimization of chemical-treated torrefaction of wheat straw to improve energy density by response surface

Recent advances in biofuels production: industrial applications   273

methodology, *Biomass Convers. Biorefinery.* (2023). https://doi.org/10.1007/s13399-023-04192-y.

[29] S.K. Thengane, K.S. Kung, A. Gomez-Barea, and A.F. Ghoniem, Advances in biomass torrefaction: Parameters, models, reactors, applications, deployment, and market, *Prog. Energy Combust. Sci.* 93 (2022) 101040. https://doi.org/https://doi.org/10.1016/j.pecs.2022.101040.

[30] S. Negi, G. Jaswal, K. Dass, K. Mazumder, S. Elumalai, and J.K. Roy, Torrefaction: A sustainable method for transforming of agri-wastes to high energy density solids (biocoal), *Rev. Environ. Sci. Bio/Technol.* 19 (2020) 463–488. https://doi.org/10.1007/s11157-020-09532-2.

[31] A.A. Adeleke, J.K. Odusote, P.P. Ikubanni, O.A. Lasode, M. Malathi, and D. Paswan, Essential basics on biomass torrefaction, densification and utilization, *Int. J. Energy Res.* 45 (2021) 1375–1395. https://doi.org/https://doi.org/10.1002/er.5884.

[32] P. Piersa, H. Unyay, S. Szufa, *et al.,* An extensive review and comparison of modern biomass torrefaction reactors vs. biomass pyrolysis—Part 1, *Energies.* 15 (2022) 2227. https://doi.org/10.3390/en15062227.

[33] D.A. Rodríguez-Alejandro, H. Nam, D. Granados-Lieberman, *et al.,* Experimental and numerical investigation on a solar-driven torrefaction reactor using woody waste (Ashe Juniper), *Energy Convers. Manag.* 288 (2023) 117114. https://doi.org/https://doi.org/10.1016/j.enconman.2023.117114.

[34] D. Chen, K. Cen, Z. Gan, X. Zhuang, and Y. Ba, Comparative study of electric-heating torrefaction and solar-driven torrefaction of biomass: Characterization of property variation and energy usage with torrefaction severity, *Appl. Energy Combust. Sci.* 9 (2022) 100051. https://doi.org/https://doi.org/10.1016/j.jaecs.2021.100051.

[35] P.N.Y. Yek, X. Chen, W. Peng, *et al.,* Microwave co-torrefaction of waste oil and biomass pellets for simultaneous recovery of waste and co-firing fuel, *Renew. Sustain. Energy Rev.* 152 (2021) 111699. https://doi.org/https://doi.org/10.1016/j.rser.2021.111699.

[36] E. Valdez, L.G. Tabil, E. Mupondwa, D. Cree, and H. Moazed, Microwave torrefaction of oat hull: Effect of temperature and residence time, *Energies.* 14 (2021) 4298. https://doi.org/10.3390/en14144298.

[37] E.A. Silveira, L.A. Macedo, K. Candelier, P. Rousset, and J.-M. Commandré, Assessment of catalytic torrefaction promoted by biomass potassium impregnation through performance indexes, *Fuel.* 304 (2021) 121353. https://doi.org/https://doi.org/10.1016/j.fuel.2021.121353.

[38] L. Chen, C. Wen, W. Wang, *et al.,* Combustion behaviour of biochars thermally pretreated via torrefaction, slow pyrolysis, or hydrothermal carbonisation and co-fired with pulverised coal, *Renew. Energy.* 161 (2020) 867–877. https://doi.org/https://doi.org/10.1016/j.renene.2020.06.148.

[39] H. Lu, Y. Gong, C. Areeprasert, *et al.,* Integration of biomass torrefaction and gasification based on biomass classification: A review, *Energy Technol.* 9 (2021) 2001108. https://doi.org/https://doi.org/10.1002/ente.202001108.

[40] Z. He, F. Zhang, R. Tu, *et al.*, The influence of torrefaction on pyrolysed biomass: The relationship of bio-oil composition with the torrefaction severity, *Bioresour. Technol.* 314 (2020) 123780. https://doi.org/https://doi.org/10.1016/j.biortech.2020.123780.

[41] S. Valizadeh, D. Oh, J. Jae, *et al.*, Effect of torrefaction and fractional condensation on the quality of bio-oil from biomass pyrolysis for fuel applications, *Fuel.* 312 (2022) 122959. https://doi.org/https://doi.org/10.1016/j.fuel.2021.122959.

[42] M. Manouchehrinejad, E.M.T. Bilek, and S. Mani, Techno-economic analysis of integrated torrefaction and pelletization systems to produce torrefied wood pellets, *Renew. Energy.* 178 (2021) 483–493. https://doi.org/https://doi.org/10.1016/j.renene.2021.06.064.

[43] D. Aboelela, H. Saleh, A.M. Attia, Y. Elhenawy, T. Majozi, and M. Bassyouni, Recent advances in biomass pyrolysis processes for bioenergy production: Optimization of operating conditions, *Sustainability.* 15 (2023) 11238. https://doi.org/10.3390/su151411238.

[44] T.K. Dada, M. Sheehan, S. Murugavelh, and E. Antunes, A review on catalytic pyrolysis for high-quality bio-oil production from biomass, *Biomass Convers. Biorefinery.* 13 (2023) 2595–2614. https://doi.org/10.1007/s13399-021-01391-3.

[45] G. Yin, F. Zhang, Y. Gao, W. He, Q. Zhang, and S. Yang, Increase of bio-char yield by adding potassium salt during biomass pyrolysis, *J. Energy Inst.* 110 (2023) 101342. https://doi.org/https://doi.org/10.1016/j.joei.2023.101342.

[46] L.M. Terry, C. Li, J.J. Chew, *et al.*, Bio-oil production from pyrolysis of oil palm biomass and the upgrading technologies: A review, *Carbon Resour. Convers.* 4 (2021) 239–250. https://doi.org/https://doi.org/10.1016/j.crcon.2021.10.002.

[47] Y. Li, B. Xing, Y. Ding, X. Han, and S. Wang, A critical review of the production and advanced utilization of biochar via selective pyrolysis of lignocellulosic biomass, *Bioresour. Technol.* 312 (2020) 123614. https://doi.org/https://doi.org/10.1016/j.biortech.2020.123614.

[48] D. V Suriapparao, R. Vinu, Biomass waste conversion into value-added products via microwave-assisted Co-pyrolysis platform, *Renew. Energy.* 170 (2021) 400–409. https://doi.org/https://doi.org/10.1016/j.renene.2021.02.010.

[49] Y. Liu, Y. Song, J. Fu, *et al.*, Co-pyrolysis of sewage sludge and lignocellulosic biomass: Synergistic effects on products characteristics and kinetics, *Energy Convers. Manag.* 268 (2022) 116061. https://doi.org/https://doi.org/10.1016/j.enconman.2022.116061.

[50] Q. Dai, Q. Liu, X. Zhang, *et al.*, Synergetic effect of co-pyrolysis of sewage sludge and lignin on biochar production and adsorption of methylene blue, *Fuel.* 324 (2022) 124587. https://doi.org/https://doi.org/10.1016/j.fuel.2022.124587.

[51] J. Bai, X. Fu, Q. Lv, *et al.*, Co-pyrolysis of sewage sludge and pinewood sawdust: The synergistic effect and bio-oil characteristics, *Biomass*

Recent advances in biofuels production: industrial applications 275

Convers. Biorefinery. 13 (2023) 9205–9212. https://doi.org/10.1007/s13399-021-01809-y.

[52] S. Li, J. Li, and J. Xu, Investigating the release behavior of biomass and coal during the co-pyrolysis process, *Int. J. Hydrogen Energy*. 46 (2021) 34652–34662. https://doi.org/https://doi.org/10.1016/j.ijhydene.2021.08.053.

[53] W. Cai, X. Wang, Z. Zhu, *et al.*, Synergetic effects in the co-pyrolysis of lignocellulosic biomass and plastic waste for renewable fuels and chemicals, *Fuel*. 353 (2023) 129210. https://doi.org/https://doi.org/10.1016/j.fuel.2023.129210.

[54] M. Cortazar, L. Santamaria, G. Lopez, *et al.*, A comprehensive review of primary strategies for tar removal in biomass gasification, *Energy Convers. Manag*. 276 (2023) 116496. https://doi.org/https://doi.org/10.1016/j.enconman.2022.116496.

[55] Z. Lian, Y. Wang, X. Zhang, *et al.*, Hydrogen production by fluidized bed reactors: A quantitative perspective using the supervised machine learning approach, *J*. 4 (2021) 266–287. https://doi.org/10.3390/j4030022.

[56] K. Abouemara, M. Shahbaz, G. Mckay, and T. Al-Ansari, The review of power generation from integrated biomass gasification and solid oxide fuel cells: Current status and future directions, *Fuel*. 360 (2024) 130511. https://doi.org/https://doi.org/10.1016/j.fuel.2023.130511.

[57] S. Pattanayak, L. Hauchhum, C. Loha, and L. Sailo, Feasibility study of biomass gasification for power generation in Northeast India, *Biomass Convers. Biorefinery*. 13 (2023) 999–1011. https://doi.org/10.1007/s13399-021-01419-8.

[58] S. Fertahi, D. Elalami, S. Tayibi, *et al.*, The current status and challenges of biomass biorefineries in Africa: A critical review and future perspectives for bioeconomy development, *Sci. Total Environ*. 870 (2023) 162001. https://doi.org/https://doi.org/10.1016/j.scitotenv.2023.162001.

[59] D.C. de Oliveira, E.E.S. Lora, O.J. Venturini, D.M.Y. Maya, and M. Garcia-Pérez, Gas cleaning systems for integrating biomass gasification with Fischer-Tropsch synthesis – A review of impurity removal processes and their sequences, *Renew. Sustain. Energy Rev*. 172 (2023) 113047. https://doi.org/https://doi.org/10.1016/j.rser.2022.113047.

[60] D. Yang, S. Li, and S. He, Zero/negative carbon emission coal and biomass staged co-gasification power generation system via biomass heating, *Appl. Energy*. 357 (2024) 122469. https://doi.org/https://doi.org/10.1016/j.apenergy.2023.122469.

[61] R. Tamilselvan, A. Immanuel Selwynraj, Enhancing biogas generation from lignocellulosic biomass through biological pretreatment: Exploring the role of ruminant microbes and anaerobic fungi, *Anaerobe*. 85 (2024) 102815. https://doi.org/https://doi.org/10.1016/j.anaerobe.2023.102815.

[62] Z.-W. He, A.-H. Li, C.-C. Tang, *et al.*, Biochar regulates anaerobic digestion: Insights to the roles of pore size, *Chem. Eng. J*. 480 (2024) 148219. https://doi.org/https://doi.org/10.1016/j.cej.2023.148219.

[63] H.-Y. Jin, L. Yang, Y.-X. Ren, *et al.*, Insights into the roles and mechanisms of a green-prepared magnetic biochar in anaerobic digestion of waste

activated sludge, *Sci. Total Environ.* 896 (2023) 165170. https://doi.org/ https://doi.org/10.1016/j.scitotenv.2023.165170.

[64] Y. Jiang, E. McAdam, Y. Zhang, S. Heaven, C. Banks, and P. Longhurst, Ammonia inhibition and toxicity in anaerobic digestion: A critical review, *J. Water Process Eng.* 32 (2019) 100899. https://doi.org/https://doi.org/10.1016/j.jwpe.2019.100899.

[65] J.L. Chen, R. Ortiz, T.W.J. Steele, and D.C. Stuckey, Toxicants inhibiting anaerobic digestion: A review, *Biotechnol. Adv.* 32 (2014) 1523–1534. https://doi.org/https://doi.org/10.1016/j.biotechadv.2014.10.005.

[66] L. Altaş, Inhibitory effect of heavy metals on methane-producing anaerobic granular sludge, *J. Hazard. Mater.* 162 (2009) 1551–1556. https://doi.org/ https://doi.org/10.1016/j.jhazmat.2008.06.048.

[67] F.R. Amin, H. Khalid, H. Zhang, *et al.*, Pretreatment methods of lig-nocellulosic biomass for anaerobic digestion, *AMB Express.* 7 (2017) 72. https://doi.org/10.1186/s13568-017-0375-4.

[68] J. Kainthola, A.S. Kalamdhad, and V. V Goud, A review on enhanced biogas production from anaerobic digestion of lignocellulosic biomass by different enhancement techniques, *Process Biochem.* 84 (2019) 81–90. https://doi.org/https://doi.org/10.1016/j.procbio.2019.05.023.

[69] B. Yang, C.E. Wyman, Pretreatment: The key to unlocking low-cost cel-lulosic ethanol, *Biofuels, Bioprod. Biorefining.* 2 (2008) 26–40. https://doi.org/https://doi.org/10.1002/bbb.49.

[70] M. Shahbaz, M. Ammar, D. Zou, R.M. Korai, and X. Li, An insight into the anaerobic co-digestion of municipal solid waste and food waste: Influence of co-substrate mixture ratio and substrate to inoculum ratio on biogas production, *Appl. Biochem. Biotechnol.* 187 (2019) 1356–1370. https://doi.org/10.1007/s12010-018-2891-3.

[71] T. Rétfalvi, P. Szabó, A.-T. Hájos, *et al.*, Effect of co-substrate feeding on methane yield of anaerobic digestion of Chlorella vulgaris, *J. Appl. Phycol.* 28 (2016) 2741–2752. https://doi.org/10.1007/s10811-016-0796-5.

[72] S. Ponsá, T. Gea, and A. Sánchez, Anaerobic co-digestion of the organic fraction of municipal solid waste with several pure organic co-substrates, *Biosyst. Eng.* 108 (2011) 352–360. https://doi.org/https://doi.org/10.1016/j.biosystemseng.2011.01.007.

[73] J. Masih-Das, W. Tao, Anaerobic co-digestion of foodwaste with liquid dairy manure or manure digestate: Co-substrate limitation and inhibition, *J. Environ. Manag.* 223 (2018) 917–924. https://doi.org/https://doi.org/10.1016/j.jenvman.2018.07.016.

[74] R.A. Alrawi, A. Ahmad, N. Ismail, and M.O.A. Kadir, Anaerobic co-digestion of palm oil mill effluent with rumen fluid as a co-substrate, *Desalination.* 269 (2011) 50–57. https://doi.org/https://doi.org/10.1016/j.desal.2010.10.041.

[75] K. Bułkowska, W. Mikucka, and T. Pokój, Enhancement of biogas pro-duction from cattle manure using glycerine phase as a co-substrate in

anaerobic digestion, *Fuel.* 317 (2022) 123456. https://doi.org/https://doi.org/10.1016/j.fuel.2022.123456.

[76] A. Thanarasu, K. Periyasamy, K. Devaraj, P. Periyaraman, S. Palaniyandi, and S. Subramanian, Tea powder waste as a potential co-substrate for enhancing the methane production in anaerobic digestion of carbon-rich organic waste, *J. Clean. Prod.* 199 (2018) 651–658. https://doi.org/https://doi.org/10.1016/j.jclepro.2018.07.225.

[77] A. Tayyab, Z. Ahmad, T. Mahmood, *et al.*, Anaerobic co-digestion of catering food waste utilizing Parthenium hysterophorus as co-substrate for biogas production, *Biomass Bioenergy.* 124 (2019) 74–82. https://doi.org/https://doi.org/10.1016/j.biombioe.2019.03.013.

[78] M.O. Fagbohungbe, B.M.J. Herbert, L. Hurst, *et al.*, The challenges of anaerobic digestion and the role of biochar in optimizing anaerobic digestion, *Waste Manag.* 61 (2017) 236–249. https://doi.org/https://doi.org/10.1016/j.wasman.2016.11.028.

[79] J. Pan, J. Ma, L. Zhai, T. Luo, Z. Mei, and H. Liu, Achievements of biochar application for enhanced anaerobic digestion: A review, *Bioresour. Technol.* 292 (2019) 122058. https://doi.org/https://doi.org/10.1016/j.biortech.2019.122058.

[80] S.O. Masebinu, E.T. Akinlabi, E. Muzenda, and A.O. Aboyade, A review of biochar properties and their roles in mitigating challenges with anaerobic digestion, *Renew. Sustain. Energy Rev.* 103 (2019) 291–307. https://doi.org/https://doi.org/10.1016/j.rser.2018.12.048.

[81] M. Chiappero, O. Norouzi, M. Hu, *et al.*, Review of biochar role as additive in anaerobic digestion processes, *Renew. Sustain. Energy Rev.* 131 (2020) 110037. https://doi.org/https://doi.org/10.1016/j.rser.2020.110037.

[82] F. Codignole Luz, S. Cordiner, A. Manni, V. Mulone, and V. Rocco, Biochar characteristics and early applications in anaerobic digestion-A review, *J. Environ. Chem. Eng.* 6 (2018) 2892–2909. https://doi.org/https://doi.org/10.1016/j.jece.2018.04.015.

[83] W. Zhao, H. Yang, S. He, Q. Zhao, and L. Wei, A review of biochar in anaerobic digestion to improve biogas production: Performances, mechanisms and economic assessments, *Bioresour. Technol.* 341 (2021) 125797. https://doi.org/https://doi.org/10.1016/j.biortech.2021.125797.

[84] L. Qiu, Y.F. Deng, F. Wang, M. Davaritouchaee, and Y.Q. Yao, A review on biochar-mediated anaerobic digestion with enhanced methane recovery, *Renew. Sustain. Energy Rev.* 115 (2019) 109373. https://doi.org/https://doi.org/10.1016/j.rser.2019.109373.

[85] C. Luo, F. Lü, L. Shao, and P. He, Corrigendum to "Application of eco-compatible biochar in anaerobic digestion to relieve acid stress and promote the selective colonization of functional microbes" [Water Res. 68 (2014) 710–718], *Water Res.* 70 (2015) 496. https://doi.org/https://doi.org/10.1016/j.watres.2014.12.046.

[86] Y. Cai, M. Zhu, X. Meng, J.L. Zhou, H. Zhang, and X. Shen, The role of biochar on alleviating ammonia toxicity in anaerobic digestion of nitrogen-

278 *Clean energy for low-income communities*

rich wastes: A review, *Bioresour. Technol.* 351 (2022) 126924. https://doi.org/https://doi.org/10.1016/j.biortech.2022.126924.

[87] T.G. Ambaye, E.R. Rene, A.-S. Nizami, C. Dupont, M. Vaccari, and E.D. van Hullebusch, Beneficial role of biochar addition on the anaerobic digestion of food waste: A systematic and critical review of the operational parameters and mechanisms, *J. Environ. Manag.* 290 (2021) 112537. https://doi.org/https://doi.org/10.1016/j.jenvman.2021.112537.

[88] M. Kumar, S. Dutta, S. You, *et al.*, A critical review on biochar for enhancing biogas production from anaerobic digestion of food waste and sludge, *J. Clean. Prod.* 305 (2021) 127143. https://doi.org/https://doi.org/10.1016/j.jclepro.2021.127143.

[89] Z. Bin Khalid, M.N.I. Siddique, A. Nayeem, T.M. Adyel, S. Bin Ismail, and M.Z. Ibrahim, Biochar application as sustainable precursors for enhanced anaerobic digestion: A systematic review, *J. Environ. Chem. Eng.* 9 (2021) 105489. https://doi.org/https://doi.org/10.1016/j.jece.2021.105489.

[90] H. Thomson, C. Liddell, The suitability of wood pellet heating for domestic households: A review of literature, *Renew. Sustain. Energy Rev.* 42 (2015) 1362–1369. https://doi.org/https://doi.org/10.1016/j.rser.2014.11.009.

[91] V. Karkania, E. Fanara, and A. Zabaniotou, Review of sustainable biomass pellets production – A study for agricultural residues pellets' market in Greece, *Renew. Sustain. Energy Rev.* 16 (2012) 1426–1436. https://doi.org/https://doi.org/10.1016/j.rser.2011.11.028.

[92] P. Daniel Ciolkosz, *Manufacturing Fuel Pellets from Biomass, PennState Extemsopm.* (2023) 1–4. https://extension.psu.edu/manufacturing-fuel-pellets-from-biomass (accessed February 19, 2024).

[93] A. Saravanakumar, P. Vijayakumar, A.T. Hoang, E.E. Kwon, and W.-H. Chen, Thermochemical conversion of large-size woody biomass for carbon neutrality: Principles, applications, and issues, *Bioresour. Technol.* 370 (2023) 128562. https://doi.org/https://doi.org/10.1016/j.biortech.2022.128562.

[94] G. Schmidt, G. Trouvé, G. Leyssens, *et al.*, Wood washing: Influence on gaseous and particulate emissions during wood combustion in a domestic pellet stove, *Fuel Process. Technol.* 174 (2018) 104–117. https://doi.org/https://doi.org/10.1016/j.fuproc.2018.02.020.

[95] L. Schwarzer, A.M. Frey, M.G. Warming-Jespersen, *Design of Low Emission Wood Stoves*. Technical Guidelines, IEA Bioenergy (2022). Available from: https://www.ieabioenergy.com/wp-content/uploads/2022/11/IEA-Bioenergy-Task-32_Report-D1-2_Low-emission-wood-stove-design_final.pdf.

[96] J. Lehto, A. Oasmaa, Y. Solantausta, M. Kytö, and D. Chiaramonti, Review of fuel oil quality and combustion of fast pyrolysis bio-oils from lignocellulosic biomass, *Appl. Energy.* 116 (2014) 178–190. https://doi.org/https://doi.org/10.1016/j.apenergy.2013.11.040.

[97] X. Hu, M. Gholizadeh, Progress of the applications of bio-oil, *Renew. Sustain. Energy Rev.* 134 (2020) 110124. https://doi.org/https://doi.org/10.1016/j.rser.2020.110124.

[98] M. Zhang, Y. Hu, H. Wang, *et al.*, A review of bio-oil upgrading by catalytic hydrotreatment: Advances, challenges, and prospects, *Mol. Catal.* 504 (2021) 111438. https://doi.org/https://doi.org/10.1016/j.mcat.2021.111438.

[99] L. Zhang, R. Liu, R. Yin, and Y. Mei, Upgrading of bio-oil from biomass fast pyrolysis in China: A review, *Renew. Sustain. Energy Rev.* 24 (2013) 66–72. https://doi.org/https://doi.org/10.1016/j.rser.2013.03.027.

[100] A. Khosravanipour Mostafazadeh, O. Solomatnikova, P. Drogui, and R.D. Tyagi, A review of recent research and developments in fast pyrolysis and bio-oil upgrading, *Biomass Convers. Biorefinery.* 8 (2018) 739–773. https://doi.org/10.1007/s13399-018-0320-z.

[101] M. Sharifzadeh, M. Sadeqzadeh, M. Guo, *et al.*, The multi-scale challenges of biomass fast pyrolysis and bio-oil upgrading: Review of the state of art and future research directions, *Prog. Energy Combust. Sci.* 71 (2019) 1–80. https://doi.org/https://doi.org/10.1016/j.pecs.2018.10.006.

[102] X. Chen, Q. Che, S. Li, *et al.*, Recent developments in lignocellulosic biomass catalytic fast pyrolysis: Strategies for the optimization of bio-oil quality and yield, *Fuel Process. Technol.* 196 (2019) 106180. https://doi.org/https://doi.org/10.1016/j.fuproc.2019.106180.

[103] A. Galadima, O. Muraza, Hydrothermal liquefaction of algae and bio-oil upgrading into liquid fuels: Role of heterogeneous catalysts, *Renew. Sustain. Energy Rev.* 81 (2018) 1037–1048. https://doi.org/https://doi.org/10.1016/j.rser.2017.07.034.

[104] O.S. Djandja, Z. Wang, L. Chen, *et al.*, Progress in hydrothermal liquefaction of algal biomass and hydrothermal upgrading of the subsequent crude bio-oil: A mini review, *Energy & Fuels.* 34 (2020) 11723–11751. https://doi.org/10.1021/acs.energyfuels.0c01973.

[105] W. Yang, X. Li, D. Zhang, and L. Feng, Catalytic upgrading of bio-oil in hydrothermal liquefaction of algae major model components over liquid acids, *Energy Convers. Manag.* 154 (2017) 336–343. https://doi.org/https://doi.org/10.1016/j.enconman.2017.11.018.

[106] R. Shakya, S. Adhikari, R. Mahadevan, E.B. Hassan, and T.A. Dempster, Catalytic upgrading of bio-oil produced from hydrothermal liquefaction of Nannochloropsis sp., *Bioresour. Technol.* 252 (2018) 28–36. https://doi.org/https://doi.org/10.1016/j.biortech.2017.12.067.

[107] S. De, B. Saha, and R. Luque, Hydrodeoxygenation processes: Advances on catalytic transformations of biomass-derived platform chemicals into hydrocarbon fuels, *Bioresour. Technol.* 178 (2015) 108–118. https://doi.org/https://doi.org/10.1016/j.biortech.2014.09.065.

[108] R. Kadam, N.L. Panwar, Recent advancement in biogas enrichment and its applications, *Renew. Sustain. Energy Rev.* 73 (2017) 892–903. https://doi.org/https://doi.org/10.1016/j.rser.2017.01.167.

[109] L. Deng, Y. Liu, D. Zheng, *et al.*, Application and development of biogas technology for the treatment of waste in China, *Renew. Sustain. Energy Rev.* 70 (2017) 845–851. https://doi.org/https://doi.org/10.1016/j.rser.2016.11.265.

Chapter 10

Modelling and forecasting the energy mix scenarios for Türkiye via LEAP analysis

Fazıl Gökgöz[1] and Fahrettin Filiz[1]

There exists a global trend characterized by the increasing adoption of renewable energy sources to overcome climate problems. This research attempts to construct a scenario-based model to undertake a quantitative analysis of the prevailing state of electricity generation and forecast the future composition of the generation mix. There is variability in future electricity demand. To account for these uncertainties, nine scenarios were developed using the Long Emissions Analysis Platform (LEAP) tool: the Low Demand case following a business-as-usual (Low-Demand-BAU), a scenario involving the implementation of a renewable plan (Low-Demand-Renewable), a scenario combining renewable and nuclear energy plans (Low-Demand-Renewable-Nuclear), the Base-Demand case, business-as-usual scenario (Base-Demand-BAU), a scenario implementing a renewable plan (Base-Demand-Renewable), a scenario integrating renewable and nuclear energy plans (Base-Demand-Renewable-Nuclear), the High Demand case business-as-usual scenario (High-Demand-BAU), a scenario adopting a renewable plan (High-Demand-Renewable), and a scenario involving a renewable and nuclear energy plan for meeting high demand. The business-as-usual scenarios are predicated on the current generation mix to meet demands. The renewable and nuclear plans aim to leverage both nuclear power and renewable energy potential. Comparative evaluations of these scenarios are conducted to assess their environmental impact, and ultimately, the results are analyzed.

Keywords: Climate; Energy mix; LEAP

10.1 Introduction

The sustainable development goals have various aspects, posing significant challenges for decision-makers. There is a conflict between environmental

---

[1]Faculty of Political Sciences, Department of Management, Ankara University, Turkey

282 *Clean energy for low-income communities*

sustainability and economic sustainability objectives. The reduction of CO_2 emissions plays a critical role in mitigating climate change. Greenhouse gases produce an increase in global temperatures. This results in adverse effects on the environment and human well-being. The electricity generation sector, being a significant contributor to global greenhouse gas emissions, holds the potential for emission reduction. This electricity generation sector involves different energy sources, such as coal, oil, and natural gas. These fossil fuels release substantial amounts of CO_2. By transitioning to renewable energy sources, as well as implementing cleaner technologies like nuclear power, the electricity generation sector can significantly reduce its carbon footprint. Additionally, improving energy efficiency, grid integration, and demand-side management can further contribute to emission reductions.

The global phenomenon of climate change presents an inevitable and enduring threat to humanity. Because of the variability in climatic patterns, various sectors, including agriculture, the economy, and public health, confront significant risks and vulnerabilities. Climate change is a global problem that continues to impact countries around the world, including Türkiye.

One notable impact of climate change on Türkiye is the increased frequency and intensity of heat waves. In recent years, the country has experienced high temperatures, which have resulted in health risks, decreased agricultural productivity, and increased demand for energy. For instance, in 2021, Türkiye faced a severe heat wave with temperatures that caused wildfires. These wildfires caused the loss of lives, damage to forests and ecosystems, and adverse effects on air quality, posing risks to human health.

Another significant consequence of climate change in Türkiye is the variation in precipitation. While some areas are experiencing heavy rainfall, others are facing drought conditions. These changes have harmful effects on agriculture, water resources, and ecosystems. Türkiye experienced a drought that affected agricultural production, particularly in regions dependent on irrigation, in recent years. Overall, the current state of climate change is having impacts on Türkiye. It highlights the urgent need for mitigation efforts, adaptation strategies, and international cooperation.

In 2015, the Ministry of Energy and Mineral Resources of Türkiye introduced the National Energy Plan for the 2015–2023 years. Türkiye submitted an Intended Nationally Determined Contribution (INDC) in 2015, before the Paris Conference of the Parties (COP 21), pledging a 21% reduction in greenhouse gas emissions compared to the projected business-as-usual scenario by 2030 [1]. Türkiye aims to achieve carbon neutrality by the year 2053 [2]. Renewable energy goals must be aligned with Türkiye's Nationally Determined Contribution (NDC) to the Paris Climate Agreement, showing the country's commitment to the international climate change agenda.

As expressed within the framework of the Paris Agreement, all nations need to decarbonize their energy systems, which remains central in the collective pursuit of constraining global warming to a threshold below 2°C. To fulfill this commitment, analysts and policymakers in numerous countries employ energy system models as

Forecasting Türkiye's energy mix scenarios with LEAP analysis   283

indispensable tools for assessing the evolution of energy and climate policies. These models serve to enhance understanding, promote a greater understanding of the intricate interplay between energy dynamics and climate imperatives, and facilitate informed decision-making processes in the realm of long-term energy planning.

There are not many quantitative studies on the relationship between electricity production and related CO_2 emissions in Türkiye. This study examines the correlation between electricity production and associated CO_2 emissions in Türkiye. Specifically, this empirical study intends to provide insight into Türkiye's power generation expansion and future model energy sustainability to reach a carbon neutrality target by 2053. Lastly, the study attempts to inform energy planning and policy formulation, thereby focusing on the context of climate change mitigation and environmental sustainability.

In Section 10.2, this study contains general information on the power sector in Türkiye. Section 10.3 explains the Long Emissions Analysis Platform (LEAP) model used for scenario development and the assumptions and data input to the LEAP model. The model results regarding scenario development are shown in Section 10.4 and the results are shown in Section 10.5. Finally, Section 10.6 summarizes the conclusion and policy recommendations.

10.2   Türkiye's power system overview

Türkiye's power sector is characterized by a competitive market structure that encompasses a diverse array of generation technologies, including coal, natural gas, hydroelectric, wind, and solar. This multifaceted landscape reflects Türkiye's commitment to a balanced and resilient energy portfolio capable of meeting the nation's evolving energy needs. Operating within this framework, the Turkish Electricity Transmission Corporation (TEİAŞ) plays a pivotal role in ensuring the seamless transmission and distribution of electricity across regions, contributing to the reliability and efficiency of the entire power network. Meanwhile, the Energy Market Regulatory Authority (EPDK) assumes the crucial role of shaping and maintaining fair competition, enforcing market rules, and safeguarding against market abuses.

The diversification of electricity generation sources improves the consistency of the electricity system against supply issues. Avoiding dependence on a single energy source is crucial for energy security. Türkiye has made progress in diversifying the energy sources used in electricity generation. Implementing strategies to minimize reliance on external energy inputs and expand domestic production also enhances the resilience and stability of the country's energy supply.

As a country that imports energy, Türkiye can reduce the import of non-renewable energies by increasing the share of renewable energy in electricity generation. Türkiye is dependent to the extent of 60% on coal, 93% on oil, and 99% on natural gas [3]. However, Türkiye has the potential to increase the share of renewable energy sources to 50% of total production by the year 2026. With the

284   Clean energy for low-income communities

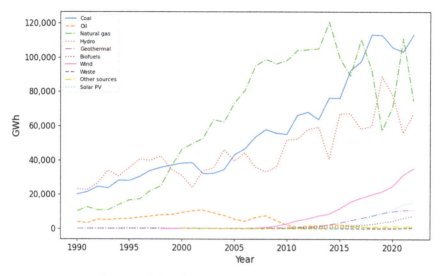

Figure 10.1   Electricity generation by source

realization of this potential, the production of imported fossil fuels will decrease [4]. This shift in energy generation sources would significantly reduce the dependency on imported fossil fuels. Figure 10.1 shows Türkiye's electricity generation by source [5].

The evaluation of Türkiye's power system transformation is based on achieving a 55% portion of renewable energy in the overall electricity generation. This was combined with enhancements in energy efficiency of 10% and the adoption of electric power in various sectors [6].

Over the years, Türkiye has been following an increasing trend in its renewable energy installed capacity. Renewable energy capacity has more than doubled in recent years, jumping from 25.6 GW in 2013 to roughly 53 GW by September 2021. The share of renewable sources in the total installed capacity, which was 40% in 2013, has risen to 53.2% as of September 2021. During the period between 2013 and September 2021, the largest contribution to the increase in installed capacity based on renewable energy sources, with a growth of 9.2 GW, comes from hydroelectric power plants. There is an increase of 7.5 GW in solar power plants, 7.4 GW in wind power plants, 1.6 GW in biomass energy plants, and 1.3 GW in geothermal energy plants [7].

10.3   Methodology

In recent years, energy system models have changed beyond their initial focus on energy security and costs to include their role in controlling climate change policies and measuring greenhouse gas emissions. The power system plays a crucial role in guiding efforts to decarbonize the energy system and facilitating transformative changes in energy-consuming sectors [8].

Scenario development has been used as an important approach to exploring uncertain environments, particularly in long-term energy production and consumption models. These scenarios provide a quantitative foundation for informed decision-making by interpreting results across alternative pathways [9]. Long-term energy planning models usually cover a wide range of factors without getting detailed in terms of time, which makes calculations simpler. Also, these models can be customized to focus on specific areas. In the power sector, for example, they aim to create a plan for generating electricity that combines different technologies to meet varying demands [10].

Stockholm Environment Institute (SEI) developed The Long Emissions Analysis Platform (LEAP) which is an energy scenario and planning software system. LEAP is a commonly used integrated modeling tool for analyzing national energy systems. Operating on an annual time step, the LEAP model has the flexibility to extend its time horizon for an indefinite number of years, commonly falling within the range of 20 to 50 years [11]. Utilizing the LEAP model offers the capability to scrutinize energy provision and greenhouse gas (GHG) emissions across local, national, and international scales. Through the application of LEAP to Japan's energy sector, diverse trajectories of CO_2 emissions are discerned. The Business-as-Usual (BAU) trajectory manifests a decline of 10% from the 2005 benchmark by 2020, followed by a 13% reduction by 2030. In the presence of a nuclear phase-out policy alongside other BAU assumptions, post-2009 sees a surge in CO_2 emissions, resulting in a 2% elevation relative to 2005 levels by 2030 [12].

In the context of the study on electricity generation and its relation to global warming potential, the LEAP model is highly relevant. It enables the construction of scenario-based models that simulate the future composition of electricity generation. Input various parameters, such as energy demand projections, renewable energy targets, policy measures, and technological advancements, into the model to assess their impacts on the electricity mix and associated greenhouse gas emissions.

Through an assessment of the Power Generation System in Ecuador's energy production system, a projection has been conducted concerning electricity supply and demand up to the year 2040. In pursuit of this objective, the LEAP methodology has been employed to scrutinize three distinct potential alternative scenarios. These scenarios, designated as the Business-as-Usual Scenario, the Energy Production Master Plan Scenario, and the Sustainable Energy Generation System Scenario, have been formulated. The sequential objectives of this investigation encompass the analysis of electricity supply and demand to transform the existing petroleum-based system into a sustainable framework, the evaluation of plausible alternatives for the prospective configuration of the equatorial energy production system, and the computation of quantities of renewable energy generation, including solar, wind, biomass, and hydroelectric, projected until the year 2040 [13].

The LEAP model was used by China to perform a simulation of the nation's energy consumption, energy composition, and CO_2 emissions spanning the period from 2019 to 2035. The outcomes of this simulation offer insights into the

286 *Clean energy for low-income communities*

evaluation of diverse policy interventions with regard to their effects on energy consumption, energy composition, and CO_2 emissions. Furthermore, these results serve as a basis for identifying the optimal and efficient approaches to evaluating CO_2 emissions [14].

LEAP models are used to create scenarios and explore potential transformations in Norway's future energy system. Scenarios are built upon assumptions, that shape the forecast. These assumptions are population growth and urbanization trends, income growth, transportation patterns, and industrial production levels. The current account scenario and 2050 projection scenario are created with assumptions and highlight the importance of renewable energy [15].

The LEAP model is studied to investigate the potential benefits of transitioning Zhangjiakou, China, from conventional to renewable energy. Utilizing a LEAP model, analyze environmental and socio economic impacts through 2050 under two scenarios: Business as usual case, which maintains current energy practices, and Integrated case, which combines three scenarios: alternatives renewable energy sources, industrial structure optimization, and energy-saving facilities. BAU models maintain current energy practices. Renewable energy alternatives promote renewable energy adaptations. Industrial structure optimization optimizes industry structures. Energy-saving facilities implement energy-saving technologies [16].

LEAP model to assess the environmental, and economic implications of three different electricity generation scenarios in Korea up to the year 2050. The analysis is grounded in the reference year of 2008, and the scenarios considered encompass the baseline, governmental strategy, and sustainable scenarios. Notably, the governmental policy scenario demonstrates higher electricity demand growth compared to the baseline scenario, whereas the sustainable society scenario indicates a growth rate lower than that of the baseline scenario [17].

The LEAP model was utilized in a comprehensive examination of the Turkish electricity sector, focusing on the evaluation of potential emissions reduction and the formulation of diverse policy-influenced scenarios. Within this analytical framework, the LEAP model has been effectively deployed to analyze both Business-as-Usual and Mitigation Scenarios. These scenarios serve as pivotal instruments for projecting energy consumption and CO_2 emissions trajectories stemming from Türkiye's electricity sector [18].

10.4 Scenario development

Energy models are commonly used in scenario development. As the scope for future applications expands, the utilization of scenario development becomes increasingly helpful for decision-makers. Various scenario development approaches, including those within LEAP models, offer diverse perspectives on scenario development [19–23].

Scenario development in the context of renewable and nuclear energy integration within Türkiye's electricity sector involves constructing well-defined pathways that outline potential trajectories for the nation's energy landscape.

These scenarios serve as important tools for exploring the relationship between energy sources, environmental considerations, and climate goals. By visualizing various future states of the energy sector, policymakers, researchers, and stakeholders can better anticipate the consequences of their decisions and strategies.

Multiple scenarios are introduced with policy choices, electricity demand trends, and emission reduction objectives. The findings indicated an increase of 9.8 points (or 21.3%) in renewable capacity compared to the Business-as-Usual (BAU) scenario, requiring an additional investment of $887 million in 2019. Wind power exhibited consistent growth in all scenarios without subsidies, while solar power's viability was heavily reliant on subsidies. Türkiye faces limitations due to technical and economic constraints, particularly in deploying higher levels of renewable energy and reducing emissions, mainly due to the need for backup sources like lignite and natural gas plants to support intermittent solar and wind energy [24].

Within the context of Türkiye's electricity system, a comprehensive analysis has been conducted, involving the development of various scenarios aimed at examining potential pathways for the Turkish electricity sector. The first scenario, commonly referred to as the BAU scenario, assumes that fossil fuel capacities will continue to be utilized at the current rate to meet the escalating electricity demand. In contrast, the renewable energy scenario focuses on decreasing greenhouse gas emissions associated with the business-as-usual scenario by diminishing the reliance on fossil fuel resources and increasing the utilization of renewable energy sources. Additionally, the renewable and nuclear energy scenario incorporates the integration of nuclear resources alongside renewable energy sources. Subsequently, these three scenarios were evaluated based on different energy demand projections, facilitating an analysis of their outcomes and implications.

Electricity demand plays the main role in electricity generation. Electricity demand is influenced by multiple factors, including social, economic, and global changes. Scenarios are primarily separated based on demand conditions, which are high-demand scenarios, base-demand scenarios, and low-demand scenarios. The high-demand scenario assumes that robust economic growth, increased population, and rapid technological advancements lead to a rise in electricity consumption. The base-demand scenario represents a continuation of current trends in electricity demand, serving as a baseline for comparison. The low-demand scenario assumes slower economic growth and demographic shifts lead to lower electricity consumption compared to the base scenario. Figure 10.2 shows different demand variations.

The selection of the nine scenarios examined in the study was based on various factors, including energy demand projections, the goal of reducing greenhouse gas emissions, and the potential of different energy sources to contribute to a sustainable energy future. Each scenario aimed to explore different pathways for meeting electricity demand while considering the reduction of greenhouse gas emissions. Figure 10.3 shows modeling scenarios. Electricity demands are categorized into high, low, and base levels, with the energy mix being designed to meet these demands in the models. Higher demands necessitate increased energy generation capacity. Utilizing the LEAP tool facilitates the calculation of energy demand and

288 *Clean energy for low-income communities*

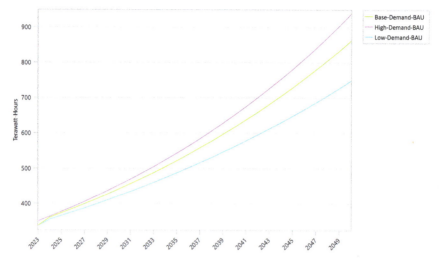

Figure 10.2 Electricity demand variations

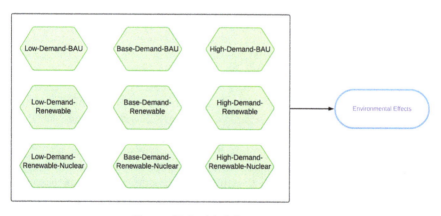

Figure 10.3 Modeling scenarios

generation requirements. The modeling process shows varying environmental impacts and CO_2 emissions in different scenarios.

10.4.1 Low-demand-BAU

This scenario represents a future where electricity demand increases at a relatively low rate, following a business-as-usual scenario. It serves as a baseline scenario to compare against other scenarios and assess the impact of different strategies on greenhouse gas emissions.

Forecasting Türkiye's energy mix scenarios with LEAP analysis 289

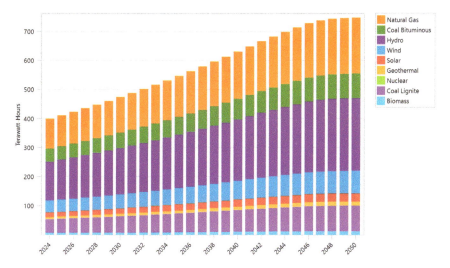

Figure 10.4 Low-demand BAU scenario

The feedstock fuel graph presented in Figure 10.4 shows the distribution of various feedstock fuels employed in the energy system scenario. The gradual decline of coal as a feedstock fuel is consistent with the transition toward cleaner energy sources. The proportion of natural gas experiences an upward trend, signifying its increasing significance in the energy landscape. Simultaneously, renewable sources such as wind and solar exhibit notable growth, aligning with the broader goals of sustainable energy generation. The feedstock fuel graph underscores the strategic shifts in energy resource allocation, reflecting policy interventions and technological advancements that shape the energy sector's evolution over the projection period.

10.4.2 Low-demand-renewable
In this scenario, a renewable plan is implemented to meet the low demand increase. The rationale behind this scenario is to explore the potential of renewable energy sources to reduce greenhouse gas emissions while meeting electricity demand. Figure 10.5 shows renewable sources exhibit growth, aligning with the goals of sustainable energy generation.

10.4.3 Low-demand-renewable-nuclear
This scenario combines renewable and nuclear energy plans to meet the electricity demand. It examines the potential synergies and trade-offs between renewable and nuclear energy in achieving low greenhouse gas emissions and meeting electricity demand. Figure 10.6 shows additional nuclear energy generation. Wind and hydro energy sources are the two main sources.

290 *Clean energy for low-income communities*

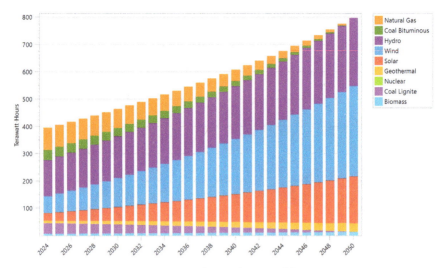

Figure 10.5   Low-demand renewable scenario

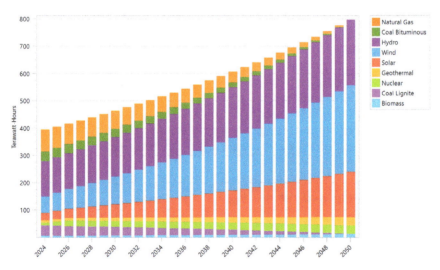

Figure 10.6   Low-demand renewable nuclear scenario

10.4.4   Base-demand-BAU

This scenario represents a future where electricity demand increases at a moderate rate, following a business-as-usual trajectory. It serves as another baseline scenario to compare against alternative strategies and assess their impact on greenhouse gas emissions. Figure 10.7 illustrates the energy composition pertaining to the base

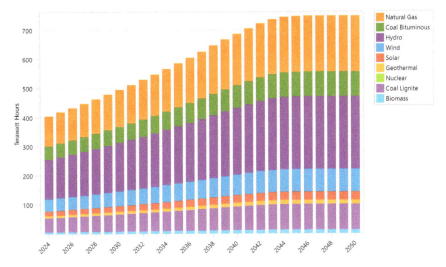

Figure 10.7 Base-demand BAU scenario

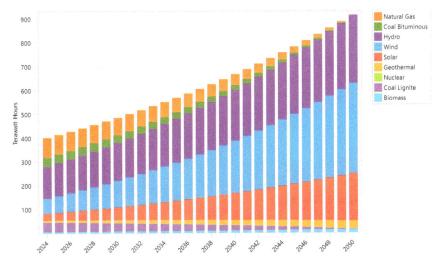

Figure 10.8 Base-demand renewable scenario

demand Business as Usual (BAU) scenario, with natural gas, coal, and hydroelectric power constituting the primary sources of energy.

10.4.5 Base-demand-renewable

In this scenario, a renewable plan is implemented to meet the base electricity demand increase. Renewable energy sources hold the potential to cut greenhouse gas emissions and pave the way for a sustainable energy system [25]. Figure 10.8 demonstrates a decline in electricity generation attributed to natural gas and coal, while conversely indicating an increase in generation associated with hydroelectric power.

10.4.6 Base-demand-renewable-nuclear

This scenario combines renewable and nuclear energy plans to meet the base electricity demand increase. It explores the potential of a diversified energy mix for reducing greenhouse gas emissions and ensuring a reliable electricity supply. Figure 10.9 shows the incorporation of nuclear energy for electricity generation. The efficacy and deployment of nuclear energy significantly rely on reactor quantity and specifications.

10.4.7 High-demand-BAU

This scenario represents a future with a high rate of electricity demand increase, following a business-as-usual trajectory. It serves as a challenging reference scenario to evaluate the implications of meeting high demand while considering greenhouse gas emissions. Figure 10.10 shows a scenario where there is a high electricity demand met primarily by the supply from natural gas, coal, and hydroelectric sources. Furthermore, it is evident that in the absence of integrating nuclear or renewable energy sources into the energy generation system, the demand for electricity remains unfulfilled.

10.4.8 High-demand-renewable

In this scenario, a renewable plan is applied to meet the increase in electricity demand. The objective is to determine the feasibility and effectiveness of renewable energy sources in a high-demand scenario. Figure 10.11 illustrates the integration of

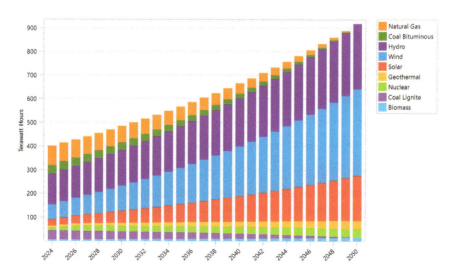

Figure 10.9 Base-demand renewable nuclear scenario

Forecasting Türkiye's energy mix scenarios with LEAP analysis 293

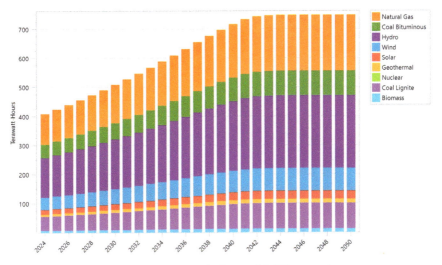

Figure 10.10   High-demand BAU scenario

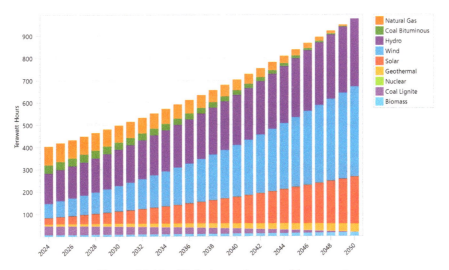

Figure 10.11   High-demand renewable scenario

renewable energy sources into electricity generation. The addition of renewable sources contributes to a more sustainable approach to meeting electricity demands.

10.4.9 High-demand-renewable-nuclear

This scenario combines renewable and nuclear energy plans to meet the increase in electricity demand. It examines the potential role of both renewable and nuclear

294 *Clean energy for low-income communities*

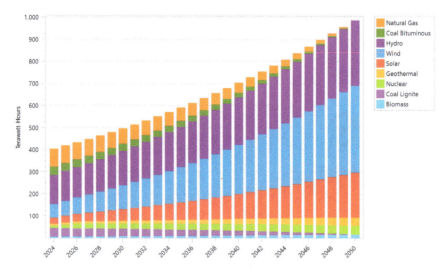

Figure 10.12 High-demand renewable nuclear scenario

energy in meeting high electricity demand while minimizing greenhouse gas emissions. Figure 10.12 provides an additional nuclear energy source to fulfill the electricity supply requirements.

This study presents a list of scenarios constructed based on a variety of criteria. The scenarios consider factors such as the continuation of renewable energy investments, the implementation of nuclear power plants, and the maintenance of the current energy mix. The results of these scenarios are directly linked to the respective criteria, and any variation in these criteria will have a direct impact on the scenario results. Additionally, the selection of these nine scenarios allows for a comprehensive analysis of different energy pathways, considering different levels of electricity demand and the integration of renewable and nuclear energy sources. By exploring these scenarios, the study can provide insights into the feasibility, benefits, and challenges of different strategies for reducing greenhouse gas emissions while ensuring a reliable and sustainable electricity generation profile.

10.5 Results

The scenario analyses reveal the potential future of Türkiye's electricity generation. Results demonstrate that scenarios with increased renewable energy integration show encouraging reductions in greenhouse gas emissions compared to business-as-usual scenarios. Renewable energy sources, when combined with nuclear power, exhibit the potential to reshape Türkiye's electricity generation mix while contributing to environmental sustainability. This study introduces nine different scenarios, each showing various environmental impacts. Figure 10.13 shows these environmental effects, particularly concerning CO_2 emissions. These outcomes

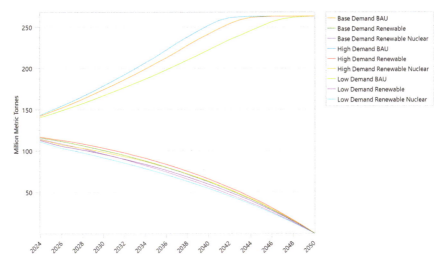

Figure 10.13 Environmental effects

depend upon electricity demand forecasting, the quantity of electricity produced, and the arrangement of the energy supply. It is observed that a significant amount of CO_2 emissions will occur from the supply of electricity if renewable energy is not utilized in the future. Remarkably, there exists an inverse correlation between emission levels and the degree of integration of renewable energy sources within the existing energy generation mix. A greater inclusion of renewables corresponds to a decrease in CO_2 emissions. By using renewable energy, Türkiye can achieve net-zero carbon emissions by 2053. To achieve this goal, there should be a significant increase in the proportion of renewable energy in the energy mix.

10.6 Conclusion

In conclusion, this study contributes an analysis of Türkiye's electricity generation mix strategies in the face of climate change challenges. Assessing diverse scenarios with different forecasted demands provides insights into achieving sustainable energy mix generation and reducing greenhouse gas emissions. The research offers policymakers, researchers, and the global community a roadmap for informed decision-making to drive a low-carbon energy future. The research findings present a comprehensive roadmap for policymakers and for informed decision-making toward the development of a low-carbon energy future. Policymakers can use these findings to develop different scenarios aimed at reducing carbon emissions. The study emphasizes the importance of integrating quantitative methodologies into energy planning and policy formulation.

As a critical component of global emissions, the electricity generation sector has a substantial responsibility to proactively address climate change and should serve as

296 *Clean energy for low-income communities*

a main point for policy interventions, technological advancements, and sustainable practices aimed at achieving a decarbonized energy system. By prioritizing clean and renewable energy sources, the power system acts as a vital enabler for sustainable end-use sector development. Its transformation unlocks the potential for resource efficiency and circularity, propelling the systemic shift toward a sustainable future. This transition not only aligns with global environmental goals but also enhances energy security, reduces dependence on fossil fuels, and fosters sustainable economic growth. Collaborative efforts on national and international levels are essential to realizing this transformative vision. As we continue to navigate the complexities of a changing climate, this study serves as an encouragement of hope, guiding Türkiye toward a more sustainable and resilient energy mix landscape.

References

[1] Alkan, A., Binatlı, A.O., and Değer, Ç.: "Achieving Turkey's INDC target: Assessments of NCCAP and INDC documents and proposing conceivable policies," *Sustainability*, 2018, 10(6), p. 1722.

[2] "Türkiye | Climate Promise," https://climatepromise.undp.org/what-we-do/where-we-work/turkiye, accessed January 2024.

[3] International Energy Agency: "Turkey 2021—Energy Policy Review" (2021).

[4] Saygin, D., Tör, O.B., Cebeci, M.E., Teimourzadeh, S., and Godron, P.: "Increasing Turkey's power system flexibility for grid integration of 50% renewable energy share," *Energy Strategy Reviews*, 2021, 34, p. 100625.

[5] https://www.iea.org/data-and-statistics/data-tools/energy-statistics-data-browser: "IEA (2023), Energy Statistics Data Browser, IEA, Paris."

[6] Acar, S., Kat, B., Rogner, M., Saygin, D., Taranto, Y., and Yeldan, A.E.: "Transforming Türkiye's power system: An assessment of economic, social, and external impacts of an energy transition by 2030,"*Cleaner Energy Systems*, 2023, 4, p. 100064.

[7] Türkiye Sınai Kalkınma Bankası (TSKB). "Enerji sektörü Görünümü" (2021). Available from: https://www.tskb.com.tr/i/assets/document/pdf/enerji-sektor-gorunumu-2021.pdf.

[8] Bistline, J.E.T., Blanford, G.J.: "The role of the power sector in net-zero energy systems," *Energy and Climate Change*, 2021, 2, p. 100045.

[9] Bhattacharyya, S.C., Timilsina, G.R.: "A review of energy system models," *International Journal of Energy Sector Management*, 2010, 4, (4), pp. 494–518.

[10] Gebremeskel, D.H., Ahlgren, E.O., and Beyene, G.B.: "Long-term electricity supply modelling in the context of developing countries: The OSeMOSYS-LEAP soft-linking approach for Ethiopia," *Energy Strategy Reviews*, 2023, 45, p. 101045.

[11] Connolly, D., Lund, H., Mathiesen, B. V., and Leahy, M.: "A review of computer tools for analysing the integration of renewable energy into various energy systems," *Applied Energy*, 2010, 87, (4), pp. 1059–1082.

[12] Takase, K., Suzuki, T.: "The Japanese energy sector: Current situation, and future paths," *Energy Policy*, 2011, 39, (11), pp. 6731–6744.

[13] Rivera-González, L., Bolonio, D., Mazadiego, L.F., and Valencia-Chapi, R.: "Long-term electricity supply and demand forecast (2018–2040): A LEAP model application towards a sustainable power generation system in Ecuador," *Sustainability*, 2019, 11(19), p. 5316.

[14] He, X., Lin, J., Xu, J., *et al.*: "Long-term planning of wind and solar power considering the technology readiness level under China's decarbonization strategy," *Applied Energy*, 2023, 348, p. 121517.

[15] Malka, L., Bidaj, F., Kuriqi, A., Jaku, A., Roçi, R., and Gebremedhin, A.: "Energy system analysis with a focus on future energy demand projections: The case of Norway," *Energy*, 2023, 272, p. 127107.

[16] Yang, D., Liu, D., Huang, A., Lin, J., and Xu, L.: "Critical transformation pathways and socio-environmental benefits of energy substitution using a LEAP scenario modeling," *Renewable and Sustainable Energy Reviews*, 2021, 135, p. 110116.

[17] Park, N.B., Yun, S.J., and Jeon, E.C.: "An analysis of long-term scenarios for the transition to renewable energy in the Korean electricity sector," *Energy Policy*, 2013, 52, pp. 288–296.

[18] Özer, B., Görgün, E., and Incecik, S.: "The scenario analysis on CO_2 emission mitigation potential in the Turkish electricity sector: 2006–2030," *Energy*, 2013, 49, (1), pp. 395–403.

[19] Wu, Q., Peng, C.: "Scenario analysis of carbon emissions of China's electric power industry up to 2030," *Energies*, 2016, 9(12), p. 988.

[20] Hernández, K.D., Fajardo, O.A.: "Estimation of industrial emissions in a Latin American megacity under power matrix scenarios projected to the year 2050 implementing the LEAP model," *Journal of Cleaner Production*, 2021, 303, p. 126921.

[21] Khan Khalil, A., Zou, X., Wang, R., Hu, G., Rong, Z., and Li, J.: "CO_2 emissions forecast and emissions peak analysis in Shanxi Province, China: An application of the LEAP model," *Sustainability*, 2022, 14(2), p. 637.

[22] El-Sayed, A.H.A., Khalil, A., and Yehia, M.: "Modeling alternative scenarios for Egypt 2050 energy mix based on LEAP analysis," *Energy*, 2023, 266, p. 126615.

[23] Cai, L., Luo, J., Wang, M., *et al.*: "Pathways for municipalities to achieve carbon emission peak and carbon neutrality: A study based on the LEAP model," *Energy*, 2023, 262, p. 125435.

[24] Kat, B.: "Clean energy transition in the Turkish power sector: A techno-economic analysis with a high-resolution power expansion model," *Utilities Policy*, 2023, 82, p. 101538.

[25] Gökgöz, F., Filiz, F.: "Electricity price forecasting: A comparative analysis with shallow-ANN and DNN," *International Journal of Energy and Power Engineering*, 2018, 12(6), pp. 421–425.

Chapter 11

Realizing clean energy for every earthling

David S-K. Ting[1] and Jacqueline A. Stagner[1]

Sharing clean energy with every fellow earth dweller is imperative for ensuring a bright tomorrow. Clearly, there is enough clean energy to go around. The holdbacks include financial viability, political will, a lack of appropriate know-how transfer, and cultural and social acceptance. Recent strivings furnish us with sound lessons concerning many dos and don'ts. The latest advancements in renewable energy technologies, accompanied by decreasing costs, make it the necessary time to forge ahead, realizing clean energy for every earthling.

Keywords: clean energy; remote; poor; community; sustainability

11.1 A tomorrow for every earthling

The only future for planet Earth is to build a beautiful tomorrow that considers every inhabitant. As cleaning the energy is not an isolated challenge, measures that aim at resolving issues such as poverty and food must be integrated into the solution approach. As eluded by Reader in Chapter 2, the largest obstacle to overcome is possibly breaking down the barriers between the haves and the have-nots. Many of the past government subsidies and incentives do not seem to resolve this challenge. In fact, most of the political intervention appears to make matters worse. Maybe it is time to provide everyone with some essential free, clean energy, and only when the usage goes beyond a life-sustaining level will a charge be imposed. If this is the solution, the road that leads to it is long and hilly. The first step is to strive to reach every earthling with a major power grid extension or a multifaceted mini-grid system that encompasses a significant amount of renewable energy. This, along with the other eight chapters and cited references, highlighted some of the outstanding issues, failures, and successes in providing clean energy for all.

11.2 Cleaner cooking

Roy and Acharya [1] evaluated the switch from highly polluting biomass-based solid fuel to liquefied petroleum gas for cooking in India. The changeover was realized for

[1]Turbulence & Energy Laboratory, University of Windsor, Canada

300 *Clean energy for low-income communities*

8 million households; however, over 90% of the poorest communities failed to make the switch to a cleaner fuel. It is no wonder that the concentration of particulates of 2.5 μm or less was negligibly improved. It should be mentioned that not everyone regards liquefied petroleum gas as a solution. This is despite its sure benefits, especially for cooking purposes. Haselip *et al.* [2] present the case for liquefied petroleum gas for the 82 million displaced people. Their study clearly illustrates that liquefied petroleum gas provides clean cooking for refugees. Liquefied petroleum gas is not only economically viable but also culturally appropriate in many rural areas.

11.3 More solar

Other than regions far from the equator, the generous and abundant sun invites our embracement. Some of the challenges associated with solar for remote communities have been spelled out in this volume. With advancing solar technologies and decreasing costs, photovoltaic systems have a promising outlook for isolated communities. It follows that solar is typically the primary source for providing clean energy for off-macro-grid communities, as presented by Abolarin *et al.* in Chapter 5 and Babaei *et al.* in Chapter 6. Proper matching is the first step. When this is met, the systems that maximize energy output can be chosen, as detailed in Chapter 4 by Sua and Balo. It is worth noting that solar is also a key factor in devising energy-efficient housing. The way to make it affordable is expounded by Shadmand and S. Arslan Selçuk in Chapter 3. Finally, as demonstrated in Chapter 8 by Khanuja *et al.*, solar can also be maximized to lower the cost of growing produce for the local community.

11.4 Microgrid

Distributed energy systems, or microgrids, consisting of multiple renewable energy sources with storage appear viable for providing clean energy far and wide. Babaei *et al.* illustrated in Chapter 6 that such an integrated system composed of photovoltaic, wind turbine, combined heat and power unit, battery, and brackish water reverse osmosis desalination can meet the essential needs of the community in Sar Goli village. By the same token, Ahmad *et al.* [3] studied a rural community in Kalam Valley powered by a distributed energy system since 2014. With the establishment of electricity, pharmacies, medical laboratories, health units, maternity and healthcare centers sprung forth. Whether micro or macrogrids, performing a quantitative analysis invoking the Long Emissions Analysis Platform tool can guide decision-making, as detailed in the Türkiye case by F. Gökgöz and F. Filiz in Chapter 10.

11.5 Green and white hydrogen

With the push for hydrogen as an energy vector, it makes sense to extend this front to implement green hydrogen for off-grid communities. A systematic review performed by Viteri *et al.* [4] indicates that proton exchange membranes and metal hydride are two suitable technologies for realizing hydrogen for remote and off-grid communities. Nevertheless, hydrogen for storing and transporting energy as it stands suffers gravely in terms of low efficiencies and high costs.

Harvesting natural or white hydrogen from the subsurface of the earth is another story. Studies on natural hydrogen exploration are starting to escalate. Maiga *et al.* [5] argued that the presence of a hydrogen field should be supplemented with the existence of a sound natural trapping system before it is worth considering a hydrogen site. In a recent literature review, Tian *et al.* [6] divided natural hydrogen into (1) free hydrogen, (2) hydrogen in inclusions, and (3) dissolved hydrogen. One not-so-recent discovery of a large volume of gas with a molecular hydrogen concentration of more than 1% is disclosed by Larin *et al.* [7]. They estimate that 21,000 to 27,000 m^3 of hydrogen seeps out at the surface in the Russian part of the European craton every day.

11.6 Other considerations

It is perplexing that the implementation of clean energy can adversely affect non-energy Sustainable Development Goals [8]. That being the case, care must be exercised so that the implementation of clean energy is not achieved at the cost of other good things. An important element of success is to involve the community, especially the educated, young minds, along with educating the young in spirit who are interested in clean energy, from the planning stage. Such a success story is conveyed by Das *et al.* [9]. Old-fashioned kerosene-based lights were replaced with solar-powered charged lights for a 1.4 million-person community in South Odisha.

In Northern Australia, Hunt *et al.* [10] evaluated renewable energy developments on aboriginal land using the Capabilities Approach, that is, determining if a development will enhance or inhibit the capabilities of the indigenous people. They investigated: (1) large-scale developments for export, (2) remote utility-owned networks, and (3) small-scale standalone off-grid applications. The last option is the only clear winner in terms of offering reliable, affordable, and culturally fitting energy services. It is worth mentioning that to be successful, measures must be taken to train the community to operate and maintain the systems. In contrast to Hunt *et al.* [10], Chipango [11] argued against the Capabilities Approach for assessing clean energy development. Utilizing Zimbabwe as a case study, she showed that relationality is the key. In plain English, the relationships among the various entities, such as the state, the market, civil society, and the citizenry are critical.

An energy source, including its conversion and usage, cannot be considered clean if it is ecologically unfriendly. That being the case, extra care must be exercised in devising clean energy for remote communities that are in ecologically sensitive areas. A disclosure on this matter is furnished in Chapter 7 by Segura-Salas.

11.7 Moving forward

The centralized power grid must be expanded to reach more people. Having said that, it may not be viable to reach some remote rural and island communities with

302   *Clean energy for low-income communities*

the macrogrid. That being the case, localized off-grid renewable energy systems, or microgrids, especially on a community scale, are the more suitable solution. Feron *et al.* [12] assessed the off-grid photovoltaic programs in Chile. Notwithstanding many successful pilot off-grid photovoltaic projects, the large-scale deployment did not go well. They attributed the impediment to social acceptance resulting from reliability issues with the photovoltaic system due to a lack of standards and maintenance. It is thus clear that knowledge transfer must accompany technology transfer.

To mention but one other essential element that leans on reliable, clean energy is healthcare. Olatomiwa *et al.* [13] considered a few rural regions in Nigeria that have abundant solar resources. Over the range of studied conditions, they found varying standalone hybrid systems consisting of photovoltaic, battery, diesel, and/ or wind, depending on the local wind and other resources, to be suitable for empowering quality healthcare delivery in those communities. It is readily apparent that there are many fringe benefits to clean energy. Clean energy also enables the attainment of many other sustainable development goals.

Without a doubt, distributed energy systems, or mini grids are a potent driving force for cleaning up energy on a global scale. To ensure success, community involvement and a viable business model are inevitable. Among other researchers, Gill-Wiehl *et al.* [14] advocate for getting the community on board early, broadening inclusion, increasing technical know-how and capacity, and establishing transparent and viable governance models. Amidst challenges, we are gaining ground, more so in some parts of the world than others. One of the most challenging regions is Sub-Saharan Africa. Almeshqab and Ustun [15] furnish us with useful details concerning the situations in eight developing countries. To this end, their paper serves as a good reference in terms of lessons learned. Kapole *et al.* [16] expound on solar mini-grid installations in Zambia. They concluded that many grids were inappropriately sized and lacked viable business models, making them unsustainable. The solar mini-grid situation in Ghana is no better. Adu-Poku *et al.* [17] attribute the less-than-desirable performance to a mismatch between demand and capacity, and many losses and inefficiencies. Castro *et al.* [18] investigate a transition to 100% renewable energy via mini grids for 208 islands in the Philippines. They simulate different scenarios and conclude that a one-size-fits-all strategy does not work. Moreover, enforcing too drastic a switch is not financially viable. Not too far away from the Philippines is Papua New Guinea. Nepal and Sofe [19] present the situation in the most linguistically diverse country in the world, highlighting the challenge from a policy and political perspective. While private investments can accelerate clean energy, full privatization poses unnecessary risks. An appropriate regulatory framework is required to mitigate these threats to the people.

Diesel is still a practical means of providing energy to many remote and isolated communities. Instead of dismissing it because it is not typically or politically perceived as clean energy, efforts should be made to further clean diesel energy and diesel energy technologies. Much cleaner "diesel fuels" are emerging, and many of these are renewable biodiesels, not to mention renewable biofuels, as disclosed in

Realizing clean energy for every earthling 303

Chapter 9 by Kang *et al.* A hot-off-the-press discussion specifically on biodiesel is given by Khan *et al.* [20]. They obtained promising results when comparing six non-edible seed oil feedstocks for biodiesel production. They suggest growing these plants on barren lands, converting wasteland into green pasture while producing clean "diesel" fuel. Another recent example is the use of teak sawdust pyrolysis oil to blend with commercial diesel. Mankeed *et al.* [21] analyze such blends in a four-stroke compression ignition engine. They find that 25% teak sawdust pyrolysis oil with diesel results in good emissions performance and, thus, conclude that such fuel blends are suitable and sustainable for agriculture and rural applications. If we are serious about helping our less privileged neighbors, let us embrace all viable solutions objectively as we continue to strive for a cleaner tomorrow. After all, how can we convince someone to adopt clean energy when they are starving for basic life-sustaining fuel to survive? Maybe we can first assist in providing our fellow homo sapiens with some essential energy and then help in transforming and increasing the supply into cleaner options. For example, a living laboratory study performed by [22] indicates that diesel generators can be complemented by solar and batteries, leading to a more than 30% reduction in diesel fuel usage annually at the Kluane Lake Research Station in Yukon Territory, Canada.

It is evident that solar and wind are possibly the most readily available energy sources. This is also the case for Queensland, Australia, where Islam *et al.* [23] estimate there is 14,448 GW of solar photovoltaic potential, along with 4,041 GW of wind power. From the many lessons learned, tapping into these vast energy resources must start with respect for the local community and culture. Starting on this footing can significantly improve the success rate, provided the other coupled factors are also carefully considered from the outset.

The good news is that there has been notable progress in supplying clean energy to low-income and isolated communities in recent years. The advancement, along with the accumulated knowledge gained, gives us hope for a clean Earth where every earthling has access to clean energy. This volume is a discourse that contains, though not exhaustive, details on how to tread forward.

References

[1] A. Roy, P. Acharya, "Energy inequality and air pollution nexus in India," *Science of the Total Environment*," 876: 162805, 2023.

[2] J. Haselip, K. Chen, H. Marwah, and E. Puzzolo, "Cooking in the margins: Exploring the role of liquefied petroleum gas for refugees in low-income countries," *Energy Research & Social Science*, 83: 102346, 2022.

[3] T. Ahmad, S. Ali, and A. Basit, "Distributed renewable energy systems for resilient and sustainable development of remote and vulnerable communities," *Philosophical Transactions of the Royal Society, A: Mathematical, Physical and Engineering Sciences*, 380: 20210143, 2022.

[4] J. P. Viteri, S. Viteri, C. Alvarez-Vasco, and F. Henao, "A systematic review on green hydrogen for off-grid communities – Technologies, advantages, and

limitations, *International Journal of Hydrogen Energy*, 48(52): 19751–19771, 2023.

[5] O. Maiga, E. Deville, J. Laval, A. Prinzhofer, and A. B. Diallo, "Trapping processes of large volumes of natural hydrogen in the subsurface: The emblematic case of the Bourakebougou H_2 field in Mali," *International Journal of Hydrogen Energy*, 50: 640–647, 2024.

[6] Q-N. Tian, S-Q. Yao, M-J. Shao, W. Zhang, and H-H. Wang, "Origin, discovery, exploration and development status and prospect of global natural hydrogen under the background of carbon neutrality," *China Geology*, 5: 722–733, 2022.

[7] N. Larin, V. Zgonnik, S. Rodina, E. Deville, A. Prinzhofer, and V. N. Larin, "Natural molecular hydrogen seepage associated with surficial, rounded depressions on the European craton in Russia, *Natural Resources Research*, 24(3): 369–383, 2015.

[8] R. E. Oghenekaro, S. Kant, "Interactions between proposed energy-mix scenarios and non-energy sustainable development goals (SDGs): a Sub-Sahara African perspective," *Environmental Research Communications*, 4: 035002, 2022.

[9] S. S. Das, D. D. Behera, and B. B. Nayak, "Use of clean energy for sustainable livelihood in rural areas: A case from South Odisha," *Materials Today: Proceedings*, 60: 765–772, 2022.

[10] J. Hunt, B. Riley, L. O'Neill, and G. Maynard, "Transition to renewable energy and Indigenous people in Northern Australia: Enhancing or inhibiting capabilities," *Journal of Human Development and Capabilities*, 22(2): 360–378, 2021.

[11] E. F. Chipango, "Why do capabilities need Ubuntu? Specifying the relational (im)morality of energy poverty," *Energy Research & Social Science*, 96: 102921, 2023.

[12] S. Feron, H. Heinrichs, and R. R. Cordero, "Sustainability of rural electrification programs based on off-grid photovoltaic (PV) systems in Chile," *Energy, Sustainability and Society*, 6: 32, 2016.

[13] L. Olatomiwa, R. Blanchard, S. Mekhilef, and D. Akinyele, "Hybrid renewable energy supply for rural healthcare facilities: An approach to quality healthcare delivery," *Sustainable Energy Technologies and Assessments*, 30: 121–138, 2018.

[14] A. Gill-Wiehl, S. Miles, J. Wu, and D. M. Kammen, "Beyond customer acquisition: A comprehensive review of community participation in mini grid projects," *Renewable and Sustainable Energy Reviews*, 153: 111778, 2022.

[15] F. Almeshqab, T. S. Ustun, "Lessons learned from rural electrification initiatives in developing countries: Insights for technical, social, financial and public policy aspects," *Renewable and Sustainable Energy Reviews*, 102: 35–53, 2019.

[16] F. Kapole, S. Mudenda, and P. Jain, "Study of major solar energy mini-grid initiatives in Zambia," *Results in Engineering*, 18: 101095, 2023.

[17] A. Adu-Poku, G. S. K. Aidam, G. A. Jackson, *et al.*, "Performance assessment and resilience of solar mini-grids for sustainable energy access in Ghana," *Energy*, 285: 129431, 2023.

[18] M. T. Castro, L. L. Delina, and J. D. Ocon, "Transition pathways to 100% renewable energy in 208 island mini-grids in the Philippines," *Energy Strategy Reviews*, 52: 101315, 2024.

[19] R. Nepal, R. Sofe, "Electricity reforms in small island developing states under changing policy contexts – Lessons for Papua New Guinea," *Energy Policy*, 186: 114012, 2024.

[20] I. U. Khan, H. Long, and Y. Yu, "Potential and comparative studies of six non-edible seed oil feedstock's for biodiesel production," *International Journal of Green Energy*, 21(4): 883–903, 2024.

[21] P. Mankeed, N. Homdoung, T. Wongsiriamnuay, and N. Tippayawong, "Biomass pyrolysis oil/diesel blends for a small agricultural engine," *Energy Exploration & Exploitation*, 42(1): 250–264, 2024.

[22] D. J. Sambor, H. Penn, and M. Z. Jacobson, "Energy optimization of a food–energy–water microgrid living laboratory in Yukon, Canada, *Energy Nexus*, 10: 100200, 2023.

[23] M. K. Islam, N. M. S. Hassan, M. G. Rasul, K. Emami, and A. A. Chowdhury, "Green and renewable resources: an assessment of sustainable energy solution for Far North Queensland, Australia," *International Journal of Energy and Environmental Engineering*, 14: 841–869, 2022.

Index

'access to electricity' indicator 59
Additive Ratio Assessment (ARAS)
  methodologies 106
affordable energy 9
  for end-use sectors 51–3
  fuel poverty benchmarks 11
    changing methodologies in the UK
    13–17
  poverty and energy challenges 23–6
  poverty measures and sustainable
    development 18
    global population growth, impact
    of 21–3
agricultural greenhouses, enhancing
    solar insolation in 231
  literature review 234–5
  model development 236
    greenhouse orientations 236–7
    greenhouse shapes 238
    roof inclinations 237–8
  solar radiation availability
    impact of greenhouse orientations
    on 241–2
    impact of greenhouse shapes on
    244–5
    impact of roof inclinations on
    242–4
  TRNSYS-18 model 238–41
air conditioning units 137–9, 142
air-fuel ratio 268
American Society of Heating,
    Refrigerating and Air Conditioning
    Engineers (ASHRAE) 125

anaerobic digestion of biomass 259–63
  pre-treatment approaches 260–1
Analytic Hierarchy Process (AHP)
    103, 106, 110, 196
Analytic Network Process (ANP) 106
Announced Pledges Case (APC) 54
authorship and co-authorship analysis
    87–9

base-demand-BAU scenario 290–1
base-demand-renewable-nuclear sce-
    nario 292
base-demand-renewable scenario
    291–2
battery sizing 132
battery storage 173–4
BAU scenario 287
bibliometrics 82
biochar 262–3
biodiesel 302–3
biofuels production 249
  anaerobic digestion of biomass
    259–63
  industrial applications of biofuels
    263
    biomass-based energy for heating
    264–6
    bio-oil as fuel oil 266–9
    use of biogas 269–70
  thermochemical conversion of
    biomass 251
    combustion 251–3
    gasification 257–9

308   *Clean energy for low-income communities*

pyrolysis 255–7
torrefaction 253–5
biogas, use of 269–70
biomass-based energy for heating 264–6
biomass-fired power plants 266
bio-oil 254–5
  as fuel oil 266–9
  upgrading 267–8
bio-syngas 257
BREDEM software 14, 40
Building Research Establishment Domestic Energy Model (BREDEM) 13
Business-as-Usual Scenario 285–7

'Can't Pay, Won't Pay' situations 11
Capabilities Approach 301
capital cost (CAPEX) 198
catalytic cracking 267
central composite design (CCD) 254
Clarivate Analytics' WoS database 82
clean cooking fuels 47–50
clean energy 26–9, 302
cleaner cooking 299–300
climate change 283
  in Türkiye 282
C/N ratio 261
CO_2 emissions 283, 295
  reduction 282
co-authorship analysis 87–9
co-digestion 261–2
combined heat and power (CHP) system 165–6, 168, 172
combustion, biomass 251–3
computational fluid dynamics (CFD) modeling 252
concentrated solar power (CSP) 252
conductive additives 262

Controlled Environmental Agriculture (CEA) 233
conventional distribution system, heuristic approach for the expansion of 193–6
converter 174–6
co-pyrolysis 255–6
cost of electricity (COE) 165
Cost of Energy (COE) 166
COVID pandemic 43, 56
Criteria Importance Through Intercriteria Correlation (CRITIC) 106

data collection 125–7
data reduction
  battery sizing 132
  energy consumption analysis 128–9
  energy efficiency improvements 129–30
  energy yield of solar PV array 131–2
  inverter sizing 132
  solar PV sizing 130–1
  solar sizing analysis 130
Democratic Republic of Congo (DRC) 24
desalination unit 169–70
diesel fuels 302–3
direct interspecies electron transfer (DIET) 263
distributed energy systems 300
domestic heating 264–5

electricity 33
  access to 29
    electricity versus energy 37–45
    energy measurement and data gathering 32–7
    household electricity access benchmarks redefined 45–7

need for electricity 29–32
demand 287–8
versus energy 37–45
electricity mix energy data 39
electric-powered systems 264
electric vehicles (EVs) 52
electrification in remote communities
185
multicriteria electrification
approaches 189
heuristic approach 193–6
microgrid formation methodology
196–7
Pantanal, Mato Grosso do Sul,
Brazil 204–16
socio-environmental criteria
characterization 192–3
stand-alone systems 197–204
prototype implementation,
electrification model and
experiences in 216
electrochemical battery application, experiences in 217–19
implementation stage 219–22
operations and maintenance plan
222–4
socio-environmental analysis 188–9
socio-environmental characterization
187–8
electrochemical battery application,
experiences in 217–19
electronic energy audit and solar sizing
(e-EASZ) tool 119, 125, 130
emissions control 252
energy consumption analysis 128–9
energy efficiency and solar PV systems
119
air conditioning units 137–9
data collection 125–7
data reduction

battery sizing 132
energy consumption analysis
128–9
energy efficiency improvements
129–30
energy yield of solar PV array
131–2
inverter sizing 132
solar PV sizing 130–1
solar sizing analysis 130
energy-efficient case
air conditioning units 142
alternative case 150–1
base case 148–50
lighting 140–1
renewable energy supply,
powering the appliances with
143–7
solar module and batteries,
correlation to predict the
number of 148
solar PV energy yield 152–3
lighting 136–7
model validation 132–4
energy efficiency improvements
129–30
energy efficiency ratio (EER) 127, 138
energy-efficient case
air conditioning units 142
alternative case 150–1
base case 148–50
lighting 140–1
renewable energy supply, powering
the appliances with 143–7
solar module and batteries, correlation to predict the number of 148
solar PV energy yield 152–3
energy-efficient housing: see low-cost
and energy-efficient housing
design

310 *Clean energy for low-income communities*

energy insecurity 10

Energy Institute (EI) 33

energy management opportunities (EMOs) 129

Energy Market Regulatory Authority (EPDK) 283

energy measurement and data gathering 32–7

energy poverty 10

energy sources and convertors 29

energy yield of solar PV array 131–2

environmental restriction indicators (ERIs) 192

environmental vulnerability index (EVI) 23–4

Evaluation Based on Distance from Average Solution (EDAS) 106

every earthling 299

Flower Pollination Algorithm (FPA) 166

fossil fuel equivalency 35

fuel poverty benchmarks 11

changing methodologies in the UK 13–17

Fuel Poverty Energy Efficiency Rating (FPEER) methodology 16

fuel poverty gap 13–14

fuel poverty in the UK 14–15

Fuzzy AHP 106

gasification, biomass 257–9

Geographic Information Systems (GIS) 189

Glasgow Financial Alliance for Net Zero (GFANZ) 31

Global Ambition Alliance 30

Global Methane Pledge (GMP) 30

global population growth, impact of 21–3

Great Point Energy 258

greenhouse gas (GHG) emissions 30, 232, 282, 285

greenhouse orientations 236–7

on solar radiation availability 241–2

greenhouse shapes 238

on solar radiation availability 244–5

green hydrogen 300–1

grid-distributed electricity 45

gross national income (GNI) 23

high-demand-BAU scenario 292

high-demand-renewable-nuclear scenario 293–4

high-demand-renewable scenario 292–3

high-income groups 81

High-Level Expert Group (HLEG) 30

High Temperature Winkler (HTW) Gasifier 258

Hills LIHC approach 15

Hills reports 13

HOMER optimization tools 165

household, defined 12

household electricity access benchmarks redefined 45–7

human assets index (HAI) 23–4

Human Development Index 24

hybrid energy system (HES) 166, 181

hydrodenitrogenation (HDN) 267

hydrodeoxygenation (HDO) 267–8

Incident Energy Approach 35

'input-equivalent' energy 34

integrated gasification combined cycle (IGCC) systems 258

Inter-agency and Expert Group on SDG Indicators (IAEG-SDGs) 28

International Electrotechnical Commission (IEC) 141

Index 311

International Energy Agency (IEA) 20, 55, 79, 164

International Renewable Energy Agency (IRENA) 28, 123

Interstate Renewable Energy Council (IREC) 108

inverter sizing 132

keyword network analysis 90–3

least developed countries (LDC) 23–6

levelized cost of energy (LCOE) 200–1, 212–13

LiFePO4 batteries 217

lighting 136–7, 140–1

lignocellulosic biomass 261

liquefied petroleum gas (LPG) 264, 300

load profile 170–2

Long Emissions Analysis Platform (LEAP) 283

  methodology 284–6

  scenario development 286

    base-demand-BAU 290–1

    base-demand-renewable 291–2

    base-demand-renewable-nuclear 292

    high-demand-BAU 292

    high-demand-renewable 292–3

    high-demand-renewable-nuclear 293–4

    low-demand-BAU 288–9

    low-demand-renewable 289

    low-demand-renewable-nuclear 289–90

Loss of Power Supply Probability (LPSP) 175–6

low-cost and energy-efficient housing design 79

  authorship and co-authorship analysis 87–9

  keyword network analysis 90–3

  most cited studies published between 2000 and 2023 93–7

  publication and citation numbers by institutions 89–90

  publication and citation numbers by year 84

  publication numbers by country 87

  publication type 86

  research areas 84–5

  results and potential areas for future studies 97–8

  sources of publications 86–7

low-cost housing 81–2

low-demand-BAU scenario 288–9

low-demand-renewable-nuclear scenario 289–90

low-demand-renewable scenario 289

low-income groups 81

'Low-Income High Cost (LIHC)' methodology 13–16

Low-Income Low Energy Efficiency (LILEE) indicator 16–17

microgrid formation methodology 196–7

microgrids 300

middle-income groups 81

mini grids 302

Minimum Energy Performance Standards (MEPS) 119, 127, 139, 142

Mitigation Scenarios 286

multi-attribute resolution analysis 105

Multi-Criteria Decision Making (MCDM) techniques 103, 105–7

multicriteria electrification approaches 189

  heuristic approach 193–6

  microgrid formation methodology 196–7

312 *Clean energy for low-income communities*

Pantanal, Mato Grosso do Sul, Brazil 204
distribution system expansion 206–9
microgrids formation 209–11
stand-alone systems 211–16
socio-environmental criteria characterization 192–3
stand-alone systems 197
capital cost (CAPEX) 198
levelized cost of energy (LCOE) 200–1
operational cost (OPEX) 198–200
Regulatory criteria 201–3
Technological criteria 203–4
multifaceted mini-grid system 299
Multi-Objective Optimization on the basis of Ratio Analysis plus full multiplicative form (MULTIMOORA) 106
Multi-Tier Framework' (MTF) approach 20, 45–7, 50

National Determined Contributions (NDC) 54
National Renewable Energy Laboratory (NREL) 108
Net Present Cost (NPC) 166
Net-Zero commitments 32
Net-Zero plan 31
Normalized Difference Water Index (NDWI) 222

off-grid communities 300
operational cost (OPEX) 198–200
operation and maintenance (O&M) optimization algorithm 216
Organization of African Unity (OAU) 26
Organization of Arab Petroleum Exporting Countries (OAPEC) 10

Organizations for Economic Co-operation and Development (OECD) countries 232
Our World in Data (OWID) 33

Pantanal, Mato Grosso do Sul, Brazil 204
distribution system expansion 206–9
microgrids formation 209–11
stand-alone systems 211–16
Paris Climate Agreement 282
photovoltaic (PV) systems 172–3, 300
pollution 1
poverty and energy challenges 23–6
poverty measures and sustainable development 18
global population growth, impact of 21–3
poverty statistics in the US 11
power grid extension 299
Preference Ranking Organization Method for Enrichment Evaluation (PROMETHEE) approach 105
prototype implementation 216
electrochemical battery application, experiences in 217–19
implementation stage 219–22
operations and maintenance plan 222–4
publication and citation numbers by institutions 89–90
PVBAT systems 217, 220
pyrolysis, biomass 255–7

'Race to Zero' campaign 30
Reliability Factor (RF) 166
renewable and nuclear energy scenario 287
Renewable Energy Sources 104

renewable energy supply, powering the appliances with 143–7

renewable resources 169

residual income 16–17

reverse osmosis (RO) 165–6, 169

Ritchie's four ways of measuring energy 36

roof inclinations 237–8
  on solar radiation availability 242–4

root mean square error (RMSE) 149

Russia-Ukraine war 43–4

SankeyMATIC 83

SAP approach 40

Saudi Arabia 43

SDG7 energy targets 55–6

sharing clean energy with the poor
  complex challenge 2
  need for 1

SketchUp software 236

socio-environmental analysis 188–9

socio-environmental characterization 187–8

socio-environmental criteria characterization 192–3

socio-environmental indices (SEIs) 192–3

solar module and batteries, correlation to predict the number of 148

solar photovoltaic systems 122

solar power at low-income communities 103, 107
  methodology 109–10
  potential funding resources for cities 109
  results and discussion 110–14
  solar energy implementation challenges 107–9

solar PV array, energy yield of 131–2

solar PV energy yield 152–3

solar PV sizing 130–1

solar radiation availability
  impact of greenhouse orientations on 241–2
  impact of greenhouse shapes on 244–5
  impact of roof inclinations on 242–4

solar sizing analysis 130

solar technologies 300

solvent addition 267

space heating and warm households 50–1

stand-alone systems 197
  levelized cost of energy (LCOE) 200–1
  regulatory criteria 201–3
  stand-alone power generation projects, cost calculation of
    capital cost (CAPEX) 198
    operational cost (OPEX) 198–200
  technological criteria 203–4

Standard Assessment Procedure (SAP) 13

stated policies scenario (STEPS) 54

Stepwise Weight Assessment Ratio Analysis (SWARA) approach 106

Stockholm Environment Institute (SEI) 285

Sub-Saharan Africa 302

substitution method 34

sun power 105

sustainable development goals 281

Sustainable Development Goals (SDGs) 185

Sustainable Energy for All (SE4ALL) 45

sustainable energy solutions for rural electrification 163
  battery storage 173–4

combined heat and power (CHP)
plant 172
converter 174–6
desalination unit 169–70
load profile 170–2
photovoltaic (PV) module 172–3
renewable resources 169
study area 168–9
wind turbine 173

Technique for Order Performance by
Similarity to Ideal Solution
(TOPSIS) techniques 106–7
thermal cracking 267
thermochemical conversion
technologies 251
tier classification for global SDG
indicators 29
torrefaction, biomass 253–5
total appliance energy consumption
(TAEC) 134
total daily energy consumption
(TDEC) 134
total net present value (TNPV) 166
TRNBuild 240
TRNSYS-18 model 238–41
TRNSYS software 231, 236
Turkish Electricity Transmission
Corporation (TEİŞ) 283
Türkiye's Nationally Determined
Contribution (NDC) 282
Türkiye's power system overview
283–4
The UK Fuel Poverty Strategy 12

Ukraine-Russia war 43–4
UNESCO (United Nations
Educational, Scientific and
Cultural Organization) 189

United Nations Children's Fund
(UNICEF) 50
United Nations Development
Programme (UNDP) 44
United Nations Food and Agriculture
Organization (UN-FAO) 232
United Nations Statistics Division
(UNSD) 19
United States Agency for International
Development (USAID) 50
United States Energy Information
Administration (USEIA) 34–5,
55
US Department of Energy (USDOE) 9
US Environmental Protection Agency
(USEPA) 26

Visual Basic for Application (VBA)
125–6
VlseKriterijumska Optimizcija I
Kaompromisno Resenje
(VIKOR) 106
VOSviewer program 83, 87–8, 90

Warm Homes and Energy
Conservation Act 2000 12
warmth, adequate 11
Web of Science 82
Weighted Aggregated Sum Product
Assessment (WASPAS) 106
welfare spending compared to poverty
rate in the US 6–7
white hydrogen 300–1
wind turbine 173
wood pellets 264
World Health Organization
(WHO) 26